W9-CAH-264

DEFIANCE

DEFIANCE

The Bielski Partisans

NECHAMA TEC

New York Oxford
OXFORD UNIVERSITY PRESS
1993

Oxford University Press

Oxford New York Toronto
Delhi Bombay Calcutta Madras Karachi
Kuala Lumpur Singapore Hong Kong Tokyo
Nairobi Dar es Salaam Cape Town
Melbourne Auckland Madrid

and associated companies in
Berlin Ibadan

Published by Oxford University Press, Inc.,
200 Madison Avenue, New York, New York 10016

Library of Congress Cataloging-in-Publication Data
Tec, Nechama.
Defiance : The Bielski partisans
Nechama Tec.
p. cm. Includes index.
ISBN 0-19-507595-1
1. Jews—Belarus—Persecutions.
2. Holocaust, Jewish (1939–1945)—Belarus.
3. World War, 1939–1945—Underground movements, Jewish—Belarus.
4. World War, 1939–1945—Jews—Rescue—Belarus.
5. Belarus—Ethnic relations. 6. Bielski, Tuvia.
I. Title. DS135.B38T33 1993
940.53'18'094765—dc20 92-33501

2 4 6 8 9 7 5 3 1

Printed in the United States of America
on acid-free paper

To the memory of
Arieh Gelblum

Preface

My research about the Nazi annihilation of European Jews alerted me to a serious omission and an equally serious distortion. The omission is the conspicuous silence about Jews who, while themselves threatened by death, were saving others. The distortion is the common description of European Jews as victims who went passively to their death.

This book is based on evidence that corrects both. It shows that under conditions of human degradation and suffering, Jews were determined to survive—they refused to become passive victims. Propelled by the desire for freedom, risking death, many escaped from the ghettos to the countryside and forests of wartime Western Belorussia. There some of them created a Jewish partisan unit, the Bielski partisan detachment, that gave protection to all Jewish fugitives.

Assuming the dual role of rebels and rescuers, this group grew into a forest community of more than 1200 that distinguished itself as the most massive rescue operation of Jews by Jews.

The history of this community—composed of fighters, rescuers, children, older men, and women—fits into my past research about personal courage, resistance, refusal to accept evil, and mutual help.

It is also tied to my personal history, that of a hidden child. I belong to a small minority of Polish Jews who survived World War II by staying illegally in the forbidden Christian world. For three years, protected by Christian Poles, I lived under an assumed name, pretending to be Catholic. At the end of the war I resumed my former identity, determined to put this past behind me, and shied away from all wartime memories.

For some unexplained reason, by 1975 these childhood experiences began to demand attention. When these demands turned into a compelling force, I decided to revisit my past by writing an autobiography.

As I was recapturing my wartime life, the same few questions kept recurring. What was it like for other Jews who tried to pass as Christians? What made some Poles defy all the dangers and risk their lives for Jews, who traditionally were regarded as "Christ killers" and who for many still unexplained reasons were blamed for every conceivable ill? Who were these rescuers? Who were the Jews who benefited from this protection?

Later, with my autobiography behind me, eager to find answers to these questions, I embarked on research that examined two groups: the rescuers and the rescued. I had assumed that each group, the Polish rescuers and the Jewish survivors, would react in different ways to their respective circumstances.

This study took me to several archival collections. At one, the Jewish Historical Institute in Warsaw, I came upon the story of Oswald Rufeisen. A Jewish youth of seventeen when World War II began, he survived by pretending to be half German and half Polish. He became an interpreter and a secretary in a German gendarmerie in Western Belorussia, wore a Nazi uniform, and in this capacity, while risking his life, saved hundreds of Christians and Jews.

I was intrigued by Oswald Rufeisen, not only because of his unusual life, but also because I was not sure how to classify him. His case presented me with a dilemma—was he a rescuer or a survivor? In the end, I classified him as a survivor, and concentrated on the help he received while passing as a non-Jew.

After I finished writing about Polish rescuers and Jewish survivors, I made Oswald Rufeisen's story the focus of my next book. Only after I concentrated on writing about him did I become aware that some Jewish survivors I had written about, although in less dramatic ways, had also helped others. Why had I overlooked their acts of altruism? Was my insensitivity to Jews as rescuers based on the assumption that one could not simultaneously be a victim and a rescuer? Did I think that as the main targets of Nazi persecution Jews would focus only on their own survival?

While I was considering these questions, in 1986, representatives of the organization of Partisans, Underground Fighters and Ghetto Rebels of Israel asked me to write a factual account of the Bielski partisan unit. Those representatives had survived World War II by fighting and hiding in the forests of Western Belorussia. They offered to help me find materials, take care of translations, and locate people for interviews, in Israel and the United States.

Prior to this request, and quite independently, I had been interested in the Bielski partisan group and its charismatic leader Tuvia Bielski. Both the unit's opposition to the Germans and its protection of Jews piqued my interest and seemed a logical extension of my previous projects. Intrigued by this special connection between fighting and rescue, I embarked on this study. I wanted to begin my research with an interview with Tuvia Bielski, the group's commander. Although I spoke to Tuvia on the phone, my efforts at meeting him were frustrated.

Each time I called for an appointment, his wife, Lilka, offered a different reason for not setting up a meeting. The first few refusals had to do with trips to Florida, later with Tuvia's failing health. I persisted and eventually was given a date.

But when I arrived at the Bielski home in Brooklyn in May, 1987, I was greeted by a distraught Lilka who told me that Tuvia had had a bad night and was not in a position to see me. Because I was leaving for Israel the next day, I was determined to get at least some kind of a personal impression of the man. Politely, but firmly, I explained that it had taken me two hours to reach their home, and that I would be very disappointed if I could not see him. I continued by promising that I would be brief.

It worked. Soon I was moving into a dining room dominated by a massive table surrounded by equally massive chairs. The hand-embroidered tablecloths reminded me of some faraway European place. The walls were covered with photographs and with what seemed like framed diplomas. These wall hangings contributed to the room's crowded feeling. Tuvia appeared in each picture, alone or in the company of others. What looked like diplomas turned out to be expressions of gratitude for his wartime achievements. The entire place had a somber, old-fashioned flavor.

My contemplation of the surroundings was interrupted by Tuvia's noiseless appearance. He was closely followed by his wife who, without actually touching him, gave the impression of holding him up. Towering over us, erect, yet with an obvious effort, the man I had waited for moved toward me. His face was covered by a tentative, sad smile. He knew why I had come and told me, in a feeble voice, how glad he was that I wanted to write about him.

Trying to sound friendly, not to offend, I explained to Lilka that I had to conduct the interview alone, without observers. Reluctantly, she consented to leave. Before she did, she pointed to one of the framed photos, taken at a recent Waldorf-Astoria dinner, in honor of her husband's eightieth birthday. She explained that this was sponsored by the Bielski partisans. She must have wanted me to say something pleasing. I did. Perhaps accustomed to this ritual, perhaps gratified, Tuvia looked on in silence, as a smile played around his mouth.

When we sat down for the interview, I told him that he could speak Yiddish or one of the other languages he knew. He settled on Yiddish and began to whisper. I had difficulties hearing and kept hoping my tape recorder had a better ear than I. But then, slowly, as Tuvia became absorbed in his past, his voice began to change. Soon the muffled sound was replaced by a vigorous voice. With this transformation came a sense of humor. And then, chuckling, Tuvia described how some of his men would shudder and hide their heads as soon as they fired a shot. Here and there, his mind would wander into some unexpected path, but would return with a minimum of prodding. Before my very eyes this weary giant became an animated and witty storyteller. When after a little over two hours I told him that the meeting was coming to an end, he objected, assuring me that he was not tired at all. I knew, however, that Lilka, who occasionally would peek into the room, was of a different opinion. Remembering my promise to her, I took leave, saying that this was only one of many more future meetings. I told him that I was going to Israel and that I would call soon after my return. With a resigned smile, he said, "By then I may not be around any more."

Acknowledgments

Help for the preparation of this book came from a variety of directions. The generosity of several research foundations freed me from most of my teaching obligations, making possible the completion of this project. The National Endowment for the Humanities awarded me a 1991–1992 Research Fellowship, which was followed by a 1992–1993 Research Grant from the Littauer Foundation and a 1992–1993 Research Grant from the Memorial Foundation for Jewish Culture. For this important support I am extremely grateful.

In 1986, representatives of the Organization of Partisans, Underground Fighters and Ghetto Rebels in Israel asked me to undertake this research. All were former partisans, either members of the Bielski partisan group or of Soviet guerilla units, familiar with the Bielski partisan detachment. They wanted me to write a historical account of the Bielski partisan unit, the single most massive rescue operation of Jews by Jews.

With this request came a promise of help that included collection of data, translations, and setting up special interviews. I asked for and was given complete freedom, both at the research and writing stages.

In addition to assisting with various tasks these partisan representatives shared their own wartime experiences. For much generous time and many efforts I am very grateful to Zorach Arluk, Moshe Bairach, Pesia Bairach, Jacov Greenstein, Raja Kaplinski-Kaganowicz, Hanan Lefkowitz, and Moshe Kalcheim, who served in a partisan unit in the Naroch forest. My gratitude and special thanks to these former partisans applies not only to their actual help but also to the warm hospitality they showed me during my frequent trips to Israel.

In addition, in Israel, I conducted lengthy interviews with Chaim Basist, Chaja Bielski, Dr. Pinchas Boldo, Cila Dworecki, Luba Dworecki, Luba Garfunk, Shmuel Geler, Perale Hirschprung, Cila Kapelowicz, Baruch Kopold, Lili Krawitz, Lazar Malbin, Jasha Mazowi, Herzl Nachumowski, Moshe Nikopajefski, Daniel Ostaszynski, Tamara Rabinowicz, Riva Reich, Luba Rudnicki, Cila Sawicki, Hersh Smolar, Dora Shubert, and Abraham Viner. I wish to thank each of them for their valuable time and efforts.

Over the years, Chaja Bielski, the wife of Asael, the second in command of the Bielski unit, shared with me important information about the history of the Bielski detachment. Her enthusiastic desire to help and her profound knowledge of that period filled many of this study's historical gaps. I am very grateful to Chaja for the many, many hours she devoted to this project.

I am indebted to Hersh Smolar for our lengthy meetings. Although in precarious health, he very generously offered his observations and insights into the history of the partisan movement, including the Bielski detachment.

Moshe Bairach helped me in locating some important materials and performed all kinds of research errands. I am very grateful for his concrete aid and, even more, for the cheerfulness and special spirit with which he fulfilled all my requests.

In the United States this research involved interviews I conducted with Fruma Gulkowitz-Berger, Motl Berger, Riva Kaganowicz-Bernstein, Berl Chafetz, Itzyk Estreich, Lea Estreich, Sulia Wolozhinski-Rubin, Esia Lewin-Shor, and Rosalia Gierszonowski-Wodakow. I am very thankful to them all for their important help.

Several other members of the Bielski family were also generous with their contributions. Tuvia Bielski's interview, conducted two weeks before his death, stands out as an important addition to this study. I would also like to thank Lilka Bielski, Tuvia's wife, for agreeing to meet with me several times and for her cooperative attitude throughout this study, from 1987 until the present. I value my meeting with Michael Bielski, the son of Tuvia and Lilka Bielski, and wish to thank him for his help.

Zus Bielski, the man who was in charge of intelligence operations in the forest, also shared with me important information as did his charming wife Sonia Bielski. I am grateful to both of them.

I am indebted to Shmuel Krakowski, chief archivist at Yad Vashem, who, from the beginning, urged me to conduct this study. For years Krakowski had been supplying me with important advice, information, and documents. He also carefully read the entire book and offered valuable comments.

I would also like to thank Katrine Finsterbush, who at that time had worked at Yad Vashem, for locating important German documents for this study.

My special thanks go to my friend, Iwo Cyprian Pogonowski, for helping me with the map.

For over fifteen years, I have come to rely on Dina Abramowicz, the reference librarian at YIVO, New York City. Very knowledgeable and extremely helpful, she had supplied answers to many of my questions and never failed to fulfill any of my requests. I would also like to thank YIVO's chief archivist, Marek Web, who has always been ready to direct me to important sources.

Arieh Gelblum, a generous friend, a gifted journalist, and an outstanding editor, enhanced the quality of this book. I am particularly grateful for his careful reading and rereading of the entire manuscript and for his detailed, insightful editorial comments.

For valuable suggestions I would like to thank Herbert Spirer, a wonderful friend and statistician. I am indebted to David Roll, my editor at Oxford University Press, for his caring and supportive attitude and for his sensitive editing. David's suggestions and comments were interesting, helpful and enlightening. Rosemary Wellner's editorial skills have also enhanced the quality of this book.

For years many of my friends have supported my research efforts and thereby reduced some of the burdensome aspects of my work. Some did it by helping me find people to interview, others by translating Hebrew publications, still others by reading and commenting on parts of this book. For their support, their involvement, and their time I would like to thank Geuli Arad, Lisl Cade, Helen Hyman, Renana Ben-Gurion Leshem, David Leshem, Marion Pritchard, and Ada Tal. For thorough proofreading I am grateful to Luba Levy and Joan Peterdi.

I wish to thank Lori Somerville for her patient and careful typing of my hard-to-read handwriting and for her insightful comments.

And for suggesting the title of this book I would like to thank my friend, Alan Fischer.

Throughout this project and for most of my professional life my family, my husband Leon and my children, Leora and Roland, have given me much support, encouragement, and comfort. I appreciate their continuous involvement with and their understanding of my work. Particularly valuable, for this book, were Leora's and Roland's editorial comments.

Although my work benefited from many generous offers of help, I alone am responsible for the shortcomings of this book.

Contents

Western Belorussia during World War II

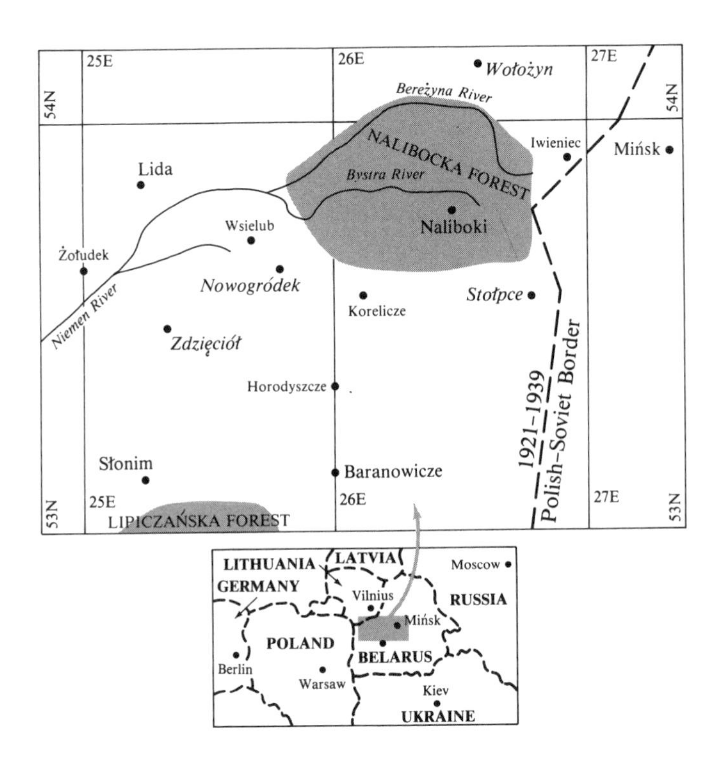

1

Before the War

An excited crowd surrounded the new arrivals. The crowd demanded to know who had come, who was left behind, what was happening to them, who else will arrive, why were they not yet here. Question upon question kept piling up. Coming all at once they were hard to follow, impossible to answer. Here and there, among the noisy haphazard efforts to tell, to connect, one could hear sudden outbursts—relatives, friends, and acquaintances were finding each other.

This was May 1943 and the center of this commotion was a group of Jewish runaways from Lida ghetto.[1] Disheveled and exhausted, most were too bewildered to talk. Passively, in silence they simply looked on.

One of the arrivals, Moshe Bairach, remembers that "A loud voice ordered: 'Arrange yourselves in rows and stand still!' Then we see a group of people coming toward us. At the head is Tuvia Bielski, followed by his brothers: Asael, Zus, Arczyk, about fourteen or so, Malbin, Grisha and others . . . Tuvia does not walk erect, not with a military bearing. His shoulders are a little stooped. He wears a leather coat, an automatic weapon dangles from his left shoulder. Yet, he salutes us in a rigid military fashion. I see that he is trying to say something but he is choking. There are tears in his eyes.

"Then he says in Russian, 'Comrades, this is the most beautiful day of my life because I lived to see such a big group come out of the ghetto! . . . I don't promise you anything, we may be killed while we try to live. But we will do all we can to save more lives. This is our way, we don't select, we don't eliminate the old, the children, the women. Life is difficult, we are in danger all the time, but if we perish, if we die, we die like human beings.' [In our group there was a man over sixty and we carried his grandchild on our shoulders.]

"What he said was simple, not flowery or showy. He talked about the difficulties in the forest. All must be quiet. All must learn how to live together, how to take orders. After this speech we were divided into small groups, each joined an already established group."[2]

About the man, Raja Kaplinski, another runaway, says, "Before I met Tuvia, I thought that a partisan commandant would be a big rough fellow, perhaps with a bushy moustache. . . . Instead, the man walking toward us was tall, graceful, with a smile that seduced all of us. The warmth and the way he received us was touching. I was all alone. My family was sent to Siberia . . . Tuvia was like a magnet that pulled us toward him. I had a feeling that with him I was totally secure, that danger could not reach me, and that around him all will be well. This is the kind of effect he had on most people; Jews, Christians, even anti-Semites. . . . This man had something that attracted and conquered people. When he got on the horse with his leather coat and the automatic gun we called him Yehuda HaMaccabee."[3]

A sixteen-year-old youth, part of a small group that had searched for the Bielski camp, also recalled: "We hear the approach of horses, then in Russian: 'Halt, password!' We give the password. Among the six riders we recognize Tuvia Bielski. He asks: 'Where are you from?'

'We are coming to you, we are from Lida.'

'Who from Lida?'

'My mother knew him. She said: 'Sima.'

'Thank God that you left, wonderful!'

"He did not ask if we had arms or anything. It was so special. I cannot forget it. It was cloudy and dark. We did not know in what world we were. I felt like a robot and suddenly this welcome! It stayed with me always."[4]

About the base itself one of the arrivals said that "When we came we saw the camp; men, women, children, all without yellow patches, smiling . . . something I had not seen in the ghetto. The first impact was deep and I am not embarrassed to admit that tears of joy ran down my cheeks. People surrounded us, friends and family members hugged and kissed, they cried. . . . Then the Bielski brothers and others from the headquarters came. Tuvia, with his tall figure, broad shoulders . . . set apart from the rest. Clearly, he was happy to see us. After a warm welcome he checked the conditions of the few weapons we brought.

"Most of us had no guns. Because of this one could notice disappointed looks on some of the faces. Then we heard loud whispering: 'Another package arrived.' These hostile remarks were meant for those of us with no weapons. I was one of them. I felt right away that because we had no arms we were a burden.

"But when I thought about the ghetto with the shameful star of David sewn into our clothes I felt free. It felt like paradise. My first day at the base passed in constant wonderment. All I saw seemed amazing. I looked at one tent, and then at another, and thought how quickly we, as a people, adjust. This is how it was in the ghetto and this is how it was in the forest. Water! I saw the two dug-up holes: wells in the forest! It was a puzzle. Everything was a puzzle. How did they do it? How?"[5]

Will another time and another place answer some of the "hows"?

For three generations Stankiewicze, a village in Western Belorussia, had been home to the Bielski family. Located between two larger towns, Lida and Nowogródek, and to the side of the main highway, Stankiewicze was a backward, isolated place. Like most surrounding villages, it had no electricity, and its simple wooden buildings linked the unpaved roads. With rain the ground would become soft and mushy, giving a sinking sensation, as if urging walkers to move on.

Each family owned a modest wooden hut with a few equally modest structures that served as barns, cowsheds, pigsties, or henhouses. Instead of their homes, the animals preferred the outdoors—on most days one could see pigs, chickens, geese, and ducks roaming the dirt roads, while cows and horses grazed in nearby meadows. At the end of the day these animals had no trouble finding their way home. The people and the animals possessed an understanding; they knew who belonged to whom.

Each household had an outhouse that for obvious reasons stood away from the main dwelling. Periodically, their contents were emptied and used as fertilizer.

Stankiewicze had no running water and no wells. A river that cut through the village supplied its water.

Individual huts stood facing each other in an uneven row. Each was surrounded by a vegetable and flower garden. Behind these gardens were farmlands and meadows. Beyond the fields were forests, extending for many, many miles. Belorussia is known for its jungle-like, huge wooded areas.

At the turn of this century the Bielskis were one of only six families who lived in Stankiewicze. All were Belorussian peasants except the Bielskis who were the only Jews. As a majority, the Belorussians had a longer history than any other group; they had their own language and culture.

The Bielskis owned a mill and farmed the land. Horses, cows, chickens, and geese also helped support the family. The Bielski mill stood at the river's edge. The barn and several sheds were close to the mill, while the living quarters were further away and partly to the side of the rest of the dwellings.

The Bielski home consisted of a wooden two-room hut. One room was only big enough to hold a medium-sized bed. Here slept the parents, Beila and David.

Each room had a small window and an unfinished clay floor. From the outside a door led into the main room that functioned as a kitchen, dining room, living room, and bedroom. To the right of this big room was a large built-in stove that served as a cooking stove, an oven, a heater, and a place to rest. During the winter months the warmth emanating from its stone surface made it a cozy sleeping place. Next to the stove a door led to the smaller room and a few wooden beds hugged the remaining walls.

The center of this room was dominated by a rectangular table with two wooden benches at opposite sides. At the ends of the table stood armchairs reserved for each parent. Although the furnishings were sparse, the room was crowded, without a free wall, devoid of a single luxury item.

David and Beila had twelve children: ten sons and two daughters. They were more fortunate than most of their neighbors because death claimed only one of their infant sons, while the rest grew into good-looking, healthy individuals. Born in 1906 Tuvia was the third oldest son. His two brothers Asael and Zus were respectively two and four years younger.

Because of limited space and the large number of children, each bed had to accommodate more than one occupant. In warm weather the barn offered additional sleeping space, easing the overcrowding.

Though casual visitors might wonder how such a large family could fit into this limited space, the Bielskis themselves gave no thought to this matter.[6] Born over a twenty-five-year span, some of the children would leave home before the birth of others. At no point did their hut accommodate the full family of fourteen.

Like the rest of their neighbors the Bielskis were poor, but the residents of Stankiewicze did not suffer from hunger nor did they see themselves as needy. Still, all were short of money and all would have welcomed extra cash.

For the Bielskis additional income would have meant more and better education for their children. Under the circumstances, they had to be satisfied with a minimum amount of schooling.

At eighty-one, Tuvia Bielski remembered well the few years he spent in Nowogródek, at the Cheder (religious school) of Rabbi Jezel. Wistfully, he spoke about his enjoyment of study and how eager he was to stay on. Tuvia's formal schooling, however, was short-lived and sporadic; it ended when he was thirteen.[7] In his case, as with his siblings, a prolonged education would have meant more money and the loss of much needed time. The Bielski children had to help in the mill and on the farm.

But Tuvia found less formal ways of learning. During World War I the German army occupied this part of the country and some soldiers came to stay in a house left unattended by its owner. Each day, for two years, the boy would spend some time with these men. He wanted to learn German. The soldiers must have liked him because not only did they tolerate his presence, but rewarded him with food and cigarettes for his father. After the Germans left, Tuvia knew how to speak their language and remembered it for the rest of his life.[8]

Not all of Tuvia's brothers and sisters were as interested in studying, but they all approved of education. They particularly valued learning that involved Jewish life and Jewish history.

The Bielskis observed Jewish dietary laws and celebrated religious holidays. Friday evening would signal an interruption of heavy work (minimal attention had to be paid to the animals). Each Friday night Beila would light the traditional candles and serve a festive meal. On Saturdays the family would walk to a nearby village or town and join other Jews for religious services. This was followed by visits with friends and relatives.

Religious observance, however, was free of fanaticism. The parents put no religious pressure on any of their children. Family loyalty and the exercise of authority were two important traditions. The father, as head of the family, was officially in charge; the mother's influence followed closely behind. Among the children, sex and age determined their importance. Because men were more valued, the sons had less trouble implementing their wishes than the daughters. Of the brothers, the oldest had to be listened to. And when the oldest son left, the one next in age would step into his position of authority.

Although firmly established, these traditions were sometimes threatened. For example, Beila was more forceful and energetic than her husband and did not always agree with him. But because the authority rested with David, she was careful not to contradict him—at least not openly. In the end, acceptance of this and other traditions prevented conflicts from bursting into the open.

As the only Jews in Stankiewicze, the Bielskis were different from their neighbors. At the same time, their healthy, attractive appearance, together with an independent spirit, defied their neighbors' prejudices about Jews. And so, within their village, the Bielskis stood out first because they were Jewish and second because they did not conform to the stereotyped images of Jews.

The Bielskis were also different from most Polish Jews. As peasants they belonged to a small minority. In prewar Poland, more than three-quarters of the Jews lived in urban settings, with less than ten percent engaged in agricultural pursuits.[9] In Belorussia, Jews were first mentioned in the fifteenth century. The majority were petty merchants and craftsmen. Only a small minority were farmers and peasants. Frequently poor, Jewish peasants had to combine farming with other occupations to make ends meet, usually different crafts. Regardless of occupation, most Jews in Belorussia had very modest incomes.[10]

As a majority, the Belorussians were involved in agricultural jobs with limited land holdings. Only a few were craftsmen or merchants, and they too, regardless of their occupation, were economically disadvantaged.[11]

Historically, this region had changed hands many times. During the second partition of Poland (1793), Belorussia fell under Russian rule. For over one hundred years a succession of czars would exploit these territories as they tried to make them an integral part of Russia.

After the 1918 rebirth of Poland, this region once more became part of Poland. From then on Poles who lived here were a privileged minority. Most identified with the so-called "intelligentsia," a label applied to school principals, government officials, the clergy, and the nobility.[12] Although Belorussia was not free of anti-Semitism, Jews found more acceptance here than in other parts of Poland.[13]

While growing up, the Bielski children were exposed to several cultures: Belorussian, German, Russian, Polish, and Jewish. They had many relatives and friends who lived in the surrounding villages and towns, including Lida and Nowogródek. The mill itself attracted a variety of customers, mostly Belorussians, Poles, a few Jews, and Russians. Although not intimate, relationships with these customers were friendly.

In Stankiewicze the Bielskis were on good terms with all their neighbors except one, a man whose land bordered the Bielski property. He had a reputation of being nasty and greedy, someone to be avoided at any cost. Each year this neighbor would mow a few inches of the Bielski's meadow and claim it as his.

With a touch of pride, Zus Bielski explains: "We grew up among the peasants, we knew them. We knew how to fight. My father was quiet, gentle. Mother was also a friendly person. . . . Father used to say that with fine people we have to be good and proper, but with bad people we have to be bad. . . . We would not let others push us around. We were never afraid, that was the kind of family we were."[14]

However, David Bielski's reaction to this particular neighbor was less belligerent. He felt that the amount of land lost was so small it was not worth bothering about and advised his sons to ignore the matter. Tuvia and his brothers disagreed—to them it was a question of principle.

One summer day when he was about fifteen, Tuvia, followed by his younger brothers, came to cut the grass. They were to start working at the edge of their property, near the neighbor's meadow. The man burst out of his hut, screaming at the top of his lungs that this was his land and they were trespassing. With a chuckle, Tuvia remembered: "He thought that if he screamed I would get scared. In his hand he held a scythe. He shouted, 'I will kill you!!' But I stood there laughing, which made him more furious. When he came closer I reached for my scythe and with it hit his. He lost his balance, landing on his back. When he was on the ground, I began to hit him with my hands. Four farmhands came to look. They stood there amused, laughing loudly at the man's misfortune.

"That day I gave him such a beating that we did not see him for two weeks. I was young then, a teenager, but I was big and strong. I was not afraid of the non-Jews and fought hard whenever one of them tried to wrong me."[15]

But neither courage nor hard work improved the Bielskis' economic circumstances. To the young it was clear that a move away might offer new opportunities and at the same time ease the family's financial burdens. In the 1920s the oldest son, Velvel, left to seek his fortune in America. Soon he was followed by another brother, Nathan. With the departure of the two older brothers Tuvia took over the position of the oldest.

Then Tuvia had to leave. In 1927 he was recruited into the Polish army. He became a sharpshooter and participated in army exercises with distinction, yet only advanced to the rank of corporal. The existing anti-Semitism offended Tuvia's sense of justice and he was quick to voice his objections.

When a Polish soldier whom Tuvia suspected of being slightly retarded called him a dirty Jew, he grabbed the man by the collar and ordered him to stop. The Pole continued by adding more anti-Semitic slogans and Tuvia reached for a knife. He hit the man over the face with the handle of the knife and let go only after the soldier, whose face was covered with blood, became silent.

Several soldiers witnessed the scene. At first they joined in the name calling. But they stopped abruptly when they saw Tuvia's handling of the man. This incident was followed by several hearings before different superior officers. Each time Tuvia defended himself saying that as a Polish citizen he could not tolerate anti-Semitic abuse. The Polish army did not expect this kind of behavior from a Jew and the case was dismissed.[16]

The army did, however, offer Tuvia some opportunities for personal growth. He was befriended by two upper-class Jewish youths from Nowogródek, the Lipner brothers. These young men helped polish Tuvia's manners and instilled in him a taste for better things.

Highly intelligent, he had a natural curiosity that was in part expressed in a facility for languages. He spoke Polish, Belorussian, and Russian each without a trace of an accent. He also knew German well. This fact in itself endeared him to many Christians.

Although he was not particularly religious, Tuvia identified strongly with Jewish traditions. He enjoyed studying the Bible and would commit entire passages to memory which, on appropriate occasions, he would quote.[17]

With his military service over, Tuvia returned to Stankiewicze. He became more aware of the surrounding poverty and tried to add to the family's income by renting another mill. When their income still proved inadequate, he continued to search for more profitable opportunities.

Eventually, Tuvia became convinced that a "proper" marriage would solve his problems. In 1929, at twenty-three, he went to Nowogródek to seek the services of a matchmaker, "shadhen," who specialized in a high-class clientele. He did not consult his parents, explaining that they were too busy and he did not want to bother them. With the help of this matchmaker, Tuvia chose Rifka, a member of a prominent Jewish family in the nearby small town Subotniki. As an owner of a general store and a big house, Rifka was rich. Without a trace of embarrassment, he explained that the store and the house rather than the bride were responsible for his selection.[18]

Tuvia and Rifka did not seem a suitable match. The bride was considered an old maid, eight to fifteen years older than her husband, depending on who tells the story. She was plain, with a stocky figure, and moved awkwardly. Although unattractive, Rifka's intelligence, education, and good upbringing compensated for her appearance.

In contrast to Rifka, Tuvia was very tall. Though broad shouldered, he was nevertheless slim and had an expressive face, a well-formed straight nose, and a clearly delineated mouth. His intelligent blue eyes had a dreamy look. Whenever Tuvia smiled, these somewhat distant eyes would give way to an animated expression of curiosity.

While men were drawn to Tuvia's independent spirit, vigor, and alertness, women found him irresistible. Rifka fell in love with her dashing suitor and gratefully took on the role of a doting and obedient wife. After the wedding Tuvia moved to her house.

Rifka continued to run the business, but tried hard to give the impression that her husband was in charge. Unobtrusively she offered him lessons in social skills and tried to share the benefits of her superior education. She was successful—Tuvia's manners, speech, and interests became more refined.

Although he knew how to work hard, he preferred the easy life and took advantage of his newly created leisure opportunities. Tuvia's position and personal charm allowed him to associate with the leading citizens of Subotniki, many of whom were Christian Poles and Belorussians. He would meet these new friends for cards, chess, and political discussions.[19]

As a teenager, Tuvia had for a short time belonged to one of the Zionist organizations that were active in the surrounding communities. Later, too independent to identify with a particular political party, he was nevertheless attracted to and curious about political conditions in the world. Unpressured by work, free of financial worries, he had time to pursue his interests. He read a lot and eventually acquired a reputation of someone worth talking to.

In the mid-1930s a teenager, Herzl Nachumowski, recalls taking a trip with his father to Subotniki, to meet Tuvia Bielski. Herzl remembers Rifka only as an "old" and unattractive woman. About her husband he says that "When I talked to him I realized that he knew Talmud, literature, and so much more. I had the feeling that the place was too small for him. All of it, the store, the petty life of a small town, was not enough. . . . On the way home I asked my father what is a man like this doing in such a primitive place? How did he end up here? I was told that he had married for money. This was not unusual. I knew that he was a person who had broad interests. . . . He lived in a fine house and no doubt made an adequate living. He dressed very well. He made a very good impression, but the place was not big enough for him."[20]

Hungry for knowledge, Tuvia subscribed to two papers: the Polish *Express Poranny* and the Yiddish *Moment.* In the late thirties, he came upon Albert Einstein's article, "Under the Shadow of Death," which made him conscious of the Nazi threat. Tuvia was convinced that Hitler would try to introduce his anti-Jewish policies to other places.

Although he was interested in the fate of the Jews and expected trouble from Germany, he had no intention of leaving. His life in Subotniki was comfortable and he was aware of his good fortune.

Despite the age and appearance difference, Rifka and Tuvia had a harmonious marriage, one dominated by Rifka's deep love and Tuvia's respect and approval.[21]

In Stankiewicze, after Tuvia's marriage, Asael stepped into the position of eldest son. He was not as tall as his older brother, but had a well-proportioned body. He had a straight nose and his deep-set blue eyes were framed by bushy eyebrows. A sensuous mouth added to his otherwise handsome appearance.

Born in 1908, Asael was a straightforward, honest man and a loyal and devoted friend. Unlike Tuvia, Asael was socially unpolished. He had difficulties expressing himself, had no intellectual curiosity, and no desire to expand his limited knowledge.

Asael was deeply attached to the countryside and its people. He had a strong love for the Jewish people, yet his Jewish identity in no way interfered with his many close Belorussian friendships.[22]

Shortly after Asael stepped into Tuvia's position as the eldest son, their father's health began to deteriorate. His mother and the rest of the family came to rely more and more on Asael. Among his new duties was the marriage arrangement of his sister Tajba, who was to wed an upper-class man. The class difference of the two families made this union unusual but Tajba's good looks were an advantage, while the bridegroom's personal limitations diminished the value of his superior social background.

Avremale, Tajba's bridegroom, was uneducated and may not have known how to read and write. A typical peasant, he had no social aspirations, few interests, and never ventured beyond his native village. Within his well-educated, well-traveled family he did not amount to much.[23]

Avremale had a sister, Chaja, who was a high school graduate, quite a distinction for this time and place. She had a reputation for being a socially conscious intellectual. Chaja tried to put her convictions into practice by educating the local peasants and easing their lot in any way she could. She heard that Asael was struggling with bookkeeping. Since he fit into her definition of the needy, Chaja offered to teach him accounting. Asael fell madly in love with his teacher. Painfully aware of the social and intellectual gap between them, he did not dare make his feelings known.

Unlike the Bielskis, Chaja's parents owned a large mill. Over the years they had also bought land from the Polish nobility who, hard pressed for funds, were willing to part with their property. Their holdings were substantial and required the hiring of farmhands.

Chaja knew that her father's employees earned two zlotys for a twelve-hour day. She also knew that these wages resulted in the laborers' abject poverty. To her it did not matter that these were the going rates and that her father treated his workers well. She also failed to consider that, without such jobs, the unemployed would be in worse shape. Chaja resented what she perceived to be the exploitation of the working class. She established contact with other communists and began to agitate against local employers, including her father.

Because pro-communist activities were then illegal in Poland, Chaja's involvement with communism could have had serious repercussions.[24] In fact, Chaja was formally charged with political subversion. Her parents succeeded in bribing the local authorities and instead of prison she was put under house arrest, in a village close to Stankiewicze. Confined to her house, she was glad to continue her lessons with Asael.

Chaja realized that her pupil was in love with her but did not give it a thought—she was preoccupied with much more important matters.

Before Chaja's brother was to marry Tajba, her mother asked for a dowry, as customary. As acting head of the Bielski family, Asael was the official negotiator. During these discussions he said that if he were to marry Chaja he would not ask for any money.[25] His remarks were treated as a joke.

Marriage between an upper-class woman and lower-class man was likely only if the woman was particularly ugly, an old maid, or both. Chaja did not fit any of these categories. She was slim, with large sparkling eyes eager to take in all the world. With boundless energy, often on the move, she had a vivacious personality and a ready, somewhat mischievous smile. Not surprisingly, Chaja had many suitors who were close to her social and intellectual level. A match between the uneducated, awkward peasant Asael and the attractive, sophisticated, politically active Chaja would have been out of the question.

Chaja and Asael continued to meet as teacher and pupil. Beyond this one joking allusion to his feelings, Asael never spoke about Chaja in a romantic way.

Already close to thirty, by local standards a ripe age, Asael should have been married. Yet no eligible young woman and no matchmaker got any sign of encouragement.

Asael's life revolved around his family. He was hard-working and had little time for leisure. More important, he avoided most social activities because he felt comfortable only with those he knew very well, usually family members or those he did not regard as his social superiors. Daughters of the Belorussian peasants fit into this last category. And because these women found Asael attractive, his contacts with them would lead to sexual liaisons. These affairs, however, did not translate into lasting attachments. Asael's deep love for his family and his equally deep involvement with Jewish traditions would not allow this to happen.

Four years younger than Asael, his brother Zus was more ready for marriage. Over six feet tall and immaculately groomed when not at work, Zus exuded an animal-like masculinity. He was easily aroused by women and they in turn welcomed his advances. Soon, a Jewish girl from Nowogródek, Cyrl Borowski, caught his eye. She would be responsible for Zus' move to Nowogródek. Later she became his wife.

With the departures of Velvel, Nathan, Tuvia, and Zus, the Bielski household had shrunk. Of the two daughters, Tajba moved to her husband's village Duża Izwa, and her younger sister Estelle, a more adventurous spirit, went to Vilna where she eventually became a bookkeeper.[26]

Except for the two sons who emigrated to America, for the rest of the Bielski children Stankiewicze acted like a magnet. For years it would be a place to come to, a place to celebrate, a place to mourn, and a place to hide.

2

The Russian Occupation

The Russians were coming.

The air was filled with enthusiastic shouts, laughter, and handclapping. Here and there a flower could be seen flying, as if searching for a safe landing. The crowd and the flowers were welcoming the Soviet soldiers who, whether walking or sitting on top of military vehicles, seemed surprised, perhaps even embarrassed by the tumultuous greetings. Some smiled, others waved tentatively. Most just looked ahead as they passed the lines of excited people.

The Russians had not expected such a reception. After all, they had come uninvited. To the beleaguered Polish government their presence was a hostile act closely identified with Germany's 1939 attack.

Stalin and Hitler were allies. Their cooperation led to the Soviet acquisition of more than half of Poland, which the USSR saw as compensation for its World War I losses.[1] For the Germans this transfer of territory represented peace on eastern borders and freedom to wage war against Western Europe. Both Germany and the USSR were winners.

The real losers were the Poles. To them it made no difference whether part of their country belonged to the Germans or to the Russians. Either occupation was a national defeat, a time of national mourning. Not surprisingly, Poles were conspicuously absent from this welcoming crowd. Not only did they refuse to greet the invading troops, but they resented those who did. They were particularly critical of the Jews, accusing them of disloyalty and communist collaboration.[2]

For Jews the arrival of the Red Army had a very different meaning.

The Russian troops shielded them from a German invasion and a threat to their lives. Jews were glad to see the Soviets because they would protect them from the German occupation.[3] Chaja, who was one of the happy greeters, concedes that "As far as the Poles were concerned, they saw that we were enthusiastic about the arrival of their enemy! They resented it. . . . They did not understand that we were happy only because we did not want the Germans. . . . The Jews were not ecstatic about the Russians. Nor were they pleased that Poland ceased to exist. If given a choice, most of us would have preferred Poland to Russia, but we were afraid of the Germans."[4]

Critical of what they interpreted as Jewish approval of the Russian occupation, the Poles did not mention that the majority of enthusiastic greeters were Belorussian. The Belorussians welcomed the Soviet takeover. They were neither Belorussian nationalists nor particularly proud of their Polish citizenship.

From 1793 to 1921 this part of Poland had been under Russian control. Unlike the Poles, who never reconciled themselves to the loss of their independence, the Belorussians were easy to govern. Consecutive czarist regimes showed their appreciation by singling out the Belorussians for special privileges. With World War I and the rebirth of Poland, this favorable treatment came to a halt.[5] At that point the Polish government showered the Poles with all kinds of social and economic advantages.

From the 1939 Russian takeover the Belorussians might have expected a return to a more favorable past. The political climate seemed ready. Stalin wanted to consolidate his grip over the newly acquired territories, and fearful of opposition from his arch enemies, the Poles, he courted the Belorussians.

But the Soviets did more than just cultivate prospective allies. With a special show of democratic procedure, they tried to politicize the newly acquired territories. They kept introducing new measures, whose real purpose was to curtail freedom, but which were presented as special benefits. Voting on all these issues was compulsory and the outcome was predictable. Over ninety percent "voted" in support of any and all changes.[6] Already in the fall of 1939, this part known as Western Belorussia was officially annexed by the USSR.[7]

Such political maneuvering was coupled with punitive measures, directed at actual or potential opponents. Elimination of the Polish leadership became a top priority. In their eagerness to be effective, the communists failed to differentiate between Polish leaders and non-leaders. Eventually, many Poles became targets of persecution, with deportations to Siberia the most common punishment. From communist-occupied Poland, including Belorussia, about a million and a half Polish citizens were forcibly transferred to the USSR.[8] According to some estimates, about fifty percent were Poles. The rest were Belorussians, Ukrainians, and Jews.[9]

Soviet persecution of Jews, although more selective, was also substantial. From the start, the Russians outlawed Zionism as a political movement. This step was followed by arrests not only of Zionists but also of non-Zionist leaders. Jewish community centers, regardless of their political coloring, were shut.

Soon Jewish businessmen were added to the "politically undesirable" category. No matter how insignificant a Jewish business establishment was, its owner automatically became a "bourgeois capitalist."[10]

Inevitably some early supporters of the Russian takeover became disenchanted. For example, Chaja experienced a conflict between her identification as a communist and as a Jew. In the end she realized that communism was not for her.[11]

Exposed to many rumors and few facts, some Jewish businessmen took fate into their hands and moved to different communities, where with changed identities they continued to live undisturbed.

As a "wealthy" store owner, Tuvia Bielski was automatically placed into the category of "bourgeois capitalists." Before the authorities got to him he ran away to Lida. There with the help of a Belorussian friend he became an executive in an enterprise that collected and transported scrap metal to foundries.

Before Tuvia left Subotniki he urged his wife, Rifka, to join him. She refused. She would not part with her house and store and continued to stay even after the Russians confiscated most of the property. Somehow, Rifka felt that her watchful presence would keep her possessions intact.

In Lida Tuvia's life was soon tied to Alter Titkin, a prosperous businessman who also had eluded Soviet authorities. Before the war, Titkin and his wife had lived in the small town Mołodeczno. A close-knit, caring family, like most traditional Jews, they valued education and sent their son and daughter to the best private schools. Lilka, the younger child, was particularly pampered with special governesses and servants.

In 1938 the family's carefree life came to an abrupt halt when Mrs. Titkin died of a heart attack. Stunned by this loss, the children rallied around their father who, stricken with pain, vowed never to remarry. When grief began to affect Titkin's health, a doctor advised him to take a prolonged vacation from home and everything that reminded him of his beloved wife.

Titkin went to Otwock, an elegant but quiet resort, not far from Warsaw. Recovery took seven months. On the way home, he met Regina Meitis, a divorcee in her mid-forties. Well dressed, attractive, and educated, she lived in the city of Vilna[12] where, as a seamstress, she supported herself and her teenage son, Grisha.

Titkin, at once charmed and intimidated by Regina's cosmopolitan manners, was flattered that she paid any attention to him. Finding her irresistible, he conveniently forgot his earlier pledge never to remarry. Nor did he pay attention to his children's pleas to weigh more carefully such a serious step. In less than two months Regina Meitis became Mrs. Titkin.

Regina was strong-willed and cold. Protective of her own family, a teenage son, a mother, and two sisters, she apparently had no affection left for anyone else. Besides, she knew that Titkin's daughter had disapproved of the marriage. Perhaps she was also envious of the girl's appearance. Lilka had a tall, willowy figure, thick chestnut hair, and large wide open, hazel eyes that looked in wonderment at the world around her.

Lilka and Regina took an immediate dislike to each other. It proved to be an unequal battle, with Regina mistreating the young girl. No one, including Lilka's father, dared interfere and Regina was rarely challenged. Quite naturally, she became the wicked stepmother and transformed Lilka into a veritable Cinderella. Some are convinced that Regina physically abused the girl.[13]

Lacking self-confidence, Lilka grew into a retiring, timid teenager. The family's move to Lida robbed her of the few friends she had. In this new place only one of the Jewish school girls paid any attention to her. Soon the two became inseparable; for Lilka her friend's home became a home away from home.

From this new friend Lilka heard that one of the rooms in their apartment was rented by an exceptionally handsome man. The girl promised to introduce this tenant to Lilka. But he was away on business, and for a long time the two friends were faced with a locked door. Both knew very little about this absent tenant. As they waited, they engaged their imaginations, weaving fantastic stories about this stranger's life. The longer he stayed away, the more mysterious and attractive he seemed.

Finally, when the door did open, Lilka met Tuvia Bielski. It did not matter that he was about twice her age—Lilka was overwhelmed. To her his looks and manners seemed a perfect mixture of intelligence and goodness. The moment she met Tuvia, this fifteen-year-old girl fell madly in love with him.[14]

When Lilka was introduced to Tuvia, she did not realize that he was in love with Sonia, her stepmother's younger sister. After the Titkins moved to Lida, Regina invited her two sisters and mother to join her. Titkin rented an apartment for them and paid for their living expenses.

Originally from Russia, Regina was born into a prosperous, well-educated family. Its women were destined for a life of leisure. During the Russian Revolution her family ran away, and shortly after they reached Vilna their father died, leaving his wife and three daughters almost destitute. In the strange city of Vilna they had to find a source of income. Regina and her sisters learned how to sew and in no time acquired an upper-class clientele. In those days, however, educated and intelligent women would not work as seamstresses since it was not a particularly valued occupation. For this reason alone, as soon as Regina became Mrs. Titkin, she stopped working. Similarly, when the rest of her family joined her in Lida, her two sisters also gave up working.

Considerably younger than Regina, Sonia was tall and slim. She moved gracefully, sliding rather than walking. She had a light complexion with a see-through quality, alert blue eyes, a small nose, thin but well-curved lips. Sonia's face was framed by light blond hair. Some remember her as "beautiful, noble looking . . . a woman of the world."[15]

Although uniformly praised for her beauty, intelligence, and style, Sonia's personality was another matter. When describing her character some felt uneasy, hesitant. Their comments follow a similar path. "I am not sure if she was nice . . . she had courage, she had guts, she was talented. She was a snob . . . people did not like her. She made others feel inferior. She did this on purpose."[16] "Very beautiful, not warm. She always kept a distance from people. But there was something about her, something classy."[17] Others emphasize that she was particularly good at cultivating those who were useful to her.[18]

In Lida, which to her had to be a provincial town, Sonia felt at ease in the company of local dignitaries. She knew how to attract men. Although in her thirties, a ripe age for a single woman, she seemed in no rush to transfer her love affairs into marriage. Tuvia was an exception. Lilka felt that Sonia pushed herself on Tuvia, that she would not let go of him.[19]

Whatever the reasons, Sonia and Tuvia's love affair became serious. Soon Tuvia wrote a letter to his wife in Subotniki. He explained that because for ten years Rifka did not become pregnant and because he wanted children he was asking for a divorce. No other woman was mentioned. He even said that he loved her, and was unhappy about his request. To Rifka's cousin who read the letter it was clear that Tuvia was concerned about his wife's feelings, trying hard not to hurt her.

According to Jewish orthodox law a man is entitled to a divorce, particularly from a childless wife. Rifka had no choice. Shaken, she went along with her husband's request. As the formalities were being finalized, she behaved in a dignified way. There were no accusations, no public tears.

If Rifka knew about her husband's mistress, she never mentioned it. As always, so in this last stage of their marriage, she remained an obedient wife.[20]

For Tuvia an escape from Soviet authorities had led to a breakup of a marriage and the beginning of a new one.[21]

When labeling businessmen like Tuvia Bielski and Alter Titkin as undesirable, the communists were objecting to their occupation, not their Jewishness. Similarly, when shutting down Jewish schools and different cultural institutions and when deporting Jews to Siberia, the Russians claimed that these moves were dictated by political expediency. But not all Jews were threatened or punished. Many remained detached from and unaffected by the political upheavals, while some even benefited from the Russian occupation.

The communists claimed that they wanted to improve the lives of the underprivileged, no matter who they were. In reality, most beneficiaries were Belorussian, some were Jewish, and practically none were Polish. The Poles were excluded in part because of the traditional Russian-Polish hostility and in part because Poles had a relatively high social and economic position in Western Belorussia.[22]

Of the different communist programs, one encouraged the participation of the poor in governmental bureaucracies and in state-owned industries. In the Bielski family, all except Tuvia were entitled to such special privileges. In Nowogródek, Zus Bielski received a well-paying administrative job. In Stankiewicze, as a poor peasant, Asael was offered a seat on the newly established village council. In his case, as in so many others, a history of poverty rather than special talent was responsible for this honor.

Other communist programs helped the underprivileged by offering them access to free education. In prewar Poland, high school and trade school attendance was costly, automatically excluding the needy. Jews with their traditionally high esteem for education were particularly impressed with the Soviet-sponsored learning opportunities. For the poor Jews in particular the offer of free education transformed dreams into reality.

Because of Poland's prewar anti-Jewish discrimination, in addition to offering concrete benefits, these Soviet-sponsored occupational and educational opportunities did away with civil rights violations.[23]

Such communist innovations compared very favorably with what Jews had to endure in Nazi-occupied Poland. But, as Germany's allies, the Soviets were carefully avoiding mention of Nazi atrocities. As a result, only some in Russian-occupied Poland knew what was happening in the rest of Poland. In Western Belorussia people received only sporadic and selective news. A certain amount of information about German terror came via Jewish refugees from Poland who, eager to elude danger, searched for escape routes.

With very limited options, Jewish refugees continued to pour into Russian-occupied Poland, hoping that they could then move to a more permanent haven. By issuing visas to remote places, a neutral country could serve as a stepping-stone for such a move. An unusual set of circumstances had made Lithuania with its ancient capital, Vilna, a neutral country. Before World War II, Vilna had been a Polish city, legally and culturally. Lithuanians, however, felt that Vilna had been taken from them illegally, that it was rightfully theirs. Their national ambition was to make Vilna their capital.[24]

When in accordance with the September 1939 Soviet-German nonaggression pact the Russians occupied Vilna, they imposed on the then neutral government of Lithuania a treaty that gave the Russians the right to station troops in that country. In return for this "privilege," the Soviets gave Vilna and parts of the surrounding territories to the Lithuanians.[25]

To Polish Jews, Lithuania's and Vilna's neutrality promised an escape route to safety.[26] At first most contacts between Russia and Lithuania were friendly and the official law that forbade movement between them was only sporadically enforced. For Jews searching for refuge such border conditions offered an opportunity first to transfer to neutral Lithuania and from there to a more permanent and safer place.[27]

Hopes were shattered when in the summer of 1940, after an artificially created incident, the USSR annexed Lithuania. No longer neutral nor independent, Lithuania stopped being a Jewish haven.[28]

Unable to move on, the majority of Jewish refugees remained in Russian-controlled Poland. Moshe Bairach was one of those refugees. Shortly after the Nazi takeover of Poland he and six Jewish youths ran toward Russia. On the border they were caught by German guards who beat them mercilessly, but this did not prevent them from reaching the Russian-occupied town of Białystok.

After many unsuccessful attempts to find a neutral haven Moshe Bairach and his friends sat at the Lida train station. Dejected, with no place to go, they waited.

A Jewish man approached them. "Why are you so unhappy? You are young, the world is big. What is the matter?"

"The world is big but it has no place for us," Bairach replied. The others nodded in agreement.

"Don't lose your spirits. There are still Jews around to help," the stranger said.

"We need a place to stay, food to eat."

"It will be done," the man said.

He brought them to the nearby town of Żołudek and placed each of them with a different Jewish family. The Jews of Żołudek were eager to help. They collected money and used it for bribing local officials who legalized these young men's stay in Żołudek. In the end, each found not only a warm home but also a job.[29] These refugees would then supply the Jews of Żołudek with information about conditions in Poland.

More news would come via other refugees from German-occupied Poland who continued to pour into different communities. Dora, a young girl from Warsaw, moved in with the Bielski family. She and Avremale, one of the Bielski sons, fell in love and married.

Dora was eager to tell about Nazi atrocities. In her adopted family she found willing listeners who accepted all she told them.

Others, however, were less receptive.

"Why," they argued, "would the Germans want to murder all the Jews for no reason?" Occasional, selective killings perhaps. Total annihilation made no sense.[30]

Whether they listened or not, believed or not, understood or not, soon all Jews were confronted with the reality of a German invasion. On June 22, 1941, over the radio, Soviet foreign minister Molotov told the Russian people that Germany had attacked their country.[31] The German onslaught was sudden and massive; it came from the air and the ground. German troops were streaming in simultaneously from several directions and Soviet authorities and civilians were taken by surprise. The Red Army issued contradictory orders or no orders at all. This resulted in a chaotic situation that prevented combat.

On the second day of the invasion Tuvia Bielski arrived promptly in the office of the military commander in charge of mobilization. He recalls:

"I was put in charge of writing notices to others who had to be mobilized. We worked until lunchtime. Suddenly about fifty planes flew over the town dropping incendiary bombs. In a very few minutes the entire place was on fire. The commander called us in, ordered us to leave the burning town and regroup in a forest about five kilometers from there. We were to continue working. We carried out his command but soon after we began our job in the forest another wave of planes flew over the area and set the woods on fire. The major called us in and said: 'Friends, you are on your own!' "[32]

With this military service behind him, Tuvia went home, only to find his house on fire. Together with his wife Sonia, he salvaged some belongings. The Germans kept up the pressure and continued to bomb the town. To escape death, people were leaving for the countryside. On June 24, Tuvia and his wife went to the Bielski home in Stankiewicze.[33]

Tuvia's two brothers, Asael and Zus, called up by their army units, were also caught in a web of confusing directives.

Zus, a sergeant, walked with five of his men until they came close to the city of Białystok. There, threatened by bands of unattached soldiers, they turned back. They looked in vain for someone to direct their next moves. In the end, frustrated, they decided to separate and each went home.

Hungry, exhausted, with swollen feet, Zus reached Nowogródek where he was reunited with his wife. She was eight months pregnant. The city was bombed; the atmosphere tense. Afraid of more destruction, the couple went to the Bielski home in Stankiewicze which seemed a relatively safe haven.

Asael too, without fighting the enemy, roamed aimlessly around the countryside and then returned to Stankiewicze.

While some Jewish soldiers unsuccessfully tried to fight, Jewish civilians were bewildered as they watched the disorderly retreat of the troops. To them this looked like total capitulation and the end of the USSR. With their initial shock behind them, the Jews realized that time was running out. Although they might not have believed that the Germans intended to annihilate all Jews, they knew that under the German occupation their lives would be in danger. Many felt it would be safer to run away with the retreating Red Army.

Of the Bielski family, Estelle and Joshua did just that; they spent the rest of the war in the USSR. The majority, however, did not make it. Of those who wanted to leave some did not because others in their family refused to join them.

Riva Reich, from a well-to-do family in Stołpce, urged her relatives to follow the retreating troops. But her mother who had been a fugitive during World War I refused to leave. Riva pleaded with her husband to run away with their baby. But her husband would not listen—the entire family continued to stay in Stołpce.[34]

But even a willingness to leave was not enough. Most Jewish efforts to run away were frustrated. Some who wanted to leave returned because the roads were being bombed. Others had reached places that were already occupied by the Germans and turned back.[35]

On the Russian border the Red Army stopped most people from entering.[36] The Russians did not want to admit Jewish fugitives. Hersh Smolar, a long-time communist and journalist who spent seven years in Polish prisons because of political activities, credits this policy to Stalin's distrust of foreigners. He thinks that "the Russians were murderers. They asked the Jews to dig trenches and then they left them there. I was left this way too. They simply ran away without us. I wrote about it to Russia, but they suppressed this kind of information."[37]

It seems that even in the disorderly conditions of the retreat the Russians favored the Belorussians. During the Soviet occupation many Belorussians enlisted into the Soviet Army while many others cooperated with communist authorities in other ways. When the Russian-German war started, most Belorussians who wanted to reach the USSR were given the opportunity to do so.

In the confusion of the massive German attack many Russian soldiers gave up fighting. No one knows how many had no choice and submitted to the enemy and how many deserted because they did not want to fight. For political reasons, successive Soviet governments have refused to disclose what must have been massive desertions.

Some of these Soviet deserters succeeded in disposing of their military clothes and in settling down as civilians. Others who originally had come from Western Belorussia went back to their homes, where they resumed their former lives.[38]

As the Red Army continued to retreat or break up, as the Jews continued to look for solutions, the Germans kept up the pressure. In some places German parachutists preceded the Russian rearguard. Local governments fell apart in the disorder of runaways, bombing, fires, and total breakdown of communication. In this sea of confusion, in some communities, the haphazard Soviet retreat created a political vacuum. In several places, even before the Germans arrived, Belorussians and Poles began to abuse and rob Jews.[39]

Anti-Semitic outbursts plus continuous bombings of cities made the countryside seem safer and more attractive. In addition to Tuvia, Zus, their wives, and Asael, other relatives and friends came to Stankiewicze to the Bielski home. Whoever reached it was welcome. The modest hut was terribly overcrowded. Instead of two, each bed had to accommodate three or four. The mill and barn were converted into temporary bedrooms. The young particularly approved of these unconventional sleeping arrangements; enjoying each other's company, they preferred not to think about the future.

One of these arrivals was a young cousin who remembers how Beila Bielski warned her guests about the possibility of attacks by local anti-Semites. She recalls her aunt's words. "We are just as strong as the Christians. We should not let them harm us. . . . They will come and try to rob us. We ought to defend ourselves."[40]

And so Stankiewicze, an overcrowded Jewish enclave, was ready to fight. During the night men took turns as guards. A few nights passed and no one came.

On July 1 a German army unit arrived. The soldiers camped on the Bielski land and established a headquarters in the backyard. These first Germans were quite friendly. They were followed by another unit.

The officer of this second group issued an order: "All Jews who are not permanent inhabitants of Stankiewicze must return to their homes. They must leave within fifteen minutes. Any Jew found after that time will be shot."

This order left no doubt. It had to be obeyed.

3

The German Invasion

On a day in July 1941, a German officer stood at the door smiling. "Will you put me up for one night?" Chaja knew enough German to understand. But what could she reply? Afraid to say yes, afraid to say no, she nodded in agreement, but made no move.

The man brushed past her. She followed, listening to him. "I like your house, it looks better than the others. My men are settled for the night. I am glad that you don't mind." Then, as an afterthought, he added, "Don't be afraid of me. I will not harm you."

The house belonged to Chaja's sister, Rachel Boldo. Like so many others, the Boldos left in a hurry for the countryside. They joined their relatives in the village Duża Izwa. Independent and ready to help, Chaja volunteered to watch the house. A girlfriend whose home was hit by a bomb moved in with her.

They had expected the Germans. After all, the Germans had so very recently forced the Red Army out of this part of Poland. And yet, when faced with a real live German, the two friends were apprehensive. The stranger must have sensed this. Reassuringly, he continued to talk. "I will prepare the food. We will eat together. You have nothing to worry about right now. All I want is to eat and rest." His words and invitation elicited only a tentative, whispered "Thank you."

The officer surveyed the kitchen. Satisfied with the lighted stove, he added wood to the fire and reached for pots and other utensils. He opened a few cans and emptied their contents into the pots. While the food was warming, he invited the two friends to join him at the table. He was in command, as if the house belonged to him.

The food looked inviting and he served generous portions. Although hungry, Chaja and her friend had a hard time swallowing. The man responsible for the tension was obviously enjoying his meal. He ate with enthusiasm and as he did he looked closely at the women before him, shook his head, and said, "What a shame, such pretty girls! A terrible shame!" Then again, to no one in particular, he repeated, "What a shame, too bad, a terrible shame." This one-way conversation came to an end only when the German excused himself and went to bed.

Left alone, the two friends sat brooding. Nothing was said. A quiet goodnight ended their silent conversation.

In the morning the man got up first. When the three met he was ready to leave. He thanked them for their hospitality and continued in a familiar vein. "It is a shame and you are so young and pretty. I feel sorry for you. You should know that difficult times are ahead of you, very difficult times."

When he was gone Chaja found a few cans of food and an unsigned note: "Be very, very careful." She welcomed the food, but those written words made her uneasy.

Why, Chaja wondered, was he so concerned about her? To be sure, this was war and war meant hardships. Besides, Chaja had heard about German mistreatment of Jews. Still, this did not necessarily mean that personally she had something to fear. She was confident she could deal with the Germans, yet this officer's strange warnings left her with a lingering anxiety.

Then through the window Chaja saw two Belorussian policemen approach the house. "How quickly they change masters!" she wondered. Without much thought, she slipped through the back door and out of their reach.

Inside the house the policemen roughly ordered Chaja's friend to come with them. After that day Chaja did not go out.

In the evening her friend returned. She looked harassed, disheveled, and distracted. All day long she was forced to clean offices that had been vacated by the Soviets. German and Belorussian policemen had watched her; any attempt to rest or slow down met with beatings.

After she finished working they had ordered her to stand on a table and sing the Jewish anthem, "Hatikva." When she refused, they grabbed the broom out of her hands and hit her with it. The broom landed on her face, head, and the rest of her body. She cried, which only made them intensify the blows. Convinced that they would kill her, she gave in and climbed on the table. Still hitting her, her tormentors now demanded that she dance and sing. She did.

Chaja kept silent. Irritated, her friend asked, "Why don't you say something?"

In a measured voice Chaja replied, "To me they will not do that. I will not allow it."

"How?" her friend wanted to know.

"I will run away. I will not serve the Germans."

From then on, whenever the police came to collect Jews, Chaja would hide.[1]

Quite independently, Riva Reich, another young upper-class woman, also refused to submit. A mother of an infant daughter, married only for a year, Riva soon faced a tragedy.

A few days after the Germans came to Stołpce they arrested a group of prominent citizens, Riva's husband among them. This happened without provocation or warning. When the police came to make the arrest the couple was still in bed. His mother admitted the policemen and brought them to her son. They took him in his pajamas.

In less than an hour Riva was at the police station with her husband's clothes. The policeman on duty would not accept the package, explaining that the prisoner was well provided for.

Among the arrested were a few Christians—an error. They were let go and they told Riva that her husband was tortured and in bad shape.

To free him Riva began to make the rounds among all kinds of officials. Some she contacted assured her that her husband was well. Others admitted that he was ill. Still others promised to arrange his release. She paid money for the news, money for the promises, but nothing happened. Among the different stories was also the truth. With the rest of the arrested Jews, he was executed in a nearby forest, only a few days after he was taken from their home. Eventually, Riva accepted this last version. When she did, she became determined not to give in and, keeping her baby with her, concentrated on avoiding the Germans.[2] This eventually made her join the defiant Bielskis.

At first ordered out of Stankiewicze, Tuvia and Zus Bielski returned to their respective homes in Lida and Nowogródek. After Tuvia settled his wife with the Titkins he went to Nowogródek. Tuvia wanted to compare the two towns, Lida and Nowogródek.

In Nowogródek Zus and Tuvia barely had time to greet each other when they were stopped by a native policeman. Wearing an armband labeled "Police," the man pushed them toward a group of about fifty Jewish men and, helped by other native policemen and Germans, led them away. They were soon joined and surrounded by a group of civilians, mainly Poles.

To get the attention of the Germans this crowd shouted in broken German and Polish: "Every Jew is a communist. We are grateful that you came to free us from the Jewish yoke." One of the German officers answered in Polish: "Under the German rule the only Jew you will see will be on the movie screen. We will settle our accounts with them." Satisfied, these civilians followed until they reached an office building.

Here the officer who had spoken to the crowd asked them to line up at the building's entrance. He supplied each with sticks or leather thongs and instructed them to hit every Jew who came out of the gate. Jewish men were ordered to carry furniture and move it to the next building. As they did they were beaten by the civilians.

Tuvia and Zus towered over the other workers. Their height together with their broad shoulders suggested strength, and the civilian tormentors avoided hitting them and concentrated on the remaining Jews.

When the furniture was transferred, the prisoners were told to arrange themselves in rows. This was in preparation for a new job: another house and more furniture to be moved.

And though the day was coming to an end, their work seemed to be continuing forever. Tuvia whispered to Zus, "I am going to run away." "Me too" came the answer. Under the cover of darkness the brothers disappeared. Before they parted each vowed never to let themselves be caught by the enemy. This decision contained no special plans, no specific steps, but it was firm.[3]

By the summer of 1941 when the Germans invaded Russian-occupied Poland, they were experienced in Jewish persecution. Not satisfied with the amount of destruction, they wanted to accelerate the process of Jewish annihilation, the so-called "Final Solution." For that purpose the Germans introduced an important innovation: separate killing units, known as the "Einsatzgruppen." These units consisted of 3000 SS men specially trained for murdering Jews and other "enemies" of the Reich. The Einsatzgruppen would follow on the heels of advancing armies and trap large Jewish population centers before the prospective victims had a chance to learn of their fate. In the mass executions that followed, the Einsatzgruppen were assisted by the German army and local collaborators.[4]

Pockets of Jewish populations, particularly in urban centers, were overlooked by these killers. These omissions were only temporary; they were followed by selective murders of the Jewish elite. Later on these policies of annihilation came to include all Jews.[5]

Though exposed to the most threatening situations, some Jews were determined not to submit to the German terror. The Bielski family fit into this category.

On one of his visits to Nowogródek in July 1941 Zus recalls seeing a public execution. He first became aware of a commotion, with people running in different directions. He hid behind a wall and saw a group of about fifty Jews, surrounded by German soldiers and policemen. "They stood in five rows, frightened, crushed, and helpless. A German officer arrived in a small car. He questioned the soldiers. They seemed to answer in the affirmative. I could not hear them. The German drew his gun and shot into the air. Immediately thereafter there was a burst of machine-gun fire and some Jews fell. There followed another shot into the air, another burst of machine-gun fire, then more Jews fell, and so on. . . . I noticed Slutzki and his son in the last row, they were distant relatives of my wife. . . . The son turned to his father, horrified, 'Father, they are killing us!' He no sooner uttered the sentence than both were dead."[6]

That day in full view of the community the Germans shot fifty men, most of them prominent leaders and professionals.[7]

Chaja only heard about these killings. While in Nowogródek she continued to play cat and mouse with the authorities. Whenever they came looking for people, she would disappear. Soon, afraid that her luck might run out, she decided to join her family in the village Duża Izwa.

In villages like Stankiewicze and Duża Izwa, where the Bielskis, Chaja, and other Jews felt safer than in towns, disturbing news and rumors began to arrive. Particularly shocking was the fate of the Jews in Horodyszcze, a nearby small town. On October 21, 1941, the town's 3000 Jews were ordered to assemble in the marketplace. Out of the 3000 the Germans selected fifty craftsmen. The rest were subdivided into two groups: first, able-bodied men and women; second, the children, the old, the weak, and the disabled. This second group was ordered to climb into trucks. The first group had to follow on foot.

Together they were all taken to the outskirts of the town to a large pit where they were put in smaller groups. Each was then told to move to the edge of the pit. On reaching the edge, each group was gunned down.

The fifty craftsmen were spared. They heard about the killings from Belorussian peasants whose job was to cover the bodies with soil. Of these fifty survivors seven were sent to Baranowicze. Yehuda Szymszonowicz was one of the seven; during the massacre he lost thirty-six relatives. In Baranowicze, Yehuda contacted members of the Judenrat and told them what happened. They shrugged in disbelief and advised him not to spread such horrible rumors.[8]

A Judenrat was a Jewish Community Council appointed by the Germans that often included prewar community leaders. The Germans assured their appointees that as Judenrat officials they would mediate between the occupational authorities and their Jewish constituents. In reality the Germans demanded total obedience to orders from these officials.

But the mass murder in Horodyszcze was not an isolated event. Rumors continued to circulate about groups of Jews who were collected from hamlets, villages, and small towns only to be executed not far from their communities.[9]

Added to these grim stories was talk about the establishment of ghettos. According to one version, these closed Jewish quarters would contain only the youths; the old and the children would be killed. Others argued more optimistically that the work of the young would protect the rest. Most Jews equated contributions to the German economy with the possibility of staying alive. There were disagreements as to whether permission to live would extend to a family's less productive members. The Germans supported all kinds of optimistic views, emphasizing the need for devoted laborers.

In Duża Izwa Chaja and her relatives, including her parents who were close to seventy, refused to submit to the Germans. But beyond a general desire to act they had no specific plans.

The change of season limited their options. As 1941 was coming to an end it took away what little warmth there was. The forest and the countryside, hospitable during the summer, became cold and uninviting. And so, apprehensive about the future, Chaja and her relatives waited.

Known for their independent spirits, members of the Bielski family soon began to defy the authorities more openly. A Belorussian, a Stankiewicze neighbor, Kushel, had joined the newly established police in Nowogródek. To curry favor with his new masters Kushel accused Asael, Tuvia, and Zus of communist collaboration. At this point the three brothers were often in their home village and a warrant was issued for their arrest.

Ironically, another Belorussian policeman warned the Bielskis about the impending danger. Without hesitation, the three disappeared in different directions for safety.

Because they were born and raised in this part of the country the Bielskis were familiar with the surrounding forests, villages, and roads and knew many local people. Some were close friends, others were friendly or hostile acquaintances.

Political upheavals would blur the distinction between friend and foe; therefore, as a special precaution, each brother pretended to be Christian. Because their changed identities would be effective only for those who did not know them, they tried to alter their appearance. Tuvia, for example, grew a bushy moustache.[10]

Familiarity with the area and the people produced more advantages than disadvantages. It often meant knowing which farm needed laborers. A shortage of farmhands had been created when many Belorussian men left with the Red Army. Tuvia and Zus benefited from this situation by offering their services in exchange for food and lodging.

Asael went to the village Dokudowo where he worked as a mason. After a few weeks informers caught up with him and he ran away. He went to the forest, close to his native village, and would visit his family only at night.

Occasionally Zus too would stay for a few days with his wife in Nowogródek. During one of those visits, the police came to arrest him. Warned of the danger, Zus slipped out of town at night. On the outskirts a ten-year-old Polish boy accosted him, saying, "You dirty Jew, how dare you use the sidewalk, you know you are forbidden!" Zus slapped him hard and escaped through the fields toward Stankiewicze.[11]

That same evening Zus cautiously approached the Bielski hut. Through a window he saw the room in total disarray, with many objects strewn all over the floor and furniture. A trusted neighbor told him that the police had just left his parents' home. They were looking for the three brothers, Asael, Tuvia, and Zus, and when they did not find them, they arrested their father instead. The rest of the family had gone in search of help.

Next day, without any explanation, David Bielski was freed.

This capricious gesture was followed by the arrest of the younger brothers, Avremale and Jacov. No reason was offered and the two were sent to prison in Nowogródek.

Beila Bielski appealed to the local priest and influential Belorussian farmers and asked them to sign a statement testifying that Avremale and Jacov were decent men who never did anything wrong. She went to the Judenrat in Nowogródek and begged them to intercede on behalf of her sons. They promised to do so, but nothing happened. She delivered food packages to the prison without ever finding out if Avremale and Jacov received them.

Through friends of friends, she was introduced to the prison administrator. In the past this man had freed Jewish prisoners. For him, the release of Jews was a business. Before Beila met him, she collected a few valuables she had and offered them. The man made it clear this was not enough. He wanted one hundred rubles in gold, an astronomical sum for Beila Bielski.

She sold everything she could. She borrowed money from relatives and friends, but was still short. Negotiations with the prison official were deadlocked. Desperately, unsuccessfully, she tried to raise more money.

Then she heard that her sons were to be transferred to another prison. Probably the two brothers saw in this move an opportunity for an escape and in October 1941 Avremale and Jacov were shot trying to run away.[12]

In Russian-occupied Poland, including Western Belorussia, Jews who evaded the initial killing waves were exposed to more gradual persecutions. In part these resembled the steps used in Western Poland. First, Jews were identified. Then their property was confiscated, and they were barred from gainful employment. Next, Jews were separated from the rest of the population and forced into ghettos. There extreme overcrowding and poverty led to the spread of disease, epidemics, starvation, and death. Each of these stages was in turn accompanied by selective killings.

Although the Jews in Western Belorussia had heard about mass murder and some even witnessed executions, they had no way of knowing what a specific German raid or "Aktion" would bring. An Aktion could lead to executions, to compulsory temporary work, or transfer to another community. Even when a raid did result in murder, one could not be sure how extensive it would be.

This uncertainty and instability applied to all aspects of Jewish life, including ghetto transfers. A move to a ghetto had shattering effects, sometimes ending with death.

Yitskhok Rudashevsky, a young teenager, captured the tragedy of a ghetto move. "The street streamed with Jews carrying bundles. . . . People fall, bundles scatter. Before me a woman bends under her bundle. From the bundle a thin string of rice keeps pouring over the street. I walk burdened and irritated. The Lithuanians (Nazi collaborators) drive us on, do not let us rest. . . . Here is the ghetto gate. I feel that I have been robbed, my freedom is being robbed from me, my home, and the familiar . . . streets I love so much . . . I find my parents and here we are in the ghetto house. . . . Besides the four of us there are eleven persons in the room. The room is a dirty and stuffy one. . . . The first ghetto night we lie three together on two doors. I do not sleep. In my ears resounds the lamentation of this day. I hear the restless breathing of people with whom I have been suddenly thrown together, people who just like me have suddenly been uprooted from their homes."[13]

For Tamara Rabinowicz the forced move to the ghetto resulted in an active defiance of Nazi orders. It has been engraved in Tamara's memory. "We packed four valises, and two knapsacks. We took a bundle of pillows and a cover and tied it to the valises and dragged it behind us. The bundle fell apart. I bent down to tie it together. Suddenly I felt that my back was burning, as if it was on fire. I raised my head and saw a German standing in front of me with a whip at the end of which were iron balls. He hit my back with it. That is why my back was burning. I did not understand what it was. I looked at him and he looked at me with cold laughing eyes and said in German: 'You dirty Jew, go to the ghetto.' I told him: 'I am a Jewess, but not dirty because I wash.' For an answer he gave me three whips over my face. Blood started pouring out of my mouth. My husband gave me a handkerchief and told me quietly to leave the bundle on the street.

"As I wiped my face I noticed in front of me a young woman with a baby in her arms. I saw a German come close to her with outstretched hands and he took the baby from her. I thought to myself, 'by chance I met a bad German, this one is trying to help this woman.' She had no packages, nothing, only the child. A beautiful baby, blue-eyed, laughing. I thought what a good German that wants to help. Before I finished my thought I saw that he took the baby, threw it into the air and caught it with the thin part of his weapon. Blood poured out of the little body. The mother jumped at the man and pushed her nails into his eyes. He had time to take out his gun. He shot her. She remained on the pavement on top of her baby. The German stood there laughing."[14]

Right then Tamara and her husband resolved not to work for the Germans or stay in the ghetto. To them, the town Stołpce, because of its small size, seemed like a haven. German brutality, they felt, might be limited to big centers.

While they put into effect their decision not to work by hiding, their escape from the ghetto had to wait. Only after three months of planning did they succeed in running away. On the outside they were faced with a journey filled with danger. Most of it they made on foot.

Penniless and exhausted, they reached the Belorussian town of Lida. In Lida they were protected by Lichtman, the head of the Judenrat. Many more refugees benefited from Lichtman's help; he would legitimate their stay by issuing special papers entitling them to jobs and food. Eventually, the Germans learned about Lichtman's "crimes." They executed him and appointed another Judenrat head.[15]

Most people had a hard time grasping that the Nazi anti-Jewish measures were part of an overall plan. Jews and those who were ready to help them often thought that some communities would be safer than others and that the Germans might overlook certain places altogether.

Anton Schmidt, a German sergeant stationed in Vilna, who risked his life to save Jews, subscribed to this last view. Schmidt was convinced that small communities would shield Jews from German atrocities. At considerable risk he would arrange illegal Jewish transfers to small Belorussian towns, away from Vilna.

A Vilna ghetto inmate, Engelstern, was assigned to Schmidt's home as a servant. After he started working he noticed bread, cheese, sausage, and other foods scattered around the house. Although hungry and eager to feed his family, he was afraid to touch any of these things. He suspected that this was a provocation and that the German wanted to catch him and accuse him of stealing. Only later did some Jews tell Engelstern how caring Schmidt was and that the food was a gift for his Jewish employee, just one of those gestures he was known for.

One way in which the Germans controlled Jewish lives was by issuing working papers that protected an individual and his or her family from raids and from death. Periodically, however, without any warnings, the Germans would issue new documents that automatically made the old papers obsolete. Those who were not covered by the new documents became "useless" people. This happened to Engelstern and he knew that as a useless person he had no right to live.

He turned to Schmidt. "My family and I are lost, I have no papers to protect us. Can you help us reach Belorussia?"

As if it were the most natural request, Schmidt replied, "No problem. I am taking a truckload of people tonight. Be ready. You and your family can come." With a wave of his hand he dismissed Engelstern's thanks.

That night a truck full of Jews left for Woronowo, a small town in Western Belorussia. The Engelsterns were among them.

For the Vilna fugitives, Woronowo proved a dangerous place. Aware that there was an influx of refugees from Vilna, the Germans were determined to eliminate them. An Aktion did the job. Most refugees were caught and executed yet Engelstern and his family managed to escape to Lida. During a raid there Mrs. Engelstern and her two daughters were taken away, never to be heard from again.

This blow only strengthened Engelstern's determination to fight for his life. While he was trying to put this resolution into effect, the selfless rescuer, Anton Schmidt, was losing his battle. Schmidt was accused of aiding Jews and a military tribunal sentenced him to death. He was executed in April 1942.[16]

The Bielski brothers belonged to the small minority of Jews who from the very beginning refused to become ghetto inmates. To elude the enemy they were constantly on the move. When they had no place for the night, they would sneak into a barn or pigsty and sleep there without the owner's knowledge. Though cautious, they had many close calls.

Tuvia recalls one of them. "I knew Kissely, the head of a particular village. I would visit him whenever I was in the vicinity, tell him where I was to be found, and ask for information of interest to me. One winter night I was visiting Kissely at home. I was sitting quietly and peacefully. Here they won't betray me. Once I had been invited to his son's baptism. My attendance at that affair had reaffirmed our friendship. Suddenly I saw two policemen outside the window. They were headed for the house. The door was opened quietly by the son of the house. He stuck his head in and whispered, 'Run, Bielski, they're after you!' The forest was not too far, I hid easily. The police actually asked for me, and Kissely said that indeed I had been there but had already gone."[17]

Despite many dangers, Asael, Tuvia, and Zus kept in touch. Frequently the connecting link was their younger brother Arczyk, who was thirteen at the time. He was familiar with the area and had a special talent for locating his brothers. Toward the end of 1941, he brought them shattering news. While Arczyk was busy feeding the animals he noticed a truck of German and Belorussian policemen drive up to their hut. He hid in the barn and saw his parents being led away.[18]

Soon the Bielski brothers learned that their parents had been taken to the prison in Nowogródek. Beila's and David's arrest coincided with the establishment of the Nowogródek ghetto. Zus' wife, Cyrl was also confined in the ghetto. He pleaded with her to leave and wanted to find her a place with Christian friends. She rejected the offer. "After all," she insisted, "thousands of Jews live in Nowogródek. They cannot all be wrong." Not wishing to upset her, he did not press; Zus did not imagine that the Germans would be so quick and efficient.[19]

Shortly after the establishment of the ghetto Nowogródek, on December 7, 1941, there was an Aktion. This raid resulted in the murder of 4000 Jews.[20] Among those who lost their lives that day were the Bielski parents, Tuvia's first wife Rifka, and Zus' wife Cyrl, as well as their baby daughter.

For the Bielski brothers, the pain that followed these losses was mixed with guilt and regret. "If only they had acted. If only they had insisted on moving them in time! If only . . . if only . . . ."

As they mourned, this tragedy became a turning point. Now more than ever they would concentrate on not submitting, on freeing themselves from the German clutches. This renewed resolution applied not only to them but also to those who were dear to them.

They knew that to succeed they needed weapons. Arms would reduce danger and increase independence. But the Germans had strict laws against the possession of arms—ownership of weapons was illegal and all civilians caught with a gun were committing a crime punishable by death. The danger surrounding the acquisition of weapons made them rare and prohibitively expensive.

Still, some people did possess arms. Usually the owners were locals who collected weapons left over by soldiers of the two retreating armies, the Polish army in 1939 and the Russian army in 1941. Frequently, these owners refused to part with their treasures and when they did, they expected to be paid exorbitant prices.

Gun transactions took place only among those who trusted each other. Occasionally, a friend would sell a gun at a moderate price as a favor. Less frequently, a gun would be offered as a gift. From a friend, the rich Belorussian peasant Vilmont, Tuvia received as a present a pistol whose trigger would occasionally stall. Although of questionable reliability, this pistol was an expression of true friendship.

Asael acquired a submachine gun from a Belorussian peasant in exchange for used clothes. Also a friend, this Belorussian was more interested in doing a favor than making a profit. Such generous transactions, however, were rare. For the next three years the Bielski brothers would devise ingenious ways for collecting arms.

In Lida the first collective execution of Jews occurred on July 5, 1941, when two hundred prominent men were murdered.[21] The realists felt that this murder was only a prelude to more Jewish destruction. Others argued that these killings were committed, in part, in the heat of battle and therefore were unique, unusual events. Soon, however, news about more mass executions challenged this last conclusion. Such news came together with rumors about the establishment of ghettos.

From a distance, Tuvia had predicted the same fate for Lida as for all the other communities: more persecution, more killings. He spoke to Alter Titkin about the need to plan for the future, but his brother-in-law refused to listen.

When confronted with the possibility of closed quarters Alter said, "There is no other alternative, we'll have to go into the ghetto. It's not too bad, somehow we'll survive. Let's not be smarter than the others. We will live just like the rest will."

"Don't you see it may mean death?" Tuvia demanded.

"They cannot kill all of us. What will be with the majority of the Jews will be with us," came the answer.

"Perhaps," Tuvia conceded, "but how do you know you will not be among the dead?"

"I will take my chances," was all Alter was prepared to say.

Tuvia's wife Sonia sided with her brother-in-law. She could not see herself living in the forest or village—she was a city person. For her even Lida was not much of a town; it was too confining.

When the Jews of Lida were ordered to move to the ghetto, Sonia, together with the rest of her family, began to pack. Tuvia forbade her to go. He had made arrangements with his friend, the rich peasant Vilmont. Sonia was to stay in his house. She was to work there as a seamstress for the entire family. Tuvia and Vilmont felt that Sonia's blond hair and blue eyes could easily support the story that she was a relative from town who had come to the village to rest.

While Sonia stayed with the Vilmonts, Tuvia would occasionally visit her. Sometimes these visits involved an overnight stay. When this happened, Vilmont's neighbors and trusted friends were on the lookout for police. Their warnings, it was hoped, would give Tuvia enough time to escape.

Once when Tuvia and the rest of the household were at breakfast they noticed two carts of policemen and gendarmes approaching. Quickly these men jumped out of their vehicles and dispersed all over the village. Then two of them came to Vilmont's yard.

The farmer's wife, pale and frightened, asked no one in particular, "What will happen now?"

Tuvia replied, "Be calm. When they ask if all of us belong to the household say yes."

Without knocking, one of the policemen burst into the room. Hardly glancing at Tuvia or anyone else he screamed: "How dare you let the dog loose!? He attacked us!" Vilmont's apologies made no impression. Fuming, the policeman began to write a report. A law was broken. The accused Vilmont tried to suppress a smile—never before was he so happy to pay a fine.[22]

Neither the war nor the events that followed diminished Asael's love for Chaja who continued to stay at her sister's house in Duża Izwa. Although he had practically no opportunity to see her, he was preoccupied with her welfare. On December 8, 1941, Asael sent Chaja a message about the mass murder in Nowogródek, urging her to run away. He explained that Jews were being collected in many villages and shot close to their dwellings. Asael's warnings led to an exodus from Duża Izwa, with various family members going in different directions. Chaja made her way into the forest with her two teenage nephews, Pinchas and Josef Boldo. Soon snow and freezing weather forced them to search for less exposed shelter. After several unsuccessful attempts they came to Chaja's friend, the Belorussian peasant Piotrus.

Chaja and Piotrus had shared an interest in the Communist Party. Financially better off than Piotrus, Chaja would supply him with books and other reading materials. Their heated discussions and work on behalf of indigent farmers created a special bond between them and Chaja knew she could count on Piotrus' friendship. He welcomed her and built a special hiding place by enclosing part of a pigsty with a wall. The area was well camouflaged, but cold.

Once a day, in the evening, Piotrus would come to the hiding place and bring potatoes, milk, and whatever else he had. He would also take away a bucket of refuse. Piotrus did not object when Chaja and her nephews were joined by her old mother and niece, Luba. More people reduced the amount of oxygen, making breathing more difficult without sufficiently raising the temperature.

One day the weather was particularly uncooperative. Snow and heavy winds made the shelter unbearably cold. Chaja was ill with a high fever but kept this news to herself so she wouldn't upset the others.

Toward evening the snow storm became more vicious. It was as if the wind and snow were waging a furious battle. On that unlikely night Asael came to see Chaja. She was feverish and her body shook. "I will get you help. I will be back," Asael said and rushed out.

He knew that he could not get a doctor since none were in the area. A trip to the pharmacy, some fifteen kilometers away, seemed the best solution.

The weather prevented him from moving fast. When he finally reached the pharmacist he skipped the greeting. "You must give me the best medicine for bringing down fever. My girlfriend's body shakes and she cannot speak." More emphatically, he added, "My brothers are in the forest. They have guns and know I came here. If anything happens to me they will take care of you."

"There is no need to threaten. I will be glad to give you whatever you need."

Without fuss, the man busied himself with the medications. He explained how to use each. As he handed over the package, he said, "Bless you. I pray that this war should be over. We must help each other in these difficult times."

More relaxed, no doubt grateful, Asael stepped into the uninviting black and white night. Only when Asael convinced himself that the medication had done its job did he leave Chaja.[23]

Now that Asael was giving rather than taking, he was less self-conscious. He gave not only to Chaja but also to her family.

In those days anyone in their mid-forties was already considered old. Older people, particularly those close to the age of Chaja's parents, were a definite liability. They were less mobile, less adaptable, and therefore burdensome. Indeed, some young people abandoned their parents. They reasoned that without older people to worry about they had a better chance of surviving.

Some parents urged their children to leave them behind. Older parents accepted the view that on the outside, in the forbidden Christian world, they would interfere with their children's safety. Selflessly, they wanted their children to live even if this hindered their own possibilities for making it.[24]

Asael treated Chaja's parents with utmost consideration and respect. It was as if some of the love he felt for their daughter spilled over to them. For that Chaja was grateful—but gratitude did not translate into love.

Chaja had a boyfriend who was her social and intellectual equal. He remained in the Nowogródek ghetto. When Chaja's Belorussian protector, Piotrus, went to town, he would smuggle her letters in and out of the ghetto. In his letters Chaja's boyfriend urged her to come and live with him. Like so many other ghetto inmates, he was convinced that if he would work hard the Germans would spare him; the same, he thought, would apply to Chaja. Why look for adventure and danger on the outside? As a city person, unused to the countryside, like Sonia, Tuvia's wife, he could not see himself living in the forest or village.

But Chaja disagreed with his assessment. To her the ghetto represented enslavement and death. She loved her boyfriend, but she also loved life.

And here was Asael. Socially and intellectually her inferior, he offered freedom and a chance to live. This promise of life, Chaja knew, included her family, even her old parents. Did she have a right to disregard it?

Although crude and unpolished, Asael was sensitive to Chaja's needs. He did not pressure. He waited. She must have given a sign, for on one of his visits he brought her a present: a pistol. Chaja knew that "This was the most expensive present that one could have imagined. With my parents' approval we decided that if we came out of this war alive I would marry him. That was our engagement. From then on we were like a couple."[25]

By 1942 the Germans accelerated the pace of Jewish destruction by coordinating their anti-Jewish measures. In addition to the increased raids on ghettos and the mass murder that followed, they became more watchful outside the ghettos. Many Jewish fugitives fell into their hands. Those caught were executed as were the Christians protecting them.

More outside raids made a stay with peasants more dangerous. Although much less convenient, the forest seemed to offer greater protection.

Still, Chaja continued to remain with Piotrus. To make her life more comfortable, Asael moved her mother and niece to the home of friendly Belorussian peasants, the Kots. Other members of Chaja's family found shelter with other Christians. Some went to the forest.

By March 1942, nine people had joined Asael in the forest near Nowogródek. Among them was his sister Tajba and her husband Avremale, Chaja's nephews Pinchas and Josef, and the Bielski brothers, Zus and Arczyk.

This group could no longer exchange work for food. It would have been hard, if not impossible, to find work for ten people. The cold weather eliminated most farm jobs and the increased persecution of Jews made an offer of the few remaining jobs unlikely. Continuing anti-Semitic propaganda had had some effect on people's attitudes. Regardless of how they felt about Jews, the local non-Jewish population became afraid of the severe punishment that inevitably followed the discovery of aid to Jews. For these reasons alone it was difficult for Jews to find work.

But Asael's group had to eat. And so, at night, a few of them would venture into a farm house, where, guns in hand, they asked for food. Those who had no weapons carried sticks in the shape of shotguns on their shoulders. Owners of these artificial "guns" stayed outside the hut but close to the windows. The idea was to make the peasants think there were many of them and that they were well armed.

For Jews on the run time had capricious and complicated effects. Time led to experience and important lessons. It taught them that in the winter the forest was more hospitable than they had thought. But time was also destructive. With time the Germans became more vigilant, more persistent. Because of such stepped-up persecutions, the Jewish fugitives had to be more ingenious. They needed to be stronger. They felt that cooperation with other German opponents, particularly partisans, could give them extra protection. Rumors about partisans, filled with images of strong and brave men, continued to circulate and inspire confidence. The Bielski brothers, particularly Tuvia, wanted to explore the possibility of cooperating with such men.

At the start of 1942, Tuvia contacted a trusted friend, a Belorussian communist Misha Radzecki. Highly intelligent and well educated, Misha came from a poor Belorussian family. In prewar Poland his communist activities got him into trouble and he spent several years in a Polish prison.

During the German occupation Misha was denounced for his communist past and became an outlaw. He avoided capture by staying in different villages and hamlets. Misha was glad to see Tuvia and felt that together they had a better chance of standing up to the Germans.

Together Misha and Tuvia continued to move from place to place. They tried to locate partisans or make contact with underground figures. They felt good in each other's company and were eager to cooperate.

Tuvia, however, noted that his friend was received with open arms in the houses they visited while he felt like an uninvited guest. As a Jew, he was regarded with suspicion. The two friends talked about it; Tuvia feared that later his friend might feel burdened by the presence of a Jew. He became convinced that the differences between them would grow and cause trouble. Tuvia knew they had to part company.[26]

He realized that a group of Jews with common needs and common expectations would be a better solution and that the strength of such a group and ultimately its survival would depend on size, organization, and weapons. These ideas and their execution developed very gradually. In the meantime the Bielski brothers continued to meet.

One day Asael and Tuvia stopped at the Kots to pay a visit to Chaja's mother. They were received warmly; generous offers of food and drinks were mixed with gossip and animated talk. Toward evening the two brothers departed, each going in a different direction.

As usual, when Chaja's mother retired for the night she put her dentures into a glass of water.

Before the village had a chance to wake up, a contingent of Belorussian and German policemen descended. They encircled the Kots' house and a few intruders burst inside shouting, "Where are the Bielskis?"

This question was accompanied by blows.

Kot answered, "The Bielskis were here, but they've gone."

"How dare you give them refuge?"

"How can I not? They're armed, they come and take what they wish!"

The police searched everywhere. One of them found the glass with the dentures.

"These are Jewish dentures. Are they Bielski's?"

"No, they belong to a relative. She left them here for safekeeping."

Pointing to Chaja's mother they asked, "Who is this old lady?"

"My godmother."

"And the girl?"

"My granddaughter."

They paid no attention to the answers—they were preoccupied with the Bielski brothers. Convinced that Kot knew where they were, the policemen arrested him. In prison Kot was tortured. He could only give them the facts: the Bielskis had been in his house and left.

After a few days he was released. Kot suffered from multiple fractures and serious wounds and was confined to bed. In less than a week he was dead.[27]

News about Kot's fate traveled fast. This news, continuous searches by the Germans, and severe punishment of those who dared to aid Jews made help to Jews more scarce. Jewish fugitives were becoming more helpless and more vulnerable. The forest now seemed to offer more safety than a stay in a Christian home.

Still, Piotrus' protection of Chaja continued. Toward the end of March 1942, Chaja left but she did so at Asael's insistence and only because he felt her hiding place was unsafe.

By this time Asael's group grew to fourteen, including Chaja's parents. Only a few were armed. But even if they had more arms, not many could have used them. Asael and Zus taught some members how to shoot. This was done without ammunition since it was too precious to be wasted on training. Besides, one had to be careful about noises that would attract attention.

This enlarged group followed the established pattern. Under the cover of darkness a few men would venture into the village for food. Intimidated by the guns, the peasants would hand over whatever provisions they had.

The Germans avoided the forest, particularly when dark. The group took advantage of this and did most of their household chores in the evening, including cooking. Their meals were made in a large pot that hung from a branch. A fire was built underneath.

They slept in tents under a tree. These tents served only as temporary shelters and were built out of branches, which were covered with all kinds of material. Although the tents offered protection from the cold and wind, they were less reliable when it came to rain or snow. No one undressed for the night—it was warmer and safer to sleep with clothes on, a habit followed by most partisans.

The group's survival depended in part on its ability to move. The peasants, their food suppliers, were poor and therefore unable to feed a large group for an extended time. Not to overburden them, the group would collect food from different farms. Besides, if these natives thought that a group of Jews had settled close to their village, they might be tempted to report them to the authorities. The Germans demanded from the local population the delivery of Jewish fugitives. To assure compliance, in addition to punishments they offered special rewards.

The group's movement was also dictated by chance meetings with civilians. One could never be sure whether these occasional encounters would lead to denouncement, and a change of location decreased the chance of discovery.

Inevitably, life in the forest led to tension. Zus argued that Chaja's parents would bring disaster on the entire group and urged Asael to send them away. Not only did Asael refuse, but he replied that if Zus didn't like it he could leave. He did and took his younger brother with him; after a few weeks both returned.[28]

Tuvia continued to roam the countryside alone. Like other places, Sonia's home at the Vilmots became more exposed. Tuvia urged her to join him, but she refused, saying that she would leave only if her family from Lida accompanied her.

Although the Lida ghetto was officially closed, for a while people were able to smuggle themselves in and out with relative ease. Because the authorities were actively looking for Tuvia, a trip to the ghetto was dangerous—even his bushy moustache was no longer an adequate disguise.

Disregarding danger, Tuvia visited the ghetto twice. Each time he met Alter Titkin and each time he tried to persuade him to take his family and leave. Tuvia promised to make all the arrangements. Titkin refused, arguing that danger would avoid them and that German killings were selective and temporary. The women were not consulted, nor did they know exactly when the meetings took place.[29]

Then on May 8, 1942, all ghetto inmates were ordered to assemble in the central section. The Titkins, Regina's mother and sister and her son, Grisha Meitis, were among them. Grisha worked as an electrician at the military base and his job entitled him to some protection. It was not clear who and how many members of Grisha's family were covered by his papers.

During this Aktion the Germans and their collaborators split the Jewish population into two groups. Grisha, Regina, Lilka, and Alter were pushed into the smaller one. Grisha's grandmother and her daughter had to join the bigger group and became two of the 5670 ghetto inmates murdered that day. Grisha's papers saved Regina, Lilka, and Alter.[30]

The May 8 mass murder, more so than the urging of Tuvia Bielski, convinced Alter Titkin that it was time to leave. He sent word of his decision. The ghetto was more closely watched and Tuvia could not enter. From the outside he orchestrated the Titkins and Grisha's escape.

For Tuvia this step marked the beginning of a long drawn-out battle to save them and many others.

4

The Beginning of the Bielski Otriad

From March 1942 Asael with thirteen followers continued to elude the Germans by staying in the forests, close to the town of Nowogródek. When collecting provisions, they made an effort not to rob poor peasants but instead took from those they thought had extra food. Familiarity with the area and with local farmers helped them decide who was an appropriate target.

Bowing to tradition, the men handled the acquisition of food while the women prepared the meals. All ingredients were cooked together. With potatoes most plentiful and with meat and fat at a premium, the mainstay of each dish was potatoes with very little fat or meat. Depending on availability, different kinds of vegetables, cereals, and flour also found their way into the same pot. How a dish tasted depended on what the men were able to collect rather than on personal preferences or the cook's skills. Even a basic item such as salt was a problem. The peasants themselves were often forced to do without it.

In mid-May 1942 Tuvia Bielski's group came to Asael's base. The new arrivals included Tuvia's wife Sonia and four members of Sonia's family: Sonia's sister Regina Titkin, Regina's son Grisha Meitis, Regina's husband Alter Titkin, and his daughter Lilka.

That evening Chaja welcomed them by serving them dinner first. Regina was unimpressed. Sure of her sister Sonia's unconditional support, she immediately requested two kitchens: one for Asael's and one for Tuvia's group. She added that the two brothers, young Arczyk and Zus, could each choose either group.

Without fuss, Chaja agreed to the division of uncooked products. Then and later, by giving in she succeeded in keeping the hostile undercurrents in check.[1] Chaja's tolerance allowed the men to stay out of these squabbles. It also gave the Bielski brothers the freedom to enjoy their reunion—with the loss of their parents and two brothers, family closeness was even more important.[2]

Following the forest custom, a tent was put up for the new arrivals. They were to share this one place. For Lilka who had fallen in love with Tuvia the moment she first saw him, sleeping in the same tent with him and Sonia had to be a painful experience. But the girl would not confide in anyone. She acted, as she always did, in a retiring, subdued way. She remembers being asked by Tuvia, "Why are you so unhappy?" To this she would answer with a vague smile and silence.[3]

For a while this enlarged group of fugitives continued to follow the habits established earlier by Asael's people. Inevitably, peasants would come to the woods to collect mushrooms, berries, or wood and some would see the Jewish refugees. Guards were posted around the Bielski base. When a stranger was noticed, this would be reported to the leaders, who had to decide whether to stay or leave.[4]

Usually it was safer to move on—one was never sure if a particular intruder would report them to the authorities. The Bielski men could kill those met by chance and thus avert potential trouble. But this was not how the group wanted to act. Their policy was to avoid people; when that was not possible they would try to get away.[5]

The group would move to a new location under the cover of darkness, traveling through side fields and side roads. On their way they sometimes had to pass through water. This they did only after the swimmers among them examined the depth of the water.[6]

All washing took place at the banks of the nearest river, toward the end of the day. The same river gave them drinking water. Their nomadic existence made digging for wells impractical.

Asael continued as the head of the enlarged unit, but he would frequently defer to his older brother Tuvia, who was always ready with advice. He was concerned about the group's vulnerability, lack of arms, the disproportionately high number of older people, and the high number of women. Their situation, he felt, could be improved by acquiring more weapons and more young men. He suggested that they recruit armed young men from the surrounding ghettos. Asael and Zus agreed and together they decided to bring some youths from ghetto Nowogródek.

This decision called for special arrangements. A messenger had to be dispatched; entering and leaving a ghetto involved serious risks. Because the Bielski group was mobile, finding the unit could be a problem and it was therefore advisable to have a stable stopover place outside the ghetto. Someone's home would be the best solution. From there ghetto runaways could be picked up by the Bielski people and transferred to the forest.

A Belorussian peasant, Konstanty Kozłowski, nicknamed Kościk, agreed to go to the ghetto and bring back a group of young men. He also offered his home as a stopover.

For years, as a youngster, Kozłowski had lived with a Jewish shoemaker from whom he had learned the trade and the Yiddish language. He spoke Yiddish fluently and had an affinity both for the Jewish culture and its people. He hated the Germans with a passion, was opposed to their murder of Jews, and welcomed any opportunity to defy them.

A widower, Kościk shared his home with his three grown sons. They lived in a "hutor," an isolated farm. Kozłowski's hutor was located not far from the village Mokrec, eleven kilometers from Nowogródek. Because of its location, away from other dwellings, a hutor offered special safety. Indeed, Kościk's home soon became a meeting place for runaway Jews and Russian partisans. He also continued to act as a courier and guide, not only for the Bielskis, but also for other Nazi opponents. His services were free of charge.

Kościk supplemented his meager income with illegal commerce that included the manufacture of vodka and the sale of all kinds of black market goods.

His family supported his anti-German position. His brother, a Belorussian policeman, supplied important information to all who were opposed to the Germans, including the Bielskis.[7]

In the summer of 1942 carrying a letter to the Bielski's cousin, Yudl Bielski, Kościk smuggled himself into the ghetto Nowogródek. This letter urged Yudl to organize a group of young men, if possible with arms, and to come with them to the forest. After Kozłowski contacted Yudl he hid in the ghetto and waited. In less than two days he and seven young men, none of them armed, were on their way to the forest.[8]

With Kościk as their guide, the seven youths reached the hutor safely. From there, well fed and rested, they went to the Bielski base. Their arrival increased the group's size to thirty. Asael, Tuvia, and Zus agreed that the time had come to organize into an official partisan detachment, an "otriad" in Russian. Other men concurred. As customary at the time, women were automatically excluded from all political decisions.

Familiar with the name of the Soviet General Zhukov, Commander in Chief of the Western Front, they decided to call themselves the Zhukov otriad. Later they were identified by another official name, yet unofficially they were always known as the Bielski otriad.[9]

The establishment of the detachment called for the creation of a few formal positions. Tuvia became the commander, responsible for running the otriad, formulating its policies, and maintaining its security. Asael, as the second in command, was in charge of the day-to-day activities of the unit and the armed men. Zus was appointed head of reconnaissance, in Russian, "razwiedka." He would collect information affecting the group's existence and safety. His job required familiarity with the area and the local population. A number of young armed men were assigned to the razwiedka.

Pesach Friedberg, one of seven young men from the recently arrived Nowogródek ghetto group, was appointed Chief of Staff. He was soon demoted to the position of Quartermaster. As Quartermaster he was in charge of the distribution and allocation of goods within the otriad; seniority rather than ability helped him retain this second position.[10]

Lazar Malbin became the Chief of Staff. He arrived from the ghetto Nowogródek soon after the group of seven. When Tuvia met Malbin in the forest he had already heard of his qualifications. He knew that Malbin belonged to a well-to-do family, was well educated, and fluent in several languages. Particularly important for his Chief of Staff position was his command of Russian and Polish. He also had had a few years of military schooling and had served in the Polish army.[11]

Malbin was a controversial figure, described by some as susceptible to alcohol and women. Others praised his organizational skills and his willingness to remain in Tuvia's shadow.[12] Initially, Malbin's job was limited to the clarification of policies and the writing of official letters to the Soviet partisan headquarters in control of the region. He would later initiate bureaucratic procedures that created a semblance of order and stability. With time Tuvia came to rely more and more on Malbin's advice and talents.

A physical impediment kept him permanently in Tuvia's shadow. Malbin was a stutterer and his speech would deteriorate greatly when he experienced any emotional stress. He was totally devoid of charisma. Still, his education, military experience, organizational skills, as well as his unbounded admiration of Tuvia, made him suitable for the position of Chief of Staff.

Toward the end of his life, in his eighties, Malbin rather immodestly said that with the exception of Tuvia's position, he should have had all the other jobs.[13] He thought highly of himself, but admired Tuvia and refused to outshine him, even by as much as a suggestion or a word.

As head of the otriad Tuvia was a strong leader—he alone would assign people to positions of power. Still, Tuvia became the commander in a special, almost casual way. When the men agreed to form an otriad they appointed Asael their leader. Asael declined the honor in favor of his older brother Tuvia.[14] At this early stage the selection of the commander did not seem very important. The decision was made mainly by Asael and Zus who felt that to avoid misunderstandings one person should be responsible. Tuvia was the oldest so they felt he should be the leader—Jewish families traditionally deferred to the older brother.[15]

Malbin was convinced that had either of the two other brothers become the leader it would have been a disaster and the otriad would not have survived.[16] Another partisan echoes this statement by saying that only Tuvia could have assumed the top leadership. "Tuvia knew how to give orders. He also knew how to be a diplomat. He knew how to deal with people strongly and gently. He would simply tell them, 'If you want to live follow me and do as I say. If you don't want to, go wherever you wish and leave me alone. If you stay with me you must obey me.'"[17]

Comparisons of the brothers suggest that Tuvia was "more sophisticated . . . Asael and Zus were peasant-like . . . Tuvia was intellectually superior to both of them, he knew how to conceptualize. He was the man of ideas, the one who made the decisions. The two would only carry out his orders."[18]

One of the Bielski partisans, not particularly grateful, at times even critical, nevertheless said that "Tuvia had an exceptionally warm Jewish heart; he was very intelligent while the other two brothers were less so."[19]

Actually no one I interviewed, not even the wives of Asael and Zus, claims that either of the younger brothers would have made a more suitable commander.[20] And while Chaja prefers to think that Asael's modesty prevented him from continuing as head of the otriad, she at the same time admitted that he was a poor speaker and awkward with people.[21]

Once Tuvia Bielski became the commander, his consistently stated goal was the group's enlargement. He argued that expansion would strengthen the unit and promote its survival. It is not clear if Tuvia was convinced that a larger otriad would result in greater safety or if he supported the group's expansion because he wanted to save more lives. Shmuel Amarant, a partisan and a historian, writes that from the beginning Tuvia was eager to accept into his unit as many Jewish fugitives as possible and that he continued to implement this policy despite vigorous internal opposition.

Indeed, as the difficulties of securing supplies increased, the tendency and desire to divide the camp into separate closed groups grew. Some wanted to take advantage of the superiority that stemmed from their possession of arms, or their experience, or their good relations with the local population, and leave behind the "useless" people. "Tuvia Bielski took no uncertain stand against these trends. His attitude carried the stamp of national responsibility and the understanding of the needs of the moment. He insisted in every case on the integration of all Jews reaching the camp, whether or not they possessed arms or were able to fight. He repeatedly said: 'Would that there were thousands of Jews who could reach our camp, we would take all of them in.'"[22]

Acknowledging that Tuvia was responsible for the policy to expand, some think it was a risky choice and by no means an obvious one. In the view of one partisan, "Post factum it seems to have been the right decision. Had we remained a small group we could not have made it. So the goal was to become a big group. Zus was against this policy. I remember he would even ask: 'What do you want to do, you want them to kill us?' My mother was also against the enlargement of the group. But the policy to expand continued due to Tuvia's pressure. . . . Tuvia insisted that if we were bigger we would have a better chance of survival and we would be more secure. I am not so sure that the idea of saving was already there. Of course we wanted to save the relatives, the close friends, but at first I don't know how important this was. At least I did not hear much about it."[23]

Toward the end of 1942 the Germans had stepped up the mass killings. This intensified German destruction of Jews coincides with Tuvia's greater involvement with saving them. He would openly insist that saving lives was more important than anything else. He would repeat over and over that it was better to save one Jew than to kill twenty Germans. This was a position he never budged from, no matter what the circumstances or the danger.

To some of the young men such views were unacceptable, many valued armed resistance and would give him little support. Only much later did these former fighters recognize the wisdom of Tuvia's attitudes.[24]

Unanimously and consistently, those who were familiar with the otriad talk about the internal opposition to saving lives. But almost all refuse to name those who were a part of this opposition.[25]

Still, the resistance to Tuvia's ideas about the enlargement of the otriad and the saving of lives had some rational validity. People had come to the forest because they wanted to live. They were convinced that a large number of children, women, and men who were unarmed might bring disaster on everyone. Also, they felt that one could not find enough food for so many people. Tuvia Bielski would not let himself be influenced by their concerns and good-naturedly answered them, "Why do you worry so much about food. Let the peasants worry. We will get what we need from peasants and let more Jews come."[26]

Tuvia must have known that there was some truth to their reasoning. "He had to . . . know that by accepting unarmed people he was taking a big burden upon himself. But it was clear to all that precisely this kind of a Jew had no other place. . . . On the other hand, too, Tuvia had to have a lot of patience and understanding toward these helpless and dependent men, women and children who were on the run and who possessed no guns. With his calm, Tuvia Bielski tried to influence those around him. At times of great danger he remained patient and quiet, and because of it we did not have many victims."[27]

During the German occupation in Poland, including Western Belorussia, there was a conspicuous absence of Jewish leaders. By 1939 some prominent members of Jewish communities had already succeeded in escaping. Abroad they failed to organize into an effective political body and could not mobilize the world's attention to the Jewish plight.[28]

Similarly, inside German-occupied Poland a variety of conditions accounted for an absence of an effective Jewish leadership. One was the German policy of murdering prominent Jews. Another, less direct blow to the emergence of effective Jewish leaders was the 1939 German order that required some members of the prewar Jewish self-governing bodies to form special councils or Judenräte.

The makeup of these councils shifted continuously. Some members were murdered, others resigned. More often than not, those who resigned were also murdered. Still others committed suicide.[29] Those who retained their jobs for any length of time were viewed by the Jewish population with suspicion. Some called them traitors. Others saw them as helpless and hence useless as leaders. Either way, German connections contaminated these leaders' reputations, depriving Polish Jewry of a potential pool of authority figures.[30]

A possible source for leadership was the Jewish underground. Most ghettos had some kind of an anti-Nazi organization. But in their case, too, powerful forces undermined the usefulness of their leaders. Jewish individuality, which had expressed itself in a diversity of political parties, worked against the establishment of a unified authority. In some ghettos an unwillingness to compromise retarded agreement while reducing the effectiveness of underground leaders. More important than Jewish individuality were German measures that deliberately destroyed Jewish communal life and isolated the Jews physically from each other and from their neighbors.[31]

Social upheavals often create leadership gaps. The two together promote the appearance of charismatic leaders; this has been true for most revolutions—the French, the Russian, the Cuban, and others.

Tuvia Bielski was this type of charismatic leader, a man who was obeyed because of his exceptional personal qualities rather than because of official or traditional rights to a position.[32]

Perhaps Tuvia's insistence on saving Jewish lives was a gut reaction, based on strong feelings rather than on rational calculations of risks. When presenting his views to the opposition, implicitly Tuvia had argued that self-preservation was intricately connected to the preservation of others. He insisted that by enlarging their group, by taking in more people, the unit would gain in strength.

Tuvia's leadership qualifications have been attributed to "a great appearance, now we call it charisma. Tuvia was the most handsome of men, very masculine. When on a horse, Tuvia really looked great. He knew how to keep a distance between himself and the people. They respected him."[33] Another partisan refers to Tuvia as a "meteor" and wonders how without any prior preparation he was able to perform a seemingly impossible task. He then continues by saying that "it was a miracle. I don't understand how he did it. . . . Somehow Tuvia sensed that we were all in the same boat. He understood that we had to help each other and that by helping each other we had a better chance to live."[34] Abraham Viner, who had the opportunity to work closely with Tuvia, feels that as a leader "Tuvia Bielski was filled with the national pain and the national love, the pain and the love for Jews. He devoted his soul, his brains, and everything else to the rescue of Jews. He saw a chance, a great opportunity, in his ability to save. He was grateful that he could save Jews. For him it was a privilege."[35]

Hersh Smolar elevates Tuvia's leadership to national prominence. Curious about the origin of Tuvia's anti-German position, Smolar inquired and was surprised by the simplicity of Tuvia's answer. "We knew what they did with the Jews in Germany, so why should they spare us? I knew that we must save ourselves."

Smolar thinks that "the ideas about arms or fighting had to come later. First he had to save. Tuvia had a strong instinct. Common sense told him that the Jews were trapped, that they were endangered."[36]

In 1987, two weeks before his death, Tuvia spoke about his wartime aims. "It was simple. The Germans caught my father, mother, and two brothers. They took them to the ghetto and from there they were taken to their death. The enemy made no distinctions. They took anyone and killed them. So would I imitate them, just by killing some Germans, any Germans?

"It did not pay. To me it made no sense. I wanted to save, not to kill . . . I saw that the Jews did not listen to each other. They were quarrelsome, unruly. They did listen to me, they had respect. So, I had to save them."[37]

Tuvia's accomplishments as a commander and the achievements of his otriad could not have been attained without internal and external cooperation. Not everyone was opposed to expansion and rescue. Asael and Malbin both strongly approved of these aims and many others were in favor. Some who supported these policies were no doubt swayed by Tuvia's arguments. People felt very strongly about him and would remark that "Tuvia had charm. He gave us all strength, we felt more secure with him. He was very close to people, he loved them . . . Tuvia was hard and soft at the same time. When anyone of us was in trouble he was always there. He would always help. One felt safe with him."[38]

Among those in favor of accepting more people were some who only wanted to save their own family and friends. Others, among them Asael and Malbin, had a sense of national responsibility and expressed their opposition to the enemy through their determination to save Jews, any Jews.

Undoubtedly, the policies of expansion and rescue were interpreted and put into effect differently at different times—inevitably many steps occurred between the ideas and their translation into reality. At first the basic motivation of those who came to the forest was "to stay alive. Now people idealize and exaggerate the motivations. Our aim was to survive. When we left for the forest we felt that it was close to the end and so we wanted to live. We did not plan to fight the Germans, we thought about staying alive."[39]

This is not an isolated view. Zorach Arluk distinguished himself as a fighter in the Russian partisan group, Iskra. He too insists that "If someone tells you that when he went to the partisans he was motivated by a desire to fight and by a desire to take revenge, that is incorrect. All of us left the ghetto in the hope of staying alive. We hoped just for a chance. And if not to survive, at least one wanted to die differently from the way most Jews were dying. Not to be shot in a mass grave and not to go to a concentration camp. I think that these motivations were similar for all who ran away from the ghetto. They did not leave to fight, they left to live."[40]

In fact, the three Bielski brothers ran away and became fugitives because they wanted to be free and protect their own lives. Only when they began to feel relatively safe did they consider helping others. At first, these others were family members or people they loved. Later, and in part as a reaction to the persistent German mass murder of Jews, their determination to save became stronger. At that point they enlarged the circle of those they wanted to save.

Tuvia, Asael, Malbin, and those who agreed with them were able to transcend both the principle of self-preservation and their desire to save only their families and close friends. They were determined to save all Jews regardless of who they were. They and others who developed similar attitudes to rescue certainly did not object to bringing in family members and friends; what they were opposed to was the exclusion of others because they lacked family and friendship ties or did not live up to other criteria such as age, sex, or the possession of arms.

In contrast, members of the Bielski otriad who objected to an unqualified acceptance of Jewish fugitives continued to focus on their early commitment to self-preservation and, in some cases, preservation of those particularly close to them. Absorbed in their own survival, they did not identify with the overall plight of the Jewish people.

In part, because the three most powerful men in the otriad, Tuvia, Asael, and Malbin, wanted to save as many Jews as possible and regardless of who they otherwise were, this policy prevailed.

Inevitably, the expansion of the otriad created internal strains. To continue to exist the group had to make internal and external adjustments. An important part of such adjustments and a key to the group's survival was mutual cooperation.

5

Escapes from the Ghetto

True to its policy, the leaders of the Bielski otriad soon invited another group from Nowogródek ghetto. This group was to join relatives and friends already in the forest. Among the invited was Sonia Boldo, an outgoing and attractive young girl, who had an uncle, an aunt, two cousins, and a close friend, Chaja Bielski, in the otriad.

When the Bielski guides entered the ghetto, word about their presence spread rapidly. Rumor had it that Sonia Boldo was about to run away. Right away she was watched by the Jewish authorities, the Judenrat. In Nowogródek most ghetto officials were opposed to illegal departures. If an illegal escape was discovered, it could mean death for those who stayed—the Germans often employed the principle of collective responsibility.[1]

The letter that Sonia received from her uncle urged her to come to the forest. It also suggested that later she would have a chance to bring her parents. Although tempted by this last possibility, Sonia was reluctant to leave. Her parents urged her to go. The final push came when in desperation her mother said, "Go! It is better to be killed by a German bullet in the forest than to be slowly stifled here."[2]

Sonia Boldo left with a group of young people on August 20, 1942. That evening her father supplied the Belorussian guards with generous quantities of vodka. Drunk and unable to attend to their duties, these guards did not turn on the searchlights that covered a large area around the ghetto.

With two guides as leaders, and a small bundle in hand, each runaway passed through a freshly made opening in the ghetto wall. Outside, after hours of silent walking, they came to the Bielski base.

It was dark. Sonia remembers being startled by the sudden appearance of two armed men. The discovery that they were Jewish made her proud. In no time the group was surrounded by a mass of people, all hungry for news. The information they received made some cheerful, others sad, still others disappointed. Then Sonia Boldo was whisked away by her friend Chaja Bielski. Excitement banished all tiredness and all desire for sleep.

When Sonia met Asael, she jokingly asked, "Do you have another commander for me? Another Bielski?"

"Yes, there is another one. He sleeps with someone, but if you have sense you will get him,"[3] Chaja answered. Wasting no time she took her friend to Zus Bielski.

Zus was tall, strong looking, well groomed, and sure of himself. He exuded masculine power. Obviously interested in women, his entire manner conveyed appreciation of the opposite sex. In the forest, for many women, Zus acted as an irresistible sexual magnet.

After she met him Sonia's reaction to Zus was very similar. "I was shaken. I never saw a man like this. I was very attracted to him, but I was also afraid of him. Somehow I suspected that he was more than I could handle."[4]

That first night Zus put his fur coat on the ground next to Sonia's sleeping place, but nothing happened between them. Sonia insists that for a long time Zus continued to court her and she refused to become his mistress. When she did consent, they became an official couple. This was the start of a marriage that has lasted until the present.

Sonia claims that she became Zus' "wife" on condition that he would help bring her parents from the ghetto. He did.

Zus liked women and was not about to give up his habits. Sonia complains that "the women pushed themselves on him. They thought if they would sleep with him they will be saved. Everyone thought so. Was I better than the others? No. I was the same. And so he would go from one woman to another, from one to another."[5]

Jealous, humiliated by Zus' infidelity, Sonia would banish him from time to time. Each time he returned she took him back. With so many women why did he bother to come back? She has an explanation. "I think that he stuck to me because I did not sleep with him right away. If I had I would have been like the others." She refers to him as her Czar and thinks that she could not give him up because he was her first man. Besides, she emphasizes that, except for his infidelities, he was good to her and to her parents whom he had saved. She had much to be thankful for.[6]

Before the war Zus could have never dreamt of marrying Sonia Boldo. Educated, beautiful, and rich, she belonged to a world beyond his reach. War and the forest narrowed the social gap that separated them. As one of the leaders, Zus gave Sonia prestige and protection. She, in turn, offered him an upper-class illusion.

In Sonia's case family concern and pressure brought her into the forest. Sometimes the most obvious source of support, the family, would prevent people from running away. Family attachments and loyalty kept some people in the ghetto because they refused to part from relatives who could not and would not even try to live on the outside. Indirectly, family closeness pushed some people toward death.[7] This had almost happened to the attractive and enterprising young girl Pesia Lewit (later Bairach).

Quite early the Germans had forced Pesia, her family, and the rest of the Jews of Żołudek, a small town close to Lida, into a ghetto, isolating them from surrounding communities. At first, the ghetto inmates were exposed only to sporadic killings. German soldiers would enter the Jewish quarters and shoot at random. The unpredictable nature of these attacks convinced many ghetto dwellers that these were the only threats they had to face and that by avoiding these visitors they would be safe.[8]

Also, at first, despite German prohibitions, contacts between the Jews and their Christian neighbors continued. Belorussian peasants would smuggle a variety of farm products into the ghetto in exchange for clothing and household articles. Not only did these transactions prevent starvation, but they also resulted in an exchange of information, usually about local rather than world news.

At the beginning of 1942, the Żołudek Jews began to feel more vulnerable. To cope with their mounting insecurities, they devised ingenious hiding places. Pesia's parents and their neighbors became owners of a well-camouflaged bunker.

Pesia's parents felt that they could not have survived on the outside. Both were in their early fifties—they thought they were too old to struggle. Pesia's sister, a polio victim, had a limp and was convinced she had no chance of making it in the Christian world. Pesia feels that these attitudes testify to the Germans' success in making them lose hope and self-assurance.[9]

While Pesia's parents were sure that as a family they could not save themselves, they wanted Pesia to try. They argued that her youth, health, and "non-Jewish looks" would help her stay alive. An otherwise obedient daughter, Pesia refused to listen; family love and loyalty held her to the ghetto.

Then came May 9, 1942, the liquidation date for the ghetto Żołudek. The raid began at dawn without any warning. Most Jews could not reach their hiding places and in the confusion families became separated. On that day most of its 1500 inhabitants, including Pesia's family, were shot on the outskirts of town. The Germans spared eighty-one men, all craftsmen and professionals. Permission to stay alive was also granted to the wives of these men. This entire group was transferred to a nearby work camp. From there, under police escort, some were brought into the ghetto to clean and sort out the remaining Jewish possessions. Officially, except for these laborers and the Belorussian policemen, the ghetto was empty. But not quite.

To this day, Pesia does not know how and why she lost sight of her entire family. Somehow she reached a bunker; it was not her own, with people to whom she was not related. In this hiding place Pesia's group found little water and even less food.

The liquidation of the ghetto population took only a few hours. It was followed by an eerie silence, punctuated by sporadic shootings. These danger signals told Pesia and her companions not to venture out.

However, in less than two days, the group had no water and no food. One person moved out, followed by another, and still another. After each departure those left behind would hear shots. They made the connection: going out meant dying. But what did staying in the bunker mean? How long could they last without water and food? Soon the difference between being shot at and staying became blurry. The size of the group continued to dwindle.

In the end only Pesia and two children, ten and twelve, were left. Dejected, this little group did not complain. None spoke. Then one day, from a small opening in the bunker, Pesia recognized a Belorussian policeman, a mere acquaintance. She crawled out of the place and approached the man. "Please, Mr. Policeman, shoot me. I don't want to suffer."

Shocked, the Belorussian said, "I don't want to. I will help you. Come, let me take you back to your hiding place!"

Pesia had no strength to argue. She turned around and the policeman followed. When they reached the bunker, he showed no surprise at seeing the children and told them to wait. He returned carrying a bucket of water and two loaves of bread. As he watched them drink and devour the bread, he talked.

From this man Pesia heard about the eighty-one Jewish men who were spared and that one of them was Moshe Bairach. Because Bairach was single, the Belorussian suggested that he could claim Pesia as his wife. Later, under the cover of darkness, he removed Pesia and the children from Żołudek. The youngsters were placed with relatives in another ghetto while Pesia became the official wife of Moshe Bairach.

Moshe was glad to take care of Pesia. This might have been his way of thanking the Jews of Żołudek for the help they gave him—he had come there as a penniless and sick refugee and they took care of him.

Pesia and Moshe knew each other, but were not romantically involved. Both were young, both overwhelmed by the destruction around them. Protected by Moshe's papers, they moved to ghetto Lida. Without quite knowing how it happened, from a fictitious couple they became lovers. Their marriage has lasted for almost fifty years.[10]

Pesia, like so many other young people, could concentrate on her own survival only when she lost her family, when she could do nothing for them. But in her case, as in most others, without outside help, she would not have made it. Her safe move from the ghetto would not have happened without the Belorussian policeman or Moshe Bairach. Later, when she and her husband left for the forest they both made it because others were helping them.

Of those who tried to save themselves, particularly if they had no help, only very few made it—most were caught trying to escape. Of those who tried again and again, fewer and fewer succeeded.

Sulia Rubin, a teenager at the time, belongs to this small minority. In Nowogródek the Germans had forced her to do manual labor without pay, like the majority of Jews. Sulia had to clean German offices and apartments. Independent and proud, she found it hard to submit and was constantly on the lookout for solutions. She established contact with other youths who were also defiant and eager to oppose the Germans.

Rumors about Russian partisans and the Bielski brothers began to reach the ghetto. Sulia and her friends would have liked to join either group and they knew about ghetto departures to the forest, but only after they happened. All such escapes were illegal, dangerous, and surrounded by secrecy. Without direct contacts to the outside, Sulia and her friends continued to dream.

Then someone gave Sulia a letter from her friend, Israel Kotler. He wrote that he had recently joined the Bielski group in the forest and urged her to do the same. In effect, this was a proposition to become his girlfriend. The letter contained instructions on how to leave the ghetto, the general location of the Bielski group, and a description of a place where she could safely stop over on her way to the forest. From this stopover address she would be picked up by the Bielski people who would then bring her to the forest. Kotler urged her to commit all this information to memory—this letter must be destroyed. Sulia did as directed.

Although she was planning to escape with a friend, because of the danger involved she did not bother to share the detailed information. Each morning, as a part of her work detail Sulia would leave the ghetto. On the designated day she and her girlfriend detached themselves from their group yet someone must have seen them leave. In less than twenty minutes they were picked up by a Belorussian policeman.

At the police station the girls were asked where they were going. Following a prearranged answer, both said they were searching for food and lost their way. The interrogators were skeptical. They suspected a partisan connection and were determined to get to the truth.

The two friends were taken to one cell. During the interrogation the policemen concentrated on Sulia. When Sulia stuck to her initial story she was beaten. She tried not to scream. Eager to see her humbled, the policemen became more violent. She was afraid she might break but then she devised a system. Each time the painful burning sensation of broken skin was about to become unbearable, she willed herself to pass out.[11] Unconscious for much of the time, Sulia's bruised body stopped pressuring her to speak. She revealed nothing.

When the Belorussians realized that they could not make her talk, they transferred the girls to the German police. There, the Nazi in charge was an older officer, Wolf. A ruthless and skilled interrogator, he soon discovered that Sulia's case was different.

During World War I, as a German soldier, Wolf had been stationed in the same region. Among Wolf's pleasant memories from those faraway days was a Jewish dentist, Sulia's mother. Despite the lapse of time Wolf was still benefiting from her excellent dentistry. When he realized the connection between his dentist and the prisoner, he wanted to help. Wolf stopped the investigation and had the two girls transferred to a special prison. From there the two were whisked away by one of Wolf's trusted assistants who returned them to the ghetto.

Taking advantage of her semi-official position, Sulia hid whenever the authorities were collecting people for work. As her body began to heal, she planned another escape. Eventually with guides sent by Bielski, as a part of a larger group, she made her way into the forest and stayed there until the arrival of the Red Army.[12]

Anyone who wanted to escape from a ghetto had to take into consideration not only guards but also the weather and time of day. Because darkness offered extra protection, ghetto departures were usually scheduled for evening. Evening and bad weather were a good combination since guards were reluctant to search in bad weather. Still, extreme weather conditions could backfire. One Jewish youth from Lida ghetto left at night in the middle of a snow storm. When his group reached the outside the storm became so severe that they could hardly move. Attacked by the wind, blinded by snow, they lost their sense of direction. After hours of useless searching they realized they were moving in circles. Afraid to be caught in the daylight, they were relieved when they finally managed to return to the ghetto. This youth, like Sulia, is part of the small minority whose initial failure to escape did not prevent a later successful move to the forest. He also survived in the Bielski otriad.[13]

An escape from a ghetto did not easily translate into a permanent stay in the forbidden Christian world. Some people knew how to escape, but not how to remain on the outside.

Motl Berger was a young man who lived in the small town of Wsielub, close to Nowogródek. Two days after the Germans took over the town they arrested him along with other young men. The Germans demanded a ransom for their release. After the Jewish leaders collected valuables and obtained the men's freedom, Motl slipped out of town.

He went to Nowogródek where he was welcomed by relatives. But, here too, shortly after his arrival, the authorities caught up with him and forced him to join a road repair crew. The work was strenuous and the pay was in the form of meager food rations. Motl was not bothered by the hard work or the inadequate pay. He resented the lack of freedom and feared the possibility of death.

His initial contacts with the Germans had made him particularly sensitive to all signs of danger. At the start of December 1941, he heard from a Belorussian peasant that ditches were being freshly dug on the outskirts of town. Suspecting the worst, Motl ran away.[14] The next day he heard about the big Aktion in which 4000 Jews from Nowogródek ghetto were murdered.[15]

Convinced that this was only the beginning, Motl roamed the countryside alone. Though he felt at home in the rural surroundings, he had no close friends among the local people. Those he turned to for help were mere acquaintances or total strangers. Although some of them took pity on him and fed him, they were reluctant to let him stay overnight.

To avoid freezing to death Motl would slip into a barn, a pigsty, or any closed-in place he thought was empty. Constantly on the move, he limited his wanderings to a small area, the locale he knew best. He felt uneasy about transferring to a less familiar territory. With time, however, those who at first were moved by Motl's plight became indifferent—he was running out of people to turn to. By March, Motl reached a dead end. Without any prospects for aid, discouraged, he attached himself to a Jewish working team and with them returned to the ghetto Nowogródek.

Once more Motl's relatives welcomed him and furnished a place to stay. Though comforted by their generosity Motl remained restless. With the approach of the summer he began to sense another disaster. It was based on a premonition, yet he was able to convince a friend that it was an imminent threat. At the beginning of August 1942, Motl escaped from the ghetto, this time with a friend.

On August 7, 1942, he heard about the second big Aktion in Nowogródek. Now the smaller and partly emptied ghetto was filled with Jewish refugees from surrounding communities.

Outside the ghetto Jewish chances for survival were also drastically reduced.[16] It was bad enough to be chased away without food—now there was the constant and real possibility of denouncement.

As if to compensate, the weather improved. With warmth and dryness, a tree, a ditch on the edge of a country road, or even an open field could serve as a resting place. Fields and gardens were also hospitable by offering food. Motl and his friend tried to take advantage of these favorable conditions. Still, they had no way of escaping from the ever-present suspicions and rejections and felt like hunted animals, fearfully avoiding all people. Soon they were overcome by dejection and futility. This time hostility from others rather than lack of food and shelter pushed Motl back into the ghetto.

While a return to the ghetto gave him the longed for acceptance, it failed to eliminate his craving for freedom and safety. Experience had taught him how hard it was to blend into the Christian environment. He was convinced that without outside help there was no point in trying. So he waited.

Help came sooner than expected—he became a part of a group of young men invited to join the Bielski group in the forest. This was Motl Berger's final ghetto departure. It transformed him into a partisan, a position he held until the arrival of the victorious Red Army in the summer of 1944.[17]

Motl's determination not to give up, though strong, was not unique. As long as there were Jews in ghettos or camps, some would try to escape.

With time the Germans increased their watchfulness over all ghettos and work camps. But considerable variations existed within this general trend toward tighter controls—the Germans in charge would decide on the specific details of supervision.[18] For example, throughout its history the Nowogródek ghetto was more closely watched than the Lida ghetto. Nonetheless, efforts to run away continued even under the most difficult circumstances.

Luba Rudnicki and her group illustrate the complexities not only in leaving but more so in staying alive outside the ghetto. From a well-to-do Nowogródek family, Luba, in 1939, at twenty-four, married a dental student, Janek Rudnicki. During the German occupation, he found employment as a dental technician. His job placed him in the category of health professionals and gave him temporary privileges.

During the first big Aktion in Nowogródek, December 1941, Luba lost her parents and all her siblings. In her case, deep mourning was followed by a total indifference to life. Around that time a Pole, Jarmałowicz, came to her saying that he would like to rescue her and her husband. The man explained that Luba's father, before he was murdered, made him promise to save Luba and her husband. He explained that he was simply trying to fulfill his promise. Though Luba had known Jarmałowicz for years, they were not close friends. Besides, the man had a reputation as an anti-Semite. Suspicious of the man's motives, numbed by the loss of her family, disinterested in life, she refused the offer.[19]

This proposition was followed by one from Mrs. Sargowicki, a Polish woman and Luba's friend. This woman was ready to save Luba and her husband. Excluded from this offer was Luba's brother-in-law Meir, who had what was considered a "typical" Jewish look. Sargowicki felt that his appearance was a giveaway and therefore would interfere with the success of her mission.[20]

Luba again refused. Her decision was unrelated to her brother-in-law, as she explains, "I was not even ready to talk about rescue. I would not leave the ghetto because I did not care to live."[21]

Without quite knowing how and why, Luba's dejection began to lift after the second big Aktion in Nowogródek, August 7, 1942. Mrs. Sargowicki was still there, willing to aid. Her husband was a prisoner of war in Germany and she was bargaining with God: if she saved Jews, God would protect her husband. This time her plan included Luba's brother-in-law and two more ghetto inmates, Dr. Tamara Zyskind and her lover Dr. Berkman. Tamara's husband, Dr. Zyskind, stayed in the ghetto.

At this point the stepped-up pace of Nazi persecution required more caution. Luba and her companions were now to stay with Mrs. Sargowicki's niece, Zosia, next to the village Chrapiniewo and near the small town of Iwje. Before they reached their destination they went through several elaborate steps. When they came to Zosia's farm they moved into the barn supposedly away from the inquisitive eyes of neighbors. But this hiding place proved too exposed—a neighbor questioned Zosia as she carried a pail of food into the barn. The food, and even more so the amount, aroused the neighbor's suspicion. Thinking she might alert the authorities, Zosia decided that her charges should spend their days in the forest and return to the barn only at night.

Anxious about safety and concerned about the approaching cold, the five fugitives decided to build a bunker in the forest. By storing food supplies in it they hoped to reduce direct contacts with Zosia and the accompanying ever-present perils.

In the forest the three men, Janek, Meir, and Dr. Berkman, took turns digging. When they were almost satisfied with the size of the hole, they hit water and had to start all over. The second attempt was successful. Once they finished digging they lined the walls with wood and on the left and right wall built bunks. With straw and covers these bunks served as comfortable beds. They also built special wooden boxes for food storage. A small stove, supplied by their protector, was placed in the middle, with an opening to the outside. The top of this hiding place was well camouflaged with leaves and branches. At first water was a problem, but this too was solved by digging a well. In the end they had a well-designed refuge.

Soon word reached the five fugitives that Dr. Zyskind, Tamara's husband, was on his way. Although aware of his wife's affair with Dr. Berkman, he still chose to join her.

Dr. Zyskind had no trouble passing as a Christian. Dressed as a peasant, no one would have recognized him as a former ghetto inmate. Yet the unlikely happened. Less than a mile from his final destination, Zyskind met a Belorussian policeman, an old public school classmate. The policeman made the arrest.

At the police station they wanted to know where the prisoner was going, whom he was meeting. Dissatisfied with Dr. Zyskind's answers, they tortured him. The painful interrogation continued, yet the prisoner refused to talk. He did not give in but his body did. In less than a week Dr. Zyskind was dead. Their protector, Zosia, brought the fugitives the news.

Through his Belorussian contacts Tuvia learned about the Rudnicki group of fugitives. One day, he and his wife Sonia paid them a visit. Tuvia admired their bunker and said that "'Only intelligent people can build this way.' He appreciated people with brains and higher education."[22]

Luba Rudnicki was a cousin of Tuvia's first wife, Rifka, and the divorce did not disturb Luba's and Tuvia's friendship. From this visit Luba remembers Tuvia's parting words: "This may not be the time for change, because our group is scattered. But in the spring you should come to us. If you decide to do that, go to Kozłowski and he will direct you to our place."[23] Tuvia's invitation came with a detailed description of the location of Kozłowski's hutor. Luba stored it in her memory.

The five settled for the winter in their new bunker. From the ghetto they had brought jewelry that Zosia exchanged for food. In addition, they stored enough firewood to last a couple of months. They had to be prepared for the snowy season when movement was restricted. Snow-covered ground left footprints—an invitation to danger. It was safe to walk only during a snow storm, so that the falling snow would cover all tracks.

During the winter the five fugitives rarely ventured out of their bunker. Their isolation was occasionally interrupted by others who also benefited from the forest's protection. One of them was a Jewish youth, Joske. A gun owner, he identified himself as an independent partisan and a fighter. Dangling casually from his shoulder, his gun looked menacing. In the forest and countryside all armed men and women called themselves partisans.

When Zosia introduced Joske to her charges, his first question was "How can I help you? What do you lack?"

With his eyes glued to Joske's gun, Dr. Berkman answered, "Weapons, we need weapons."

His four companions nodded in agreement.

"If you have money, I may be able to get some."

With Joske as their go-between, the five soon acquired three guns. Without any discussion, all assumed that these weapons belonged to the three men, not the women. But this did not prevent the entire group from feeling stronger and more independent. Now they also felt like partisans.

Zosia thought that as gun owners her charges might wish to meet other armed men. She told them of a partisan she knew who had been trying to assemble a large group of fighters. "After all," she said, "you may want to unite forces. One never knows what can happen." The five agreed that there was no harm in meeting. Luba would have preferred if these people did not know the location of their bunker. Her companions laughed at her excessive caution.

It turned out that there were only three armed men, all Russians. One had married a local girl. It was not clear whether they were former POWs or Red Army deserters. Although they spoke about fighting the Germans, they only visited homes of Belorussian peasants who, afraid of their weapons, gave them food and drink.

The three Russians were glad to meet the five fugitives. They were particularly interested in Dr. Berkman. In this part of the country, especially in the forest, a doctor's skills were rare and highly valued.[24]

One day, Luba and her friends woke up to shooting sounds. From their Russian contacts they heard that the Germans had attacked the Bielski group. These men did not know how many had died, only that Zosia's mother was among those killed and that her farm had been burned down. The Bielski partisans had stayed at her farm.

Immediately after Zosia disappeared and was never heard from again; Luba and her companions lost an important link to the outside. Only Joske and the three Russian men continued to visit. Soon Joske also disappeared without a trace—he was probably killed.

The three Russians invited Janek and Dr. Berkman to join them in their nightly expeditions in search of food. Because of his "typically" Jewish appearance, Meir stayed in the bunker with the women. The two men became a regular part of these nightly ventures. On a cold winter day two of the Russians came to the bunker explaining that the third had been hurt in an exchange of fire with Germans. They needed Dr. Berkman's help, but asked Janek to come along. The two men, each carrying a gun, followed their Russian friends.

Luba, Tamara, and Meir stayed in the bunker. In less than an hour they heard a shot and then another.

Luba's immediate reaction was, "The Russians killed our men!"

"Why do you say this?" Tamara wanted to know.

"Because if it was a fight with the Germans there would be more shooting."

Luba turned to her brother-in-law, "Go out, see what happened."

In silence Meir reached for the third gun, a defective weapon. It made a lot of noise without hitting any targets.

The women waited. A shot was heard, this time quite close to their bunker. Luba jumped up. "They killed Meir. Let's run away!"

On the outside the two women faced a blanket of snow that came up to their waists. Movement was difficult. Then they heard voices. The Russian partisans were next to the bunker, calling for the women to come up. Their voices were becoming more threatening, more insistent. Clearly, the protection offered by the darkness could not last. The women decided to make their presence known. Luba remembers, "I don't know what got into me. I am retiring and shy as a woman. They were very drunk. They smelled. I began to stroke one of the men saying, 'Vanja what happened? We are friends.'"

"Today is the Red Army Day. We have an order to shoot all the Jews. We killed one of your men. One ran away, we killed also the brother. We came to take the one that ran away."

I understood that if one was alive he would come here and I had to be rid of them. I went over to the other Russian and said, "We are women alone. You know we cannot stay here. But it is not good for you to be here at this late hour. Go rest, then come back later and we will be with you."

When Tamara started to talk they were rough with her, "You are only the doctor's whore. You shut up!"[25]

Luba continued to speak to the two men gently, seductively, promising that they will all have a good time, later on. The men left.

Now they knew that at least one of their men was alive. Whoever it was was bound to come back and they had to wait for him. It turned out to be Janek. Luba recalls, "My husband came but how I don't know. The snow was deep, without knowing the direction too well, he came. He was all confused, yet asked where Meir was. I told him that he was not here. He did not even hear me. Then the three of us left. We walked in such a way that the footprints should not be clear. We moved about a kilometer from our place and then we sat down. We heard the Russians come back. We heard their shots. We heard their curses. Then they left."[26]

Although the Russians left, they would be back. Now the bunker was useless and the three fugitives' only hope was Tuvia Bielski's group. The road to Bielski led through his contact, Kościk (Kozłowski). They walked in the direction of his place and it was still very dark when they reached it.

Though friendly, Kozłowski insisted that he could not keep them. The Germans were searching for Jews, they had just left his hut, and might be back. All he could do was direct them to another contact, another Belorussian peasant.

This second man was also afraid. He told them that the Beilski unit had been attacked but had already regrouped in the forest of Zabiełowo. He was ready to take them to the forest and from there direct them to the Bielski base.

At dawn, between branches of huge pine trees, Janek, Luba, and Tamara saw a clearing. On the snow, scattered over a large area, were hastily erected tents. Coming closer they could distinguish men, women, and children. They wore tattered, messy clothes—the Jewish fugitive uniform. These people seemed busy. They carried what might have been food or water.

The three then heard a loud "Who is there?" The voice belonged to a man with a gun, a guard. He spoke Russian, but Luba answered in Yiddish. He seemed satisfied that he had not stumbled on enemies. They were joined by another man, another guard. The first one volunteered to bring them into the camp.

Word about the new arrivals traveled fast. Entering the clearing, they were surrounded by a curious, but not particularly friendly crowd. Then the three began to recognize a few familiar faces from their town Nowogródek. Others were strangers. Unaccustomed to seeing so many Jews all at once, a bewildered Luba thought, "What next?"

Next Tuvia, his brothers, and other men came to greet the newcomers. All were armed, with some weapons dangling from their shoulders or attached to their belts. Only now did Luba remember that in addition to losing Meir and Dr. Berkman they had also lost their guns.

Her thoughts were interrrupted by Tuvia's friendly greetings. "Welcome, welcome. It is so good to have you with us. Together, we have a chance to make it. You will see things will work out."

He must have noticed their anxiety. He must have remembered from his past visit to their bunker that of their group two were missing. Tuvia asked no questions. With special warmth he continued to say how glad he was that they came.[27]

Enveloped by this hospitality the three sighed with relief. Some of their most pressing apprehensions began to recede. Then as a final welcoming gesture Tuvia placed Janek in charge of a group of twenty-five people.

In Zabiełowo, in the winter of 1943, the Bielski group had over 200 people: many were older, some were women and children. In a roundabout, painful journey, Luba and her two companions became a part of this Bielski detachment. A refuge in the sea of Jewish destruction, the Bielski group continued to grow.[28]

6

The Partisan Network

After being exposed to German terror, the Jews in Western Belorussia remembered the Soviet occupation as being the "good old days." They would ask, "Since the Russians were good to us in the past why should the Soviet partisans not be good to us now? Now that we have a common enemy, the USSR has even more reason to protect us." Jews expected the Russians to help.

Even independent and apolitical people like Tuvia Bielski waited for this to happen. For quite some time, however, no one in his otriad had met these likely protectors. In the meantime, as the group grew, it had to contend with different kinds of dangers.

Of the newcomers who joined them only a fraction had weapons. This meant proportionately fewer guns. Soon ammunition even for this small number of weapons began to dwindle. When the supply became dangerously low, the Bielski brothers decided to act. They thought that by interrogating friendly peasants they would find out who among them had hidden ammunition.

On a hot summer day in 1942, a day the Bielski scouts expected to be free of raids, the three brothers set out for the countryside in search of ammunition. After a few unsuccessful attempts, the scorching sun made them stop for a swim, near their native village, Stankiewicze. Refreshed, with their guns close by, they rested on the banks of the river.

Familiarity with the area and the people made this place at once safe and dangerous. Safe, because they knew who their friends were and where to go. Dangerous, because they were known and could be easily recognized by a foe. Alert, their surveying eyes spotted three people who turned out to be two men and a boy—all strangers. With guns in hand the brothers considered moving away. But then they heard the boy shout, "Bielski wait!" They did not recognize the youngster and hesitated. Should they retreat or stay? They waited.

The two adults approached cautiously. They had friendly, yet tentative smiles. They were armed and introduced themselves as Russian partisans. One, Gramov, explained that he was a leader of a larger guerilla group. Neither the unit's location nor its size was mentioned. Such sensitive information could not be shared with strangers.

Tuvia mentioned his and his brothers' names. Then he, too, explained that they were part of a bigger detachment. These brief exchanges were cordial, free of tension. The Russians' openness and manner made Tuvia test his luck. He told them about their search for ammunition and asked if they could help.

Gramov nodded. "Sure. I know a farmer who has bullets—a lot, a lot. He will give it to you because he is my friend. We will bring you to the place."

It happened so quickly and with so little effort—in no time the Bielskis had a generous supply of bullets.

As they were concluding this transaction, the boy came shouting, "The policeman Kuznicki is here, he's just arrived at his father's house!"

A son of a wealthy Belorussian peasant, Kuznicki was an enthusiastic Nazi collaborator. He would denounce and deliver to the authorities both Jews and non-Jews and had been responsible for many deaths. Those who had any reason to fear the Germans tried to stay out of his way.

Tuvia knew about Kuznicki and told his new friends about him, suggesting that the man ought to be eliminated. "Let me do the job. I will get him and you will back me up," Zus called out eagerly. Agreed. This was to be his first partisan deed.

With Zus at the head, they marched to the policeman's hut.

Zus burst into the kitchen, leaving the others behind, close to the door. The policeman was at a table, eating with his father and mother. Within easy reach and leaning against the wall was a rifle.

No one uttered a word. Zus quickly put his gun under the man's chin. "Stay still!" Zus ordered and with his free hand he grabbed the rifle.

Those behind him ran into the room. Pointing their guns at the older Kuznickis they screamed, "Don't make a move!" No one even tried.

Turning to Gramov, Tuvia said, "Take the prisoner and his rifle." The two Russians went out with the policeman between them and the rifle in Gramov's hand. Through the open door, the others saw them enter the field and disappear into the nearby forest.

A shot rang through the air, and the two partisans returned without the policeman. Nothing was said; the parents sat vacantly watching the fields. They did not change positions or the direction of their gaze as their uninvited guests left the house.

The six men took advantage of the forest's hospitality where the protection of the trees relaxed them. They were eager to talk, eager to share. Gramov related his POW experiences. The Germans had made him work on a Belorussian farm. There he had heard about other POWs who, like him, were farmhands. Only after a friend, a Belorussian peasant, gave him a loaded gun did he contact these other men. It was winter and together they decided to organize an escape in the spring of 1942. There were six of them—two men had guns, the rest were armed with daggers. They kept moving around, close to the woods. Friendly natives were supplying them with food and important information, which led to an attack on a small police station that yielded guns and ammunition. It also resulted in the deaths of several policemen.

Encouraged, Gramov continued to collect young men and hoped to establish a large fighting unit. Tuvia admired the man's enterprising spirit and told him about their otriad and how it continued to grow by taking in all Jewish fugitives.

Then Tuvia asked, "Why don't you join us? We are three brothers, brave and strong, as we proved to you just now. Let's work together."

"Together?" Gramov was surprised.

Then with a touch of irritation he added, "You'll work by yourselves. Gather more people around you and lead them! You want to enjoy the fruits of my labors?"

"We haven't got enough arms to even make a start," Tuvia said.

"That's something else. Here is Kuznicki's rifle and take more!"

The Russians were armed young men capable of defending themselves—they were not about to unite with a group consisting of dependent civilians who had to be protected. Still, Gramov gave the brothers six rifles and a seventh that lacked a trigger.

It was indeed a generous gift. After this transaction, still friendly, but somehow uneasy, the two groups parted. They never met again.[1]

Kuznicki's rifle actually belonged to Zus. Whoever took a weapon away from an enemy became its rightful owner; this was an unwritten partisan law. But partisans were guilty of breaking this law and many others.

Independence, freedom, mobility, and violence were all a part of Russian partisan life.[2] How much there was of each depended on who these partisans were.[3] At different times different kinds of people were drawn into the woods.

After June 22, 1941, following the outbreak of the Russian-German war, entire Red Army divisions collapsed under the massive onslaught of the German Army.[4] Thousands of Soviet soldiers moved to the forests before they were caught by the enemy.[5] Others, an overwhelming majority, surrendered to the enemy and became POWs.

German treatment of these POWs was ruthless. Many fell victim to mass executions, others died a slower death, as concentration-camp slave laborers. Estimates for Russian POWs who perished in German captivity run into the millions.[6]

During the Soviet occupation some Belorussian men enlisted in the Red Army, while others cooperated closely with communist authorities. With the outbreak of the Russian-German war, the Soviets either took with them or were followed by many of their Belorussian co-workers. Many farms were left without male workers.

At the initial stages of that war, spectacular victories assured the Germans of their military superiority. Feeling invincible, the Germans had little reason to fear the humiliated and defenseless Russian POWs and saw in these captured men a solution to their labor shortages. The Germans felt secure enough to "release" some of these prisoners of war in order "to satisfy the demands for farmhands, skilled laborers and others."[7] Highly controlled and supervised, such selective releases of prisoners in no way implied a changed policy. Rather, these were exceptions motivated by self-interest. At this initial stage, the Germans recognized the economic potential of these Russian prisoners of war and used them to their own advantage.

The Soviet government was also quick to see the benefits that could accrue from their former soldiers. The USSR felt that these men could help them fight the enemy from within. As early as July 1941, the Central Committee of the Soviet Communist Party urged the formation of an anti-German partisan movement. With headquarters in Moscow, one of the first steps of this organization was the creation of a school for saboteurs.[8]

The central staff of the Partisan Movement was established in spring 1942. Marshal Voroshilov became Commander in Chief and Ponomarenko, the first secretary of the Communist Party in Western Belorussia, was appointed Chief of Staff.[9]

Specially trained men were sent by Moscow to organize a guerilla movement in the thick Belorussian forests. These government-sponsored organizers found many more former Soviet soldiers than they had anticipated but had a hard time controlling these partisans. Scattered throughout different forests, consisting of small splinter groups, these early Russian partisans lacked weapons, leaders, and discipline. Avoiding confrontations with a superior enemy, they limited their activities to finding food and shelter. At first, only very rarely would they attack Germans and only if confronted by an easy target. The main inducements in such cases were weapons.

Instead of fighting the Germans, these early partisans would rob each other of arms and anything they considered valuable. Rivalry and greed would sometimes lead to the murder of unarmed civilians.

Partisan encounters with local peasants were limited to food collection; peasants reluctant to supply them with food were robbed.[10]

News about peasant-partisan contacts circulated widely, and occasional partisan military actions led to exaggerated ideas about their actual and potential strength. In the absence of real evidence, rumors about the partisans' heroic escapades continued to multiply. Because of this wide publicity, from the outside, Russian forest dwellers appeared a formidable force.

Talk about guerillas alerted the Germans to the threat of these forest people. The Germans became concerned that more POWs would be joining those already there. To prevent this, they began to collect POWs who worked on farms close to the woods. The Nazi authorities explained these moves as selective transfers to new jobs—in reality these men were sent either to concentration camps or directly to their deaths. But the POWs were not fooled by such official assurances. Suspecting the truth, many escaped to the forests where they established contacts with the remnants of the Soviet army and other POWs.

In the forests they were soon joined by Belorussian men.

At first because the Belorussians lacked a strong sense of nationalism and had no intention of fighting the Germans, they assumed a friendly posture toward their new masters.[11] The Nazis had been initially successful in recruiting Belorussian volunteers for work in Germany. Only gradually, and only after Belorussians had learned about Nazi atrocities against the Jews and the Russian POWs and news about abuses of foreign workers by German employers reached Belorussia, did the supply of these volunteers begin to dwindle. Belorussian men refused to enlist for work in Germany.

By 1942, short of labor, the Germans began to use force. Work deportations were first limited to the urban population, then extended to the countryside. Belorussian young men retaliated by running away to the forests where they connected with already existing partisan groups or formed their own special units.[12]

About the same time, Jews added variety to the forest dwellers. Almost without exception they were ghetto escapees. Some ran away before or during an Aktion. Unlike the Russian and Belorussian partisans, not all these Jewish fugitives were young men. Some were older men, some were women, with or without children. Most did not even know how to handle weapons. Moreover, in prewar Poland most Jews lived in towns and cities and could not easily adjust to the life of a forest fighter.[13]

Jewish fugitives became easy targets. Not only were they at the mercy of German raiders, but they also had to rely on the goodwill of other forest dwellers much better equipped for this kind of life. Frequently Jews would reach the woods only to encounter new forms of persecution. Some were robbed, abused, and even murdered. Robberies were sometimes motivated by greed rather than anti-Semitism yet, more often than not, those who expected to find treasures were disappointed.[14]

The Soviet government made no effort to protect the Jews. At the early stages of the partisan movement Stalin wanted to establish control over the unruly forest bands. Communist leaders were concerned with what, to them, were more pressing partisan problems: heavy drinking, haphazard robberies, and chaotic violence.

The situation was therefore unpredictable. Some Russian otriads would welcome Jewish fugitives; others would persecute or reject them. But no matter how they were treated by other forest dwellers, ghetto runaways had to overcome many difficulties. German terror loomed everywhere and because this terror pushed them into the forests, they continued to come.

At first the Belorussian woods became a haven for Soviet deserters, former POWs, Belorussian men, and Jewish ghetto runaways. Later these different groups were joined by other types of people. Although diverse in terms of nationality and ethnic origin, these early forest dwellers did have a few things in common. None were making an ideological statement and very few were motivated by the desire to fight. They came to the forest because they wanted to live.[15]

Isolated from meaningful contacts with the outside, within the ghettos many Jews continued to cling to the idea that Russian partisans would protect them.[16] Only very slowly did reality force them to change their ideas about this protection. Some came to realize that only young men with guns could count on membership in a Russian otriad. Most did not know, or preferred not to know, that in some cases even armed Jewish men could not join a Russian otriad.

For many young men the road from the ghetto to a Russian unit was long, with unexpected, hazardous detours. Jashke Mazowi's experiences illustrate this long and unusual path. Originally from a small town, Mazowi lost his entire family at sixteen, during the liquidation of the ghetto. As part of the remnant of Jews considered by the Germans a temporary economic asset, Mazowi was transferred to Lida ghetto. In Lida, in the summer of 1942, he heard about Russian partisans. Although this information was vague, Mazowi was determined to escape from the ghetto. He explains what happened. "We organized a group of six youths. Our goal was to run away, go to the forest, build a bunker there, hide in it, and at night go for food to the peasants."

The six owned one rifle, one revolver, and ammunition. On the outside, after a safe escape, these few weapons made them feel free and powerful. Elated, they entered a village. When they came close to the police station they were suddenly fired at. Each ran in a different direction. Three of the youths were hit by bullets and fell. Two were caught. In the confusion, all the guns were lost.

Only Mazowi escaped. Aimlessly he roamed the countryside. With darkness came fear; to him every sound signaled disaster. Even the noise of his footsteps became a serious threat.

"I saw a hutor. I knocked at the door, trying to decide what language to use, Russian or Polish. Poles hate the Russians. A man opened the door. I spoke Polish and told him I was a Jew and that I ran away from the ghetto.

What should I do? He obviously knew Polish and looked at me, then turned around. I thought that this will be the end. . . . He moved to a cupboard, took out vodka, gave me half a glass, a cutlet, and said, 'Here, eat and drink.' This was the first vodka I ever had."

He spoke. "You are young. The Russians wanted to kill me. I am a settler—I moved around and they did not catch me. Move around. Don't go to the ghetto and you will survive, you have a chance."

"Where should I go?"

"I will tell you which hutor to go to. They will give you food. Make yourself a place in the forest, a bunker . . . at night people will give you food and you will live."

"He gave me courage, food for the road, and I left. During the day I stayed in a small forest. The day seemed to last forever. How will I make it? At night I went into the hutor he told me to go to. They gave me food. The next day it was easier. It did not seem so long. The day after even easier. . . .

"This is how it was, a week or two, I don't remember. . . . Still, I was afraid. It was cold. I knew already where to go. I had special places where they would give me food.

"Once on a very dark night I walked and walked. I lost the way. The hutors and all. I saw only unknown places. In the dark I came upon a small poor-looking hut. The man gave me cold potatoes, no doubt the only thing he had. He said, 'Listen, there is a group of men here that moves around. They are Russians and say that they are partisans. Go to them."

"What are you telling him?" the man's son interrupted. "They kill all the Jews. Don't go!"

I wanted to know more and was told that they were "Russian officers who stayed behind after the Soviets left. They move from place to place, get food, don't fight, just eat."

When Mazowi told his hosts he wanted to meet these Russians the son agreed to take him there in his horse-drawn wagon. Mazowi was brought to a lighted hut. When he approached, he noticed a man with a rifle. Without any greeting this man ordered him inside. There, sitting on benches around a table, were eight men busy talking, drinking, and eating. Close by, leaning against the wall in a straight row, like soldiers, stood eight rifles.

Continuing to drink and eat, the men turned toward Mazowi. One asked, "Who are you? What are you looking for?"

"I am a Jew. I ran away from the ghetto. I want to join Russian partisans and fight the Germans," Mazowi said.

"Don't you know that we shoot Jews?" the man asked.

"How come? I am young, I want to fight. Why don't you kill Nazi collaborators instead?"

Pointing to a space next to him, the man simply said, "Sit." He poured vodka into a glass and placed it in front of the youth.

Another man's pocket produced what turned out to be a document from the Polish underground. This too was placed in front of Mazowi, with an order to read it aloud. He did. The men seemed satisfied with Mazowi's knowledge of Polish, but said nothing.

When they were ready to leave, they asked Mazowi to join them. Although he was afraid they might kill him, he followed. On the way one of the Russians volunteered an explanation. "We took a Jew into our otriad, we trusted him and treated him well. He ran away on the sly, without telling us. Because of what he did we shoot all the Jews we meet."

This was not encouraging news, but Mazowi decided to wait and see. Only when these men gave him a gun and tried to show him how to use it did he know they had no intention of killing him.

Acceptance into the otriad gave Mazowi a feeling of power, and with this feeling came a desire to fight.

Mazowi became a member of the Iskra (Spark) otriad. The group started with less than ten men and grew to 300. Mazowi stayed with them until the Red Army came in the summer of 1944.[17] Later, as a member of Iskra, he would have some unusual encounters with the Bielski otriad.

News that five of Mazowi's companions were killed reached the Lida ghetto but it was not clear what had happened to him. Most people concluded that he too had died but under more mysterious circumstances. Those who had no intention of leaving felt that their decision was justified by this incident and used it as an example of what Jewish runaways can expect on the outside. Those planning to run away did not change their minds but became more cautious and took more time to prepare. None lost their sense of humor. Ironically, they nicknamed Mazowi "Jashke, the Savior."[18]

Among those who heard the story and continued to plan a ghetto escape was Zorach Arluk. Born in 1914, in Lida, Arluk had served in two armies: the Polish and the Russian. His initial decision to escape from the ghetto was triggered by a direct exposure to mass murder, the May 8, 1942, Aktion in Lida.

Arluk was spared because he was employed as an electrician in an SS construction agency but he knew this was only a temporary exemption. Arluk and some thirty youths decided to organize an underground group. On Sundays they would meet in hothouses maintained for the high German official, the "Gebietskommissar," Hanweg. Here, away from prying eyes, they would discuss the "hows," "when," and "ifs" connected to their future partisan life.[19] Beyond a general idea that there were Russian partisans, they knew little else. They had heard enough to realize that without weapons their chances of making it on the outside were slim.

Most ghetto inmates were poor and the young in particular had no funds. Arluk's group decided to collect money from the few who had a little to spare. Lack of funds was only one of their problems. Internal opposition was another. The Judenrat and some ghetto inmates objected to all illegal activities. They were convinced that hard work would result in their survival and any resistance would lead to death. They argued that the entire community would perish because of a few "sinners."[20]

Without money and with this strong opposition, Arluk and his group gave up the idea of a unified underground and settled for smaller, partly independent, loosely connected groups.

Eventually he and some friends succeeded in buying arms from local peasants. As they familiarized themselves with these treasures, they tried to establish contact with Russian partisans.

Then a special event propelled Arluk into an escape.

He and one of his co-workers, also an electrician, were asked to do repairs on a train that stopped at the Lida station. When they finished and the train pulled out, Arluk's partner realized he had forgotten to remove his toolbox. He reported this to the SS man in charge. In a measured voice, the German warned, "If by tomorrow morning you do not bring the tools I will kill you."

In the ghetto Arluk, his partner, and others spent many hours frantically looking for replacement tools. Next morning the young man showed his boss the new box and the German methodically checked its contents. The Jewish youth stood waiting. Without warning, without any visible anger, the SS man began to hit him. Powerful blows landed on the youth's head, face, and the rest of his body. As the German continued, it was clear he was seeking out the victim's hands that so desperately tried to protect sensitive areas.

The blood-covered youth tried to suppress all expressions of pain. The assault produced its own eerie sounds. The blows were coming faster. Staggering, trying to stand up, the victim fell. The SS man waited. On the ground, the crumbled figure did not move. With an expressionless face, without uttering a word, the German turned and left.

Arluk's eyes were glued to this scene. He recalls, "I decided that I had had enough and that I would not work for them any more. I climbed down the ladder and told the German supervisor that I was not feeling well. He ordered me to go to the doctor in the ghetto and bring a letter saying that I was sick. I walked some five kilometers to the ghetto. I went to the doctor, a Russian woman. I told her the truth. 'I am healthy, but I decided to leave and look for partisans and I must have two days off, give me a letter that I am sick.' She gave it to me."

This incident occurred on Friday—Arluk was determined to leave by Sunday. He contacted three friends: Melzer, Wilenski, and the physician Dr. Gordon who had all been planning to escape together. They agreed to leave Sunday morning at four. At two, Wilenski told Arluk he had decided to stay. He had a young cousin, an orphan, whom he felt he could not abandon.

Of the four Wilenski knew the area best. He also had access to Belorussians on whose friendship he could count. Arluk reached for his gun and pointed it at Wilenski. "If the Germans won't kill you, I will. We decided to leave today and we will!"

Wilenski began to cry. Still weeping he said, "I will come with you."

The move out of the ghetto was uneventful. A little later the four were sitting in a Belorussian hut that belonged to Wilenski's friend. The owner, a woman, was happy to see him alive. Right away she placed in front of them generous amounts of meat, cheese, bread, and milk—for ghetto escapees a feast. As they enthusiastically devoured these delicacies, they told her about their plans to join Russian partisans.

"This can be done," the woman said. "Not far from here a group of Russian partisans stopped over tonight. I will send my farmhand to see if they are still there."

Eager to establish contact, the four friends followed the peasant. When he brought them to the place, two men came out of the darkness and ordered: "Hands up! Put down your guns and all bundles." They did. Among the items was Dr. Gordon's medical bag with instruments and medication.

The two men gathered up these things and asked the four to come inside. In the room a huge table was covered with drink and food, and several men were eating, drinking, and talking. Their guns leaned against the wall. In contrast to the two guards who had brought them in, the men seemed relaxed and friendly.

Worried about their confiscated weapons yet pleased with the warm reception, Arluk was overcome by a mixture of trust and fear. Asked to explain their appearance, the four told the truth: they were ghetto escapees eager to join an otriad.

The Russians wanted to know if the four could shoot. Did they know how to ride a horse? Were they familiar with the countryside and the forests? Satisfied with the answers, they seemed particularly delighted that Gordon was a physician.

The four were accepted into Iskra, the same unit Mazowi had joined a few months earlier. The men they met were only a fraction of the group. Since Mazowi's acceptance Iskra had grown to 100 members.[21]

When Mazowi and Arluk joined Iskra they did not know about Jewish partisans. Like most Russian detachments, Iskra was mobile; since it was not burdened with older people and children, it could travel with relative ease. Its members also had more opportunities to stay over in villages because the local population would welcome them much more readily than they would Jews. When Russian partisans traveled they would stop in villages and farms to eat.[22]

For the Bielski otriad, meals had to be prepared at the base in the forest. Young men would go on food expeditions while older people, women, and children were excused from such jobs. This task alone put extra burdens on the young men. Feeding the entire Bielski otriad was much more difficult than feeding a group of young men.

If one had to rank the priorities of partisan life, food was second only to safety. Like safety measures, the acquisition and distribution of food also changed constantly to adjust to changing internal and external conditions. Because provisions had to be confiscated, often by force, this created problems for the takers as well as the suppliers. Some partisans were bothered by the moral issues created by these transactions.

One of them feels that "a partisan was something between a hero and a robber. We had to live and we had to deprive the peasants of their meager belongings. These natives were punished by the Germans and by us. . . . At least if they were all pro-German it would have been easier. This usually was not the case. Often we took by force from poor peasants who were not even pro-Nazi."[23]

"At times Russian partisans would take a horse in one village and then sell it for vodka in another village. I would have understood had they taken a horse in one place and sold it for wheat in another place. But most of the time this was not the case. These men were fond of vodka."[24]

Less sensitive to the peasants' plight, another forest fighter emphasizes that partisans tried to take provisions from those whom they suspected of being pro-German. Having said this, he adds, "We also took advantage of some of the special situations. Peasants were required to supply the authorities with a portion of their farm products. In every village we had our people who spied for us. When a transport of food was going to be delivered to the Germans in town, we would attack it, confiscate the provisions and bring them to our camp. This way we would divert the goods that were supposed to go to the Germans."[25]

Although occasionally the confiscated food would have gone to the Germans, as a rule, food collection by guerilla fighters was a form of plunder. "The peasants were robbed anyway by the Germans. They were poor . . . they were deprived. . . . Yet we had to do it. They would not have given us on their own. We were in a predicament. . . . Sometimes we would take away their last cow, or their last horse."[26]

Responsible for feeding his people, Tuvia was aware of these inherent problems. The partisans had to eat. The Belorussian peasants could ill afford to feed themselves, the Germans, and the different forest dwellers. The partisans and the natives were caught between these different requirements and Tuvia had to balance these needs. In the end, he doubted that he had successfully resolved the difficulties.

"Although I tried very hard to prevent robberies," Tuvia recalled, "in truth the line distinguishing robbing from 'taking,' upon which our very existence depended, was very thin. The partisan ethics dictated that taking 'essentials' was permissible, whereas taking anything that could be termed 'luxuries' was robbery. However, there was a vast difference in the concept as defined by us and that of the farmer who complained about the 'Zhidi' [Jews]. There was also a gap between their idea of right and wrong and that of the partisans. . . . There was also a distinction between the attitudes of the one who commands and the one who executes the command."[27]

These dilemmas and contradictions would appear, disappear, and reappear, under different guises. Never completely resolved, they lasted as long as there were partisans.

With Russian, Jewish, and Belorussian partisans all competing for the peasants' meager resources, what happened next was probably inevitable. In the summer of 1942 news reached Tuvia that a Russian otriad headed by Victor Panchenko had decided to annihilate the Bielski camp. The Jews were accused of robberies and the Russians thought the complaints were justified—their own confiscation of goods they did not define as robberies.

Tuvia called for a meeting of his people and explained the situation. He suggested that before the Russians acted on their threat the two sides should have a chance to talk. Tuvia and his advisors called on a peasant who was sensitive to their problems and also a friend of the Russian commandant Victor Panchenko. This Belorussian arranged a meeting at which Tuvia was to speak for his group and Victor was to act as the accuser.

Victor, formerly a lieutenant in the Red Army, was the head of the powerful otriad Octiaber, named after the October revolution. Courageous and enterprising, he was able to collect a group of ex-Soviet soldiers and disenchanted Belorussian peasants. Under Victor's direction, attacks on Belorussian and German police stations yielded arms and ammunition. The villagers liked the easy-going and dashing Victor and were happy to supply him with goods and important information. With his charm Victor knew how to neutralize hostile bands. He had established his authority around the towns of Lida, Nowogródek, and Zdzienciół.

On a prearranged day and place the two commanders met, each surrounded by his armed men. Tuvia admits to being apprehensive about the outcome. Much depended on the interpretation of facts and goodwill.

Of the two, the Russian had more authority and power and Victor spoke first. He described several robbery accusations against the Bielskis and concluded, "We've decided to shoot you!"

Calmly Tuvia took over. "I am the commander of the Soviet company of partisans named for Marshall Zhukov. We are not robbers. If you are a true Soviet leader you should know that it is in the interest of our homeland to fight the German enemy, together. Our homeland does not differentiate between Jews and non-Jews, it only separates the loyal disciplined citizens from harmful destructive bands."

"But we were told that you rob villages!" Victor said.

Tuvia continued. "We can clarify this, and should your accusations prove untrue, we are duty bound to fight together."[28]

Panchenko nodded in agreement and listened as Tuvia added that as natives his partisans were familiar with the area and spoke the different dialects peculiar to this part of the country. All this could be an asset for the Russian partisans and the struggle against the enemy.

It was clear, however, that first they had to settle the problem of robberies. Jointly they devised a system. For a few days neither otriad would go to the village that accused the Jews of robbery. After that they would together approach one of the peasants who complained the most.

On the appointed night Victor knocked on the window of the hut asking for bread, and the owner's weepy voice answered, "I have no bread, dear ones, last night the Jews took from us all we had."

Victor was furious and drew his gun, but Tuvia stopped him. "It's never too late to do this, first ask him what did the Jews take."

When Victor did, the farmer called a girl who supposedly was in the house at the time. She said, "The Jews were here past midnight, they took bread, shortening, onions, salt, butter, eggs, and even the cloth from the table."

Because of their initial agreement to avoid the village for a few days, Victor knew that the Jews had not been there. The peasants were clearly lying.

Fuming, Victor burst inside and confronted the peasant. "From now on, should a fighter-partisan come, whether he be Russian, Polish, or Jewish, you are to give him everything he asks for, food, boots, even furs. You were as good as dead a minute ago and my friend prevented me from killing you. Why did you tell lies about the partisans?"[29]

To this there was no answer. And when Tuvia introduced himself as Bielski, the peasant paled and began to shake. Partisan justice demanded an execution—the peasant had to be aware of this fact. The peasant was indeed lucky that Tuvia Bielski valued life and not death.

This incident ended with an understanding. For the purpose of food collection Victor divided the area into two parts. The Bielski people could gather provisions from farms close to and around the towns Lida and Nowogródek. The Octiaber otriad, Panchenko's unit, would take foodstuffs from and around the town Zdzienciół.[30] With this agreement in place the people were less likely to falsely accuse either otriad.

News about the Bielski-Panchenko cooperation traveled fast. This news not only helped legitimate the existence of the Bielski group, but indirectly made life easier for other Jewish fugitives. Local peasants and Russian partisans were less likely to rob Jewish fugitives. Now they would direct them to the Bielski otriad or tell the Jewish partisans about them. Altogether this made for an easier road to the Bielski camp.

A personal though guarded friendship developed between the two commanders. Thirty-six years old, less impressionable, Tuvia had a gift for anticipating the future. Good-natured and generous, not yet twenty-five, Victor would seek advice from the older man. Like so many others, Tuvia was drawn to this energetic young man. He was also keenly aware that their friendship gave a protective cover to the Bielski otriad. The younger man became a frequent visitor to the Bielski base and Tuvia recalls one such seemingly casual call. "Victor told me that in an isolated hamlet on a farm there was a group of armed Jews who robbed at night and did nothing during the day. The peasants in the surrounding area were angry and complained. They hadn't handed them over to the Germans yet because they were afraid of reprisals. They wanted to know if Panchenko was their protector as well. If he was not, then they would get rid of them quickly. Victor thought that this Jewish group should join the Bielski otriad. If they didn't, he would have to liquidate them."[31]

Tuvia liked the idea of incorporating this group into his unit. While giving him a chance to save lives, it would improve the fighting power of his otriad. With a few armed men he set out for their place.

The Belorussian owner of the farm, though friendly, claimed total ignorance about Jews. Assurances that no harm would come to them had no effect. Only when Tuvia drew his gun did the man say, "Look around, maybe you will find them."

"Where?"

"In the granary," came the answer.

On entering the hayloft the men called out in Yiddish, urging the people to come out. They were greeted with silence. After several futile attempts they began to search the place. Then one of the Bielski men, Ben Zion Gulkowitz, saw a wooden door in a corner in the ground. When he lifted it, the door revealed a hole. They guessed this to be the hideout entrance and again they spoke Yiddish. This too produced no reaction. Ben Zion slipped into the pit. Inside, to the side of the opening, he saw a group of men, each holding a rifle. One of them stepped forward, introducing himself as Israel Kesler, the group's leader.[32]

A native of the small town Naliboki, Kesler was a professional thief. Before the war he spent several years in prison. When the Germans occupied Western Belorussia, Kesler had refused to submit to their rule. Like the Bielskis, he never became a ghetto inmate. Instead, he collected a group of Jews from Naliboki and from the work camp Dworzec. Through his connections to Belorussian peasants he acquired guns and ammunition and a place to stay. On this farm he had fifteen Jews with him. Only a few were women and children.

At the beginning of the war Israel had lost his wife and children. To this hiding place he brought Rachel, a young, attractive woman who regarded him as her savior.

Kesler had heard about Tuvia and others in the group knew him personally. Bielski explained the situation and finished by saying that they would have to join his otriad, otherwise they would be shot by Victor's men.

Kesler laughed. "We can shoot too."

When Tuvia insisted he said, "Give us a chance to discuss this."

Tuvia agreed. "I'll give you half an hour, but your answer must be positive. I endangered myself in coming here, I am not a child, I have only the final goal in mind."[33]

Kesler and several men went out to confer privately. They came back with a suggestion that four of their men would visit the Bielski otriad and check out if they indeed wanted to make the change.

Tuvia lost his patience. "You will come and join us. If not, I am ready to divest you of your arms right now!"

The Bielski partisans made a belligerent gesture. One of the Kesler men interfered. "Now, now, there is no need for that. We will come with you."[34]

No one objected.

Then someone said, "We will go with you, but before we do, we have to settle some business with Albelkiewicz, a head of a village."

"What is the problem?" Tuvia inquired.

"He is the leader of a gang that opposes partisans and Jews. He has already handed over to the Germans twenty Jews, women, and children. He lives about half a kilometer from this village. He has arms."

The idea appealed to Tuvia. It would give him a chance to stop some future murders and check out the skills of his new partners. The Kesler and Bielski men went to this place and posted guards around the collaborator's hut. Five Jewish partisans entered playing the part of an anti-semitic band that was eager to deliver Jews to the authorities. Pesach Friedberg was selected as the speaker of the group. Neither his looks nor his speech could identify him as a Jew.

Ostensibly, they were looking for runaway Jews. The owner of the house greeted them like kindred souls. Relaxed, he returned to his bed. The rest of his family continued to sleep on beds that hugged the rest of the walls. Pesach sat on the collaborator's bed and asked pleasantly, "So what is new with the Jewish business?"

"We make a living and the Jews are being caught," came a half-joking answer.

"Then you must be one of us. How many Jews have you killed?" Pesach wanted to know.

"I don't kill, it is not worth soiling my hands. I just turn them over to the police and they put an end to them," the man explained.

Pesach continued. "How many Jews have you caught so far?"

"Four days ago a woman, two children, and two men, several weeks ago eleven people, I don't remember exactly. Two more several days ago. I chased them on horseback. We took a gun away from one of them, gave it to the police." Then he added, "Oh, I put the women, the children in the storage room. Kept them there all night. In the morning they were almost frozen. I tied them up like sheep and took them to the police station, in Dworzec."

Tuvia could not resist saying, "Man, you're a human being and have a conscience. How can you catch live people and turn them over to certain death at the hands of the Germans? Doesn't your conscience bother you?"

"What are you talking about? Hitler decreed and we must carry out his decree." He sounded surprised. But as yet he did not guess the true identity of his visitors and continued to tell them that he was the leader of a special guard, responsible for clearing the area of Jews.

Now Pesach lost his temper and slapped the man in the face with all the power he could muster. "I am a Jew too! Enough of playing around!"

The Belorussian knew he was trapped. He pleaded for mercy, then argued that after all he did only what the authorities wanted him to do. He did not make the laws—it was not his fault.

Quickly his visitors shot him and the entire family. They searched the hut and found arms. Before they left they made a large sign and attached it to the door. In Russian it said: "This family was annihilated because it cooperated with the Germans and pursued Jews, signed The Bielski Company."

On the way to the Bielski otriad they eliminated another family also guilty of Nazi collaboration and the murder of Jews. There too they left an announcement that these people were shot because they had cooperated in the death of Jews.[35]

The success of these undertakings led to other anti-collaborator moves. One involved the Marciniewskis. This was a large Belorussian family who for years had worked in the mill that belonged to Chaja's parents in Duża Izwa. When the Germans came they appointed the older Marciniewski manager of the mill. This new boss was determined to become the mill's legitimate owner. Greedy and anti-Semitic, the Marciniewski family wanted to kill all who in the future could claim the mill. They were also actively pursuing other ghetto runaways.

One of the Marciniewski sons became a policeman. Together with his brother-in-law, a forester, he had been catching Jews, transforming the entire area in and around Duża Izwa into a Jewish deathtrap. Whenever they heard about a Jewish hideout, they would tell the authorities. Chaja recalls, "The Bielski otriad reacted . . . one day we waited for Marciniewski, the policeman, to come home. We watched the house. When he returned we killed the entire family. . . . This policeman's brother-in-law was a forester who searched the woods for Jews. He bragged that he could recognize the Bielski brothers by their footsteps and was confident that he would catch them. This forester and his family were also eliminated."[36]

A similar fate befell the Stichkos, a Belorussian family who had worked in the Bielski mill for many years. The Stichkos had also been destroying Jews and trying to eliminate all the Bielskis so that they could become the legitimate owners of the mill. The Bielski partisans shot the entire Stichko family and left their usual notice.

Tuvia was convinced that "when the peasants realized that a group of Jewish partisans was tough, then they would try and help us as much as they could . . . this happened only after they realized that we meant business."[37]

Indeed, the natives were impressed with the efficiency of the Jewish partisans and with their courage to stand up to Nazi collaborators. They also became fearful of reprisals and many refrained from harming Jewish fugitives.

Fewer denouncements and less dangerous roads offered more and better opportunities for escapes. Of those who took advantage of these improvements only some ended up in the Bielski otriad. An unknown number joined Russian partisans, roamed the forests in small groups, or found shelter with local peasants.[38]

7

Rescue or Resistance

Tuvia was not interested in military glory. To live, to keep his people alive, to bring more Jews into the otriad, these were his goals.[1]

He would avoid talking to the non-Jewish partisans about his main preoccupation: the saving of lives. To them he presented himself as a guerilla fighter and continually repeated that he had come to the forest to wage war against the enemies of the Soviet Union, the Germans. And when Panchenko suggested that they cooperate in anti-Nazi moves he readily agreed.

Their joint military ventures began in the last quarter of 1942 and continued into the second half of 1943. Although such anti-German moves were initiated by Panchenko, the two otriads each carried the same burden. Publicly Tuvia continued to emphasize his personal commitment to anti-German activities. In reality he and his group were under pressure to participate. A refusal could have endangered the very existence of the Bielski otriad. Russian partisans would not have tolerated an unwillingness to fight, especially not from Jews whom they suspected of cowardice.

At this early stage, all forest dwellers were united in their hatred toward the Germans and their collaborators. These feelings of hostility were supported by equally strong ideas that it was important to fight their common enemy, the Germans.

Russian partisans were very belligerent when they talked about their enemy—killing Germans was equated with patriotism. Hero worship was common. How much of a hero a person was depended on how daring and life-threatening the person's opposition was to the Nazis. But this high value placed on fighting the enemy did not correspond to actual combat; their belligerence was merely verbal.

Inevitably, however, these favorable ideas about guerilla fighting came with a devaluation of those who could not wage war. Non-fighters were subjected to ridicule and contempt. The disheveled and hungry ghetto runaways in particular were sometimes greeted by Russian partisans with a sneering, "Why did you work for the Germans instead of fighting?"[2]

The forest was pervaded with ideas that fighting and causing damage to the Germans and their collaborators was good. Even young ghetto runaways after they were accepted into a Russian otriad would soon become strong advocates of the "wage a war ethic." In their case, as with others, the admiration of anti-German fighters came with a disapproval of non-fighters.

One of these young Jewish partisans admits, "I looked at the Bielski otriad with its 1000 Jews, most of whom could not fight, and thought: 'So what? But they do not fight!' I believed that they should have fought the Germans. As a Russian partisan I felt self-important."[3]

The idea that waging war against the Germans was more important than saving lives was implicitly accepted by many others. And while the degree of such acceptance varied, to some extent it was always there. Indirectly, support for this ideology sometimes had tragic consequences.

The actions of known Jewish partisan fighters reflect this attitude. Dr. Icheskel Atlas, Alter Dworecki, and Hirsz Kaplinski each distinguished himself as an outstanding partisan leader. Each courageously battled the enemy. By the end of December 1942, they had all been killed in action.[4]

During their short careers as partisan leaders, the three operated in and around the huge Lipiczańska forest. With its jungle-like growths, islands of swampy terrain, and irregular, poorly built country roads, this forest promised safety to many of the persecuted. Lipiczańska forest became home for both Jewish and Russian partisans. It also became a haven for ghetto runaways, many of whom were older people, women, and children. Small family clusters or units of unattached fugitives were scattered all over this forest. Disorganized and unprotected, these groups lived in primitive bunkers. Some would beg for food from surrounding farms; the few who had guns would get their provisions by using force. Still, the very few who brought some money or valuables with them would exchange these for food. Unaccustomed to life in the forest, many fugitives were attacked by unruly partisan bands and robbed of their meager belongings and some were murdered in the process. Without the support of a large group like the Bielski otriad many died from cold, starvation, and epidemics.[5]

A survivor of a family camp tells how Russian partisans robbed them of their few weapons. For them, no arms meant no food. They lived in a poorly constructed bunker, with water seeping through every corner. The few children with them were barely dressed, ravaged by hunger and fever. The adults helplessly awaited the children's and their own deaths. Only a fraction survived.

The three Jewish partisan leaders, Atlas, Dworecki, and Kaplinski, had witnessed the plight of these Jewish civilians. Concerned, each in a different way tried to help these poor defenseless people—with warnings about danger, with food supplies and moral support.

Dr. Atlas became very depressed during one of his visits to such a half-starved and defenseless family camp. He told the people he could take ten of them with him, but they all had to be fit for combat. He explained that all his men were fighters and were taking revenge for the suffering of the Jewish people. He is quoted as saying, "We are lost, but we must fight."[6]

Like so many other Jewish leaders who fought the Germans, Dr. Atlas identified with the Jewish plight. Yet, hardly any of these leaders devoted themselves to saving Jews. The help they offered was sporadic, not organized, and hence not very effective.

Their commitment to wage war interfered with their desire to curtail Jewish destruction. It was as if their preoccupation with fighting the enemy made no room for saving Jewish lives. They might also have thought that in the long run fighting would save more people. In comparing Tuvia Bielski and Dr. Icheskel Atlas, Hersh Smolar believes that they are the two most important symbols of Jewish resistance: the Jewish fight for existence and the Jewish fight for revenge. Atlas stood for revenge and Bielski for life, for the biological existence of the Jews. Faced with a choice between saving and fighting—a choice that seemed clearer in hindsight—Atlas and others like him opted for the latter.[7]

The great value placed on fighting the Germans appeared under different guises and penetrated the most unexpected places. In some ways it reached even the Bielski otriad. It was in part expressed in the relatively high prestige enjoyed by those who had arms. The bulk of the otriad, the unarmed, were placed into a less valued category disdainfully called "malbush." Malbush is a Hebrew term for clothes. No one seems to know how the term acquired its negative meaning. Of course this did not necessarily mean that in the Bielski otriad fighting took precedence over saving lives. It only meant that fighting carried with it prestige and status.

An astute leader, Tuvia had to be aware of the value most others placed on fighting. And so, only in the privacy of his inner circle, he would repeat, "Don't rush to fight and die. So few of us are left, we have to save lives. To save a Jew is much more important than to kill Germans."[8]

In the Bielski otriad saving lives became intricately connected to a moderate involvement with fighting. As one young partisan from the Bielski otriad asked, "If we perished, who would protect our families?"[9]

Disputes over the relative value of saving a life or fighting the enemy have continued until this day. Raja, who is convinced that saving lives was more important, tells of her exchange with a Jewish partisan who accusingly said, "'You in the Bielski otriad sat in the forest without fighting!' I asked, 'Tell me how many Germans did you kill?' 'Two.' Then I asked, 'How many Jews did you save?'"

This last question was met with silence. Raja continues, "They have been writing about Jewish resistance, uprisings of different kinds, but not about Jewish rescue by Jews. All the time I have been arguing about why no one writes about the heroism that was involved in the rescue of Jews. As far as I am concerned this is a holy thing."[10]

Only recently have others come to similar conclusions. One former forest fighter told me, "For forty years we had discussions about what was more important, fighting the Germans or saving Jews. We came to the conclusion that our heroism was not heroism. When I was fighting with guns together with other partisans, this was not heroism. Heroism was to save a child, a woman, a human being. To keep Jews in the forest for two years and save them, this was heroism."[11]

Among all partisans a discrepancy existed between the value placed on fighting and the actual number of attacks on the Germans. It was as if the pressure to oppose the Germans lost its momentum somewhere on the way to the real strikes. This lack of correspondence between ideas and actions was probably responsible for some of the exaggerated claims about extraordinary anti-Nazi escapades.

Until 1943, in Western Belorussia, attacks on the Germans by Russian partisans were rare. The few anti-enemy moves that did occur were in the form of sabotage that included the destruction of trains, bridges, telephone lines, and other installations.

At first infrequent and sporadic, such ventures became more coordinated and more common after the arrival of specially trained men from the USSR. With time, more partisans were parachuted into the area. Others had come in planes that landed at secretly constructed airports, close to the forests. Before 1943, however, only a handful of this new breed of partisans reached the forest.

On balance, throughout their entire history, partisan combat activities have been "highly exaggerated. Many of the anti-German expeditions that were planned and discussed by Russian partisans never materialized."[12]

Whatever military encounters between the partisans and the enemy did take place, they reflect the overall characteristics of guerilla warfare, particularly its avoidance of direct enemy confrontations.

And so, even when attacked, partisans preferred not to fight. Instead, they tried to disperse, thinking they would regroup later. Only when these tactics were not possible, only when the partisans had no choice, did they fight. As in Western Belorussia, so in other places, moves initiated by guerilla fighters were usually in the form of sabotage. On rare occasions, and not by design, sabotage did lead to direct confrontation.[13]

Tuvia had a particularly warm feeling for the youthful Jews who fought within the Russian otriads. Moreover, the Bielski otriad participated in armed resistance to the Germans. The joint military moves of the Bielski and Octiaber otriads might have satisfied Tuvia's need to stand up to the Germans; in some way, Tuvia's armed resistance supported his sense of freedom and independence.

One important anti-German move happened shortly after the 1942 harvest. The German law required each peasant to deliver a portion of his farm products to the authorities. In contrast to the treatment of the farmers, estate owners were divested of their entire holdings. Some estate owners were murdered; the rest were banished or sent to concentration camps. Confiscated estates were then run with cheaply paid or unpaid slave labor. What an estate produced went directly to the authorities.

The 1942 harvest was rich and the granaries were filled with wheat. Unable to confiscate this large amount of food, Victor and Tuvia, in a jointly planned operation, decided to burn the fruits of the harvest. The plan was to set fire to all the estates at once, under the cover of darkness. At each location partisans were posted to prevent people from extinguishing the flames.

The authorities and the local population knew that the partisans were responsible for the job.

But then something surprising happened.

"Just as the fires were at their peak planes flew over the area and bombed the flames. Evidently these were Russian planes that were on their way back from an enemy attack. The pilots probably understood the meaning of the fires and decided to 'lend a hand.' The Germans were overcome with horror. It looked like there was a link between the partisans and the Russian Command."[14] In reality this partisan-army cooperation happened by chance.

Convinced that the Bielski otriad had had a hand in these fires, the Germans were even more determined to eliminate the three brothers, particularly Tuvia. They promised a substantial reward for Tuvia's capture or for information leading to his arrest. Depending on who tells the story, the sum offered differs, ranging from 50,000 to a million rubles. A realistic figure probably comes closer to the 50,000 level.[15]

This eagerness to catch Tuvia only underlined his importance. The local people got the message. Their fear and respect grew as did their contempt for the "ineffectual" enemy.

Now that the partisans had destroyed an important food supply, they wanted to prevent the Germans from recuperating their losses and collecting food from the farmers. Consequently, the purpose of their next operation was to hurt the enemy and silence the peasants who grumbled, "You take, the Germans take, how shall we survive?" The partisans decided to deprive the Germans rather than the natives.

The plan was to lie in ambush, wait for cars filled with provisions, and attack. They selected the road leading from Nowogródek to Nowojelnie.

Loyal natives supplied the partisans with information about the enemy's whereabouts. Although they saw it coming, the partisans missed the first small vehicle. Success came only with the appearance of bigger trucks and much more important loot. Tuvia remembers. "Boom! Boom! The first vehicle stopped. We had hit the wheels and the driver. From our side came a hailstorm of bullets. Eight German and White Russian police jumped from the vehicle, fell to the ground and began to return machine-gun fire. Suddenly a 'Hurray!' from Panchenko's people. The enemy is overwhelmed. I don't know how many of them were running away screaming and shooting. . . . The enemy was retreating. We got up and pursued them, but most of them disappeared among the trees in the forest. . . . We decided not to miss the opportunity to gather the loot. We swooped down on the machine guns, the boots and the fowl as well as on the boxes of ammunition. All this took only a few minutes. Our booty on hand: two machine guns, four rifles and thousands of bullets. Trembling with excitement we left the area and went into the forest. The first armed victory and it was ours!"[16]

A less successful move is remembered by the Bielski partisan, Pinchas Boldo. Initiated by Victor, the idea was to destroy the Yatzuki train station and kill the Germans who guarded the place.

"We arrived at the rails in the dark. There was a small building in which there were Germans. From our positions we could see them well. We even had a machine gun. Actually it was in our way and it contributed to our defeat. Instead of going into the rooms of the Germans' house all bullets went over the roof. The enemy had realized that they were being attacked and they opened fire on us . . . not one German was killed and neither were any of us. Somehow they did not imagine that we were so close to them. Their bullets went far past us into the forest. They thought that we were in the forest, some distance away from them. None of us died. But we did not succeed."[17]

On balance, the Bielski-Panchenko cooperation gave a measure of security to the Bielski otriad and led to its expansion. But expansion also meant older people, women, and children and eventually resulted in a curtailment of movement. Tuvia felt that by dividing his people into coordinated subgroups he would improve the otriad's mobility. And so, he created subunits of twenty to twenty-five people.

Usually a person's group affiliation was based on family relations and friendships. People were free to choose those they wanted to join. Tuvia stepped in when these preferences violated the principle of safety or denied a place to those who were alone and whom no one wanted.[18]

Each group had to be properly balanced. It had to include young armed men, older people, women, and children. The correct mix was determined by the commander. Because each person belonged to a group, evacuation could be more efficiently accomplished. Each group had a leader and several other positions with specially designed functions. For example, Tuvia insisted that military protection and food distribution not be in the same hands. He felt the separation of these two basic functions would avoid abuses.[19]

Each unit was assigned to a particular space. Each had a special pot in which to cook its meals. Similarly, provisions were assigned to each group. The expansion of the otriad required more frequent food expeditions, which meant more danger. The authorities were eager to cut off all partisan food supplies and especially those of the Bielski otriad.

Partisan food collections, known as "bambioshka," took place at night. From the Bielski otriad, "Every night one or two groups were sent out to bring food. A group consisted of ten to twelve armed men. One of these men acted as the leader. Some of the participants had to be familiar with the side roads and the particular villages. Of course, one had to select people that first of all were not afraid and second of all to whom the peasants would give food."[20]

When a group reached a village it would first collect provisions from the richest farms. As one partisan explains, this was possible because "In each village we had a peasant, usually himself poor, he would give us information about the other peasants. This way we knew what each had, how many horses, cows, etc. Such a peasant we called "legalshchyk." We took nothing from him. Sometimes we would give him some of the booty. Some of the rich peasants tried to hide their products . . . we would search and if this was the case, we took more from them."[21]

Toward the end of 1942 horse-drawn wagons, confiscated from farmers, were used for food expeditions. When a group left a village, it had to subdivide and prepare the goods on the way back to the camp. For example, cows had to be killed and cut into manageable portions. All this had to be done quickly. At dawn a group was expected to be back at the base—daylight was the partisans' enemy.

Food expeditions were both risky and unpredictable. One never knew if all those who set out would return. Nor did one know what they would bring back. Some items were chronically in short supply, especially bread. It was impossible to find enough; people longed for bread and continuously talked about it.

When a bambioshka was about to return, "people would run out to see what it brought . . . they would touch the goods and if they came upon bread they would not leave. They would form a line. Those at the head would get some bread, but those behind would not get any."[22] This would lead to resentment and envy.

By October 1942, the Bielski otriad, which by that time numbered about 200 people, established a base in the two connecting forests, Perełaź and Zabiełowo. Because of cold weather and the threat of snow they decided to build more permanent quarters. The otriad tried to collect more food to last for part of the winter.

A more permanent forest home or bunker was called a "ziemlanka," a name derived from the word soil. In part it was a dug-up rectangular space. When finished, two-thirds of the bunker was in the ground and one-third above. The upper part consisted of wooden sticks or branches that touched each other, forming a pyramid. These were in turn covered by sheets of bark, leafy branches, or other materials that kept out extreme temperatures and helped camouflage the place.

If constructed properly, the inside walls of such a ziemlanka were lined with wood. This in turn prevented the soil from crumbling. Instead of beds, around the walls were wooden bunks covered with straw or leaves and, if available, flour sacks.

In the middle, between the two rows of bunks, there was an iron stove, with a pipe leading to the ceiling and outside. In the winter months this stove would cook the food and heat the place. Depending on size and need, a ziemlanka could accommodate up to forty people.

In preparation for their first winter, 1942–1943, Tuvia wanted these ziemlankas to cover a wide area. He felt that by scattering them he would be spreading the risks. If the enemy discovered one bunker, the rest had a chance to remain unnoticed. Each bunker was large, accommodating thirty to forty people. One was to serve as a hospital.

For most of the people this was their first forest winter. This alone created anxieties and fears. People were highstrung, and emotional imbalance led to much grumbling. Complaints were frequently directed at the most vulnerable among them: the newcomers.

A few made arrangements to stay in more protected places, with peasants. This method was used by part of the Boldo family. The Bielski brothers and their families could have found more comfortable quarters with some of their Belorussian and Polish friends, but they did not. They would not abandon the people. Not only did they themselves stay in the forest, but they insisted that those close to them should do the same. Others, the majority, had no choice and remained in the forest.

For those who stayed, tensions mounted as the temperature dropped. New people continued to come. With so many new arrivals, "not everyone wanted to take the old and the children. After all they were special burdens. Each bunker had a leader. Each tried to have fighting men that would be easy to run away with. There were conflicts. Tuvia made the assignments and arrangements."[23]

One day Tuvia was confronted by a representative from a group of twelve fighters that included Motl Berger, the Lubczanski brothers, Pesach Friedberg, and eight others. Attached to this group were also twelve non-fighters; among them were wives and sweethearts.

This representative came to ask permission to dig separately because his group did not want to be detained by those slower than they. On the surface this seemed like a reasonable request and it was granted.

At first it turned out well; not to be shamed, other diggers worked faster. But this became a prelude to something else. When the digging was nearly finished a representative of the same group, this time one of the Lubczanski brothers, came with new requests. His group's fighters wanted to decide themselves whom to include and exclude from their bunker. This clearly could lead to the elimination of hard-to-place individuals, usually non-fighters. The group also wanted to control their own food supply. They wanted more autonomy and were trying to create an otriad within an otriad.[24]

"And what if I will not go along with your demands?" Tuvia asked.

"You will," Lubczanski replied in a threatening voice.

Asael jumped up. "Will you fight us?"

"We will if we have to," came the answer.

Furious, Asael pointed to his submachine gun. To this Lubczanski said, "Our rifles shoot too!"[25]

Asael and Zus were ready to attack the man. Tuvia stepped in and prevented a fist fight. Instead, he called for a general assembly and presented the case for a review. He summarized the situation and then asked for reactions.

The first partisan to speak up said that permission to dig separately was a divisive step that should not have been granted. Therefore, the entire otriad was to blame for what had happened. No sooner did he finish when Victor Panchenko made an unexpected appearance. Naturally, he was curious about the meeting. Tuvia filled him in and asked for advice. He knew that Victor would condemn the rebels and that his opinion carried weight.

Unhesitatingly, Victor said, "You took them out of the ghetto, you trained them, you spend all of your time with them, and now they dare rise up against their commander? Shoot them, here and now!"[26]

Tuvia turned toward the assembly. "Who would like to express an opinion?"

"You're the commander, do whatever you please," came the answer.

Tuvia asked Panchenko to help him divest the rebels of their arms.

Having received Victor's consent, Tuvia announced that the Lubczanski group should turn in their arms immediately. They did not move. The three brothers approached them carrying their submachine guns, followed by Panchenko with a revolver in hand. Tuvia grabbed each man's rifle as Panchenko placed the revolver to the man's head.

All twelve rifles were collected.

This happened quickly, in total silence. When it was over, the crowd just stood there stunned. Above them they heard their commander's voice. "All twelve members are exiled from the company because of insubordination and are to leave the area within twelve hours; should they not leave, they will be shot. There will be discipline, we cannot exist without it. If we cannot achieve order by ordinary means, it will be done by force."[27] With this he formally dismissed the assembly.

Almost immediately word reached the Bielski brothers that the rebels were in disarray. Their wives and non-fighting relatives were shocked. They were accusing the rebels of insensitivity to their own welfare and that of the otriad. There were quarrels and name-callings. Then a few of the rebellious twelve, with Pesach Friedberg among them, came to plead with Tuvia. Clearly Pesach had changed his mind when he said, "Let them go to hell, but I do not want to be with them!"

Tuvia yelled at him, "You're big and fat like an oak, but where was your head? Couldn't you slap your friends down before they brought this trouble down upon themselves?"[28]

The verdict remained intact; the commander was unmoved. People continued to come and plead and the circle of those who tried to intervene widened.

Then a delegation of seven, representing the entire otriad, came with a message. "The entire company wishes to beg your pardon."

Once more Tuvia reassembled his people and made the following announcement. "If all of you will take on the responsibility for the future behavior of the transgressors, I'll agree to take them back."[29]

The crowd murmured its approval. The dispute was over. But only two weeks later were the rifles returned, gradually, one by one. There were no more objections. Order and discipline were restored.[30]

The Bielskis continued to remain in the two adjacent forests, Perełaź and Zabiełowo. Neither was very big or very dense. In terms of size and promises of safety these wooded areas did not stand up to the huge Lipiczańska forest.

In November 1942, a group of Jewish fugitives went from the Bielski otriad to the Lipiczańska forest. The group consisted of fugitives from Nowogródek ghetto, non-fighters, with pregnant women and children among them. They arrived at their destination at the beginning of December while a large raid was in progress. Most perished. This group was escorted by six fighters from the Bielski otriad, among them Lazar Malbin and Motl Berger. After about a month, disheveled and hungry, the six escorts returned to the otriad without the civilians.

Different versions of this event appear when one tries to move beyond the known facts and into the "whys." Tuvia says that a group of runaways from Nowogródek ghetto came to their otriad when winter preparations were in progress. The Bielskis fed them and let them stay over. Next day, another group of Jewish fighters, part of the Russian Orlanski otriad, also passed. This second group just completed a military mission and was returning to their base in the Lipiczańska forest. The fugitives from the Nowogródek ghetto decided to go with this second group. Tuvia continues, "We did not oppose their decision, we even sent an escort to accompany them. This group came to a bad end. Most of them were killed in hunts, escapes and wanderings."[31]

But why did they need an escort if they were going with Jewish partisans? Chaja Bielski says that they were, indeed, fugitives from the Nowogródek ghetto. She then explains that because they came "with children and old people it was decided to bring them to Lipiczańska forest. There were many family camps there. We did not know that there would be a raid. At the time we were in small forests and Lipiczańska was very big. We thought that they may be better off there than here in the small forests. . . . We did not see it as a bad thing." But then she adds, "Each bunker had a leader and none of them wanted to receive them. There were too many people at that time. I am not saying that it was right what we did. We should have built more bunkers and settled them with us."[32]

Motl Berger, who seems to dislike Malbin, blames the incident on him.

Motl feels that Malbin talked Tuvia into sending the forty people away from their camp, most of whom were unarmed women and children. "These people were told that it would be safer for them in the Lipiczańska. They had no choice, they had to leave . . . they arrived during a raid . . . I don't know what happened to them."[33]

Esia Lewin-Shor, a Bielski partisan, cries when she talks about them. Like Berger, she thinks that Tuvia was pressured into sending this group away. Although she refers to this event as "heartbreaking," she nevertheless says that "they felt that in Lipiczańska they would not be as easily detected."[34]

When I brought up this incident with Sonia Bielski, Zus' wife, she hesitated at first and then said, "They had relatives there. . . . They wanted to leave because we had too many people, men and women. I don't remember the reason. . . . They did not like the Bielskis. They must have left on their own, we never sent away anyone. . . . This was a tragedy. When they came, there was an 'obława' [a raid]."[35]

The news about the raid had a shattering effect on the Bielski people. Some blamed themselves for not preventing the move from happening. The entire otriad, and especially Tuvia Bielski, had learned a hard lesson.

News about this raid came in December 1942, along with a warning that the Bielski camp was to be attacked. Hurriedly, in the middle of the night, they all left and went to the Chrapiniewo forest, close to the town of Iwje.

In this new location they could erect only a few primitive tents. The soil was frozen, which made digging particularly difficult. As soon as they were settled, word reached them that the Germans were about to liquidate Iwje ghetto.

Before 1939, Iwje had been a small town with a Jewish population of less than 4000. On May 12, 1942, the Germans murdered 2000 Jews and forced the remainder into a ghetto.[36] Most of these ghetto inmates were convinced that they too would soon be killed. Here, with support from the Judenrat, the people began to prepare for an eventual ghetto breakout. They managed to buy a few arms. Baruch Kopold, a teenager at the time, remembers, "We kept guard in the attics at the edges of the ghetto to see what was happening. We were all young, the Judenrat helped us as did the Jewish police."[37]

If they saw any suspicious moves by the enemy, they were supposed to cut the wire that surrounded the ghetto and run. Kopold belonged to an underground group of fifty; his was one of several such groups who cooperated loosely with each other.[38]

Although at the end of 1942 the group was in a precarious position, Tuvia decided to save the Iwje Jews. He sent four armed men to the ghetto to warn about the impending end of their community. The four were to urge the people to come to the forest. In the meantime, instead of leaving with his people, Tuvia and about twenty of his men stayed to prepare the place for the Jews from Iwje and show them how to live in the forest.[39] As planned, as soon as the Iwje Jewish guards saw the Germans and their Belorussian helpers come from one side they cut the ghetto wires from the other side and escaped. When the Germans reached the ghetto they found very few people.[40]

Tuvia and more than twenty of his men received the Iwje runaways in the Zabiełowo forest. These fugitives wanted to stay together. Tuvia recalls, "We instructed them on life in the forest, guided them, divided them into sections and helped them choose leaders from their own group. They had only four rifles so we chose armed people of our own to accompany them on their first trip to forage for food."[41]

As for the rest of the Bielski otriad, the rumored raid on Perełaź and Zabiełowo never materialized, so most returned to their abandoned bunkers.

Regina Titkin had a flu and fever. She and a few others asked Tuvia's permission to stay over in two hutors, close to the Chrapiniewo village. Although uneasy about their safety, Tuvia consented because he felt sorry for them. For protection, he sent a few fighters with them, one with a machine gun.[42] In all there were twelve people.

Chaja describes the event. "These were Belorussian homes. They gave us permission to come because we had arms—they knew that we could force them to. We entered the two houses. I prepared breakfast. We began to clean the guns. Only a few stayed with us. Israel Kotler, Lova Volkin, Itzhak Batch, and Sonia (Zus' wife) were in one house. In the second were Sonia Bielski (Tuvia's wife), Regina her sister, Grisha Meitis, and a few other fighting men. . . . The two houses were close to each other. The name of the village and the forest was Chrapiniewo. We went to rest, and did not post guards. Everyone wanted to rest. No one gave orders."[43]

The four men who went to warn the Iwje ghetto returned, hoping the people were already on their way. Chaja was anxious. She reasoned that if the ghetto runaways would come by their place, they might be followed by policemen. Therefore, from time to time, she went to check for any suspicious signs. Each time she left the house, the freezing weather forced her back in after a few minutes.

But then she saw "hundreds of Germans and policemen, in white fur coats surrounding the place. They wanted to take us alive. There was probably a denouncement that the Bielski brothers were there. I burst into the house and screamed: 'We are surrounded!' Sonia (Zus' wife) slept in her boots. She jumped off the bed and screamed, 'I am lost, I am lost.' I dragged her with me, saying that it is better to die from a bullet than to be taken alive. Only Kotler had time to run away with us. Outside Sonia fell into a hole filled with potatoes. It was covered with snow. We did not see that it was a hole. She screamed: 'Go save yourself, leave me!'

"The Germans still don't shoot. They want to take us alive. I heard the screams, 'Hurrah Bielski, Hurrah Bielski!'

"Then Israel Kotler began to shoot at them. I dragged Sonia out of the hole and took her with me. We moved on, realizing that the Germans were coming closer. I began to shoot as well. We were trying to keep them at a distance. This way we came close to the forest and entered it. Then they began to shoot. At that point it seems that they attacked the two houses."[44]

Grisha Meitis, Regina's son, Lova Volkin, and the others continued to shoot until they ran out of bullets. When the enemy took over they found only one partisan who was alive: Lova Volkin. After the Germans took him they set fire to the two farms.

In Zabiełowo, while training the Iwje people, Tuvia and his men learned of the attack and quickly left for Chrapiniewo. On the way Tuvia and his fighters heard shots. Soon the shots became less frequent and then stopped. At the edge of the forest the Bielski partisans saw the two burning houses and witnessed the departure of the Germans and Belorussians. Then the Jewish partisans approached the burning ruins. Their search was disappointing; it yielded only parts of dead bodies.[45]

Lova, the partisan manning the machine gun, was injured. "The Germans brought him to a hospital. They tortured him to make him divulge our secrets and the secrets of other partisans. In spite of the horrible pain, he told them nothing. They abused him, cut his flesh, but he did not surrender. After three weeks they hung him in the middle of the town of Nowogródek. A note attached to his body warned that this would be the fate of all Jewish partisans."[46]

During the Chrapiniewo attack nine members of the Bielski otriad lost their lives. The three who survived were Chaja Bielski, Israel Kotler, and Zus' wife, Sonia.[47]

This attack interrupted the training of the fugitives from Iwje ghetto. The Germans announced that whoever returned to that ghetto would be admitted without any punishment, and some of the fugitives did go back. The Germans kept their promise for less than two days—after that, most of the inmates were murdered and a few moved to Lida ghetto.[48]

In the Bielski camp the death of the nine partisans shook up the entire otriad. Particularly affected were the relatives of the dead, especially Tuvia Bielski who lost his wife. Some say that Tuvia was a "broken man."[49] "He was in a deep depression, because he loved her very much."[50]

This loss pushed him out of the otriad. Zus went with him. More than forty years later Sonia Bielski, Zus' wife, is still hurt when she remembers. "Zus told me that I should go to my parents, that he could not leave Tuvia. One week the two of them were riding around. I went to my parents. I said that this was the end. If he can leave me to be with his brother, then I don't need him. Six weeks I stayed with my parents."[51]

Tuvia used to drink before the death of his wife and he knew how to hold his liquor. In the forest most people found solace in hard liquor. Vodka, most partisans felt, kept them going.

When Tuvia returned to the otriad he was deeply depressed and his drinking became heavy. People were concerned about his sadness as well as his drinking. Lilka Titkin and her father, Alter, were particularly worried. Regina's and Sonia's deaths had cut their official family link and Tuvia remembers Alter's pleading: "Do not become estranged from us."[52]

He reassured his former brother-in-law by moving back to the ziemlanka they used to share before the tragedy happened.

In love with Tuvia from the day she met him, Lilka insists that not she but he asked her to be his "wife." Still she admits that she "grabbed" the offer. Seventeen at the time, innocent and very beautiful, Lilka recalls her father's warning. "'I see that your relationship with Tuvia is not innocent. If you open the door you will go from hand to hand.' He did not believe that Tuvia will stick with me."[53]

Alter Titkin was both right and wrong. Lilka remained the official wife, but she did have a lot of competition. Tuvia was very successful with women.[54] Most of the women were after him and he tried not to disappoint them. Lilka admits to much suffering. "The women used to come to my bunker and proposition him. He was not a person to tell me or to humiliate me. I am sure he did it. He never came to me and said: 'I slept with this one or that one.' Sometimes I knew and I reproached him: 'Is this fair. She is supposed to be my best friend?' He used to say, 'I don't talk about these things. You want to find out? Not from me.' Sure it was terrible because I wanted to be the only one. . . . It was killing me."[55]

Yet, when asked how her husband felt about her, she says, "I think that he was crazy about me, till the last minute of his life. . . . Before he died he could not talk but communicated to the children that the only obligation they had was to watch over mother. He was a kind, fine man, very generous. He taught me a lot."[56]

Although many young women in the otriad would have been happy to become Tuvia's mistress, he did not limit his sexual escapades to the camp. In the surrounding villages he had all kinds of women, some Belorussian, some Polish.

In the forest, with so many women throwing themselves at him, he seemed not to have even tried to move his relationship with Lilka to another plane. He did what he liked. And he liked women.

A tragedy brought Lilka and Tuvia together. Other events continued to cement their relationship, and their marriage endured for more than four decades.

8

Eluding the Enemy

The twenty-five guerilla fighters of the Iskra otriad arranged themselves in a circle. The man responsible for this assembly stood in the middle and seemed to tower over them, not because of actual height, but some intangible inner strength.

His self-possessed manner matched his vigorous voice. "I bring you greetings from our fatherland. To continue defeating the enemy we must make changes. From now on I will be in charge of this otriad. Your commandant will be second in command."

Some of the men seemed to be shifting in their positions, others bowed their heads. Did they search the ground for the words so unexpectedly thrown at them? There was movement, but the distance between the man inside the ring and those around him remained the same.

Casually, as if it was the most natural thing, Sinichkin reached for his belt. Then, with revolver in hand, he continued, "Whoever is against this change let him come forward!"

All motion came to an abrupt stop. The men stood still, very still. Embarrassed, avoiding each other's eyes, they listened to the new commander's order to disperse.[1]

This was the beginning of 1943, right after the German debacle at Stalingrad, a turning point in the German-Russian war.

On February 23, 1943, Stalin issued a special appeal to the Russian partisans urging them to continue their courageous battles.[2] Also about that time the Soviets increased their efforts to utilize the partisan movement in the war against the Germans.

Sinichkin was a part of this renewed effort. In a group of about twenty men he had parachuted into Western Belorussia. Capable and well trained, he quickly reorganized Iskra. Sure of his effectiveness, he turned to another partisan group and then to another, bringing them together into more effective fighting forces. With the rank of general, he became a commander of a partisan brigade. Along with other partisan commanders Tuvia was invited to meet Sinichkin. At this gathering their host emphasized how proud the Soviets were of the partisans' achievements. To make their successes more effective, Stalin had sent special organizers.

Tuvia heard that "The great commander Platon had arrived. He and Sinichkin had come to reorder and unite different companies and make them part of brigades that would all work together according to a unified plan. Every disturbance caused to the enemy would be helpful to the front. It was necessary to cut rails, burn bridges, cut telephone and telegraph lines. It was also necessary to make the availability of food difficult for the Germans."[3]

Tuvia was struck by the differences between this and other assemblies. Here he noticed definite order. Each commander gave a report about his detachment while a secretary kept records of everything that was said—definite new bureaucratic ways for conducting meetings. Sinichkin showed a special interest in the Bielski otriad and assigned it to his own brigade.[4]

Similarly, the arrival of General Platon, who was in chage of Russian partisans in the Baranowicze region, further emphasized how much importance Moscow placed on an integrated partisan movement.[5]

The meeting in the Sinichkin brigade was followed by a get-together at General Platon's headquarters that involved a few hundred commanders, Tuvia among them. Many had a chance to speak with the General. When Tuvia's turn came he jokingly said that an important man like Platon should have clothes to fit his position and that he could take care of this matter. Platon laughed. "Done. Let me see you do it!"

Toward the end of his life, when he was eighty-one years old, Tuvia concluded this story with a chuckle. "Our Jewish laws say 'do not take bribes,' so I did not take. But the laws do not say 'don't give.' So I gave."[6]

Winning the war, although a top priority, was only one of Stalin's objectives. Beyond the defeat of Germany, the USSR had other political agendas, some tied to Poland's destiny.

Once Stalin felt more secure about the outcome of the war, he began to press his allies, the United States and England, for recognition of the Polish-Russian borders as specified in the Ribbentrop-Molotov Agreement. This called for a return to the USSR of the Polish land occupied by the Red Army in the fall of 1939. Stalin's aspirations, however, went beyond that. He also wanted a Moscow-sponsored government for all of postwar Poland.[7]

Determined to win both militarily and politically, Stalin set out to bureaucratize and politicize the Russian partisan movement, and Western

Belorussia with its extensive forests and thousands of forest dwellers became an important part of his plan. As the Soviets consolidated their control over the partisan movement, they established two separate centers of power: one military, the other political.[8]

With time, the Russian partisans gained control not only of other partisan groups but also actual sections of the forests. This control spilled over to adjacent towns and villages. By mid-1943 it was not unusual for local authorities to avoid these partisan enclaves. Such territorial takeovers were uneven, frequently punctuated by exceptions and changes.[9]

Although they were losing ground against the Red Army and were often hesitant about confrontations with guerilla fighters, the Germans never faltered in their dealings with civilians. German military losses went hand in hand with crueler treatment of defenseless civilians. Themselves suffering from defeats, the Germans would strike out and hurt those who were close by and too weak to openly wage war. Often without a shred of evidence the authorities would accuse some peasants of partisan cooperation and destroy entire communities as a reprisal.

When the Germans discovered that partisans had received food or shelter in a village, or if a German was attacked in the vicinity of a village, its inhabitants were quickly punished by death or deportation.

Through swift and vigorous punitive measures the Germans had hoped to stop or reduce partisan operations. Actually, these retaliations seemed to have the opposite effect. Whenever an anti-German incident happened, and if people found out about it in time, some young men would flee into the woods to join the partisans. More men in the forests only helped expand the anti-German forces.[10]

In 1943, the Germans stepped up their attacks on local villages, murdering and forcibly transporting Belorussians and Poles for slave labor. As a result, the Moscow head of the partisan movement, Voroshilov, and his Chief of Staff, Ponomarenko, issued a directive to save all civilians.[11]

With this, as with other directives, there was no guarantee if, when, and how it would be implemented. Until the very end, the partisans would comply only partly with Moscow's orders. A gap between what the government wanted and what the partisans were willing to do continued. Still, a directive, even if partly enforced, offered some benefits.

Jews who roamed the countryside and the forests were supposed to be helped by this directive. Ironically, only after the Germans stepped up their persecutions of non-Jews were the Jewish fugitives entitled to some protection. For the Bielski otriad, the Soviet call to help civilian refugees left some unsolved problems. Tuvia still had to convince the Russians that he and his people were useful supporters of the USSR.

He did this in several ways. The Bielski-Panchenko cooperation in anti-Nazi moves helped project their image as fighters. In addition, the Bielski partisans served as guides for their Russian counterparts. Familiar with the area, the Jewish youths knew how to avoid dangerous roads and were glad to share this knowledge with their fellow partisans. Also, the special relationship the Bielskis had to many of the local people helped them collect crucial information that they often passed on to Russian detachments.

While the USSR was winning the war, the Jews were losing it. By the beginning of 1943 most Jewish communities in the region had ceased to exist, among them Mir, Nieświeź, Iwje, Żołudek, Zdzienciół, and others.[12]

The Lida and Nowogródek ghettos, although still in existence, had lost most of their people.[13] Tuvia knew that it was only a question of time before all of them would be murdered.

Struggling hard to protect the people in his otriad, Tuvia fought equally hard to save Jews who were not a part of his group. Executions of Nazi informers by the Bielski partisans led to a drop in the number of denunciations. Fewer denouncements made the roads less dangerous—not only increasing the chances of survival of those who escaped from the ghetto, but also giving courage to others to try and come to the forest.[14]

The Bielski otriad also continued to send guides to the remaining ghettos. In addition to Kozłowski and other locals, there were also Jewish partisan guides, among them Orzechowski, the Oppenheim brothers, Gulkowitz, Druk, and others. Sometimes these guides took people out according to lists. Those whose names were on the lists were usually relatives and friends of people already in the forest.

Tuvia, for example, sent an invitation to his friend Chaim Dworecki, his wife, and his two teenage daughters. The Dworeckis were forced into the Lida ghetto during the liquidation of the Iwje ghetto. In his letter Tuvia wrote, "It does not interest me if you have a gun or not, you are coming with your wife and children. Nothing is important. Just come." They did, penniless and without guns.[15]

If a person on a list was not available, guides had instructions to take a substitute. At first, Tuvia would have preferred young men with arms. This specification, however, was never enforced. Soon it was altogether dropped. Sometimes guides were told to bring anyone willing to come.

Although most guides were honorable, courageous young men, the situation left room for corruption. For example, at the suggestion of the partisan Abraham Viner, who was a student of the historian Shmuel Amarant, Tuvia invited Amarant and his wife. The guide who was entrusted with bringing them took someone else instead because they paid him.[16] No matter how hard Tuvia tried he could not fully control his guides from the forest.

Many of those who ended up in the Bielski otriad did so only after long and painful wanderings. While searching for shelter, some had lost their families and met with rejections from Russian partisans. Some stumbled on the Bielski otriad by chance; others were directed to it.

To make it easier for these roaming fugitives, Tuvia Bielski sent special scouts out to look for them. And whenever these scouts heard about Jews in need of protection they would try to find them. Their searches included people who were hiding alone or in small groups in the forest, or fugitives who had overstayed their welcome with peasants.[17]

One example was Hana Berkowitz. During an Aktion in Lida her husband had urged her to run away with their baby daughter. At first she was kept by a Christian woman, then because of outside threats she left. Friendly peasants told her about the Bielski otriad and that she would be accepted there. But finding the otriad was a problem.

After much suffering and many rejections, she met, by chance, the Bielski scout and guide, Jacov Druk. The man was very friendly and took her to the Bielski camp. She recalls, "Tuvia received me with tears in his eyes. 'Where are you coming from,' he wanted to know. I told him my story. 'So few of them saved themselves!' he said. He accepted us. He was so moved."[18]

At one point they were searching the forest for Abraham Viner because someone reported that he was in need. They brought him to the camp with a teenage girl. Viner remembers, "I arrived in torn clothes without shoes. I had rags around my feet. The girl too was in rags; we both looked very disheveled. People around us said, 'Here we have more malbushim.' They were annoyed. . . . But Asael's and Tuvia's attitude was very warm. Tuvia turned to me with: 'You will live in our place.' This made a deep impression and stayed with me all the time. What did I think? I thought that I had met the Messiah. How can one think in any other way?"[19]

But not all those who joined made an effort to adjust. Chaja remembers a grandmother who came with a two-year-old grandchild from the Lida ghetto. At the beginning of 1943, "the grandmother found life too hard in our place. She did not have a brassiere and said that she could not live without it. She was going back to the ghetto. Asael approached her, this I remember, and screamed at her. 'How can you think about a brassiere at this time! At least don't take the child! Leave it here!' She left with the girl and both were murdered."[20]

Although some Bielski partisans moved to Russian otriads they knew they would always be welcomed back. Indeed, the Panchenko-Bielski cooperation was soon cemented by a romantic tie that developed between Victor and Bela, a Bielski partisan, a young woman from a nearby ghetto.

In the forest Bela's good looks, winning smile, boundless energy, and generous, giving disposition made her a favorite of many. Men in particular found her charming. No one was surprised when the equally attractive Victor Panchenko was smitten with Bela. More than forty years later her voice becomes a bit softer when she describes him as a fine man, good-looking, good-natured, with a "Russian soul." During his visits to the Bielski otriad he fell in love with her, and Bela reciprocated his feelings. In less than a month she moved to the Octiaber otriad and became Victor's official mistress.

Some feel that their love affair resembled the ancient story of Esther, the Jewish heroine who, through her attachment to the king, saved her people. On several occasions, Bela intervened with Victor on behalf of the

Jews. Sometimes this intervention led to an acquisition of arms, medication, and special privileges. Tuvia was aware of these benefits and hinted that Bela became Victor's mistress because she wanted to help her fellow Jews. Highly principled, warm and outgoing, now over sixty and married, Bela denies that her love affair with Victor was motivated by anything but love.

Sometime in February 1943, Bela came to talk to Tuvia. Distressed, she asked, "Will you take me back? I want to leave Victor."

"What do you mean? Of course. This is your home. It is a pity you left in the first place. You belong here, you should have stayed with us."[21]

Although the Bielski otriad had profited from Bela's love affair with Victor Panchenko, Tuvia did not hesitate for a moment. He asked for no explanations and Bela was too upset to offer any.

Political rather than personal changes were responsible for the breakup of this love affair. For a few months their happiness flourished. Then, around Christmas, two Ukrainians joined the otriad and when Bela listened to their talk she heard that as avid anti-Semites they were attacking defenseless Jews. More often than not, their assaults ended in the victims' deaths.

Bela recalls, "When I heard this I was hurting. I was appalled. I told Victor. He tried to get out of it, minimized the importance, but he did not stop it. Then the Ukrainians began to talk against me. I heard that too. Victor did not take a firm position. He might have been afraid of them. Or maybe they influenced him? . . . He hesitated. Then they said that he should get rid of me. They threatened to elect another leader if he did not."[22]

When Victor failed to act, Bela told him she was going back to the Bielski otriad. Victor agreed to this and promised her that in her absence he would take care of the situation and then she would return. They remained friends and Victor would come to visit Bela, but she never rejoined him.[23]

The arrival of the two Ukrainians demonstrated some of the changes taking place in the Russian partisan movement. Before 1943 Russian partisans would shoot anyone suspected of Nazi collaboration. But then a new directive came from Moscow: All who want to join the Russian partisans should be accepted. By taking all the able-bodied men, the Soviets wanted to deprive the Germans of their much needed manpower and speed up Germany's defeat.[24]

Although not officially sanctioned, anti-Semitism among Russian partisans affected the fate of Jewish fugitives. To many Jews, even the potential fighters, it denied membership in Russian otriads. For those who were already part of Russian detachments it made life more difficult.

Indirectly, anti-Semitism led to an expansion of the Bielski otriad. Some who were rejected by or mistreated as members of Russian detachments came to the Bielski camp. Tuvia explained how at different campsites he tried to cope with large size and limited movement. "Our camp was spread out, in sections, over an area of over ten kilometers. I used to think of the Biblical phrase: 'Should one part of the camp be attacked and overcome, the other part will remain.' This strategy was used by our forefathers, I always remembered it, and used it in planning our activities."[25]

Tuvia was busy visiting these scattered groups. He felt that his personal appearances helped reduce existing tensions. In fact, while making the rounds he dealt with many accusations about favoritism and unfair treatment by different group leaders.

Patiently listening to the people, he would smooth over the rough edges of these conflicts. Through his show of concern Tuvia earned his people's loyalty. This loyalty and approval, in turn, helped him establish a firmer grip over the otriad. Tuvia knew that an orderly and well-run unit prepared him for all kinds of eventualities and dangers, ranging from enemy attack to health epidemics.[26]

Indeed, one day, at the start of 1943, Dr. Isler, the camp physician, diagnosed the first case of typhoid fever. Before he had time to do anything about it, a delegation came to Tuvia, demanding that the patient be shot. They argued that the sick man was useless and dangerous to the rest.

Dr. Isler objected to this kind of "treatment." He explained to Tuvia that if there is one case many more might already be infected. If they will shoot this man, who knows how many more will have to be killed? Tuvia understood and gave Isler a free hand to act as he saw fit.

The doctor decided to put the sick man into an empty ziemlanka. As he and his helpers were transporting the patient on a primitively constructed stretcher, their path was blocked by a group of fighters. At the head was Srulik Salanter from Iwje.

Known for his arrogance, Srulik spoke up. "You are not going to take this man anywhere. I will not allow you to save his life and risk ours!"

In a split second, Dr. Isler reached for his gun. Directing it at Srulik's head he said, "If you dare to go for your gun I will shoot!"

The effect was immediate. Salanter moved away and his supporters followed. As if from nowhere a crowd appeared. Somehow they knew what it was all about and most sided with the doctor.

The patient recovered. Later, and for as long as they stayed in Zabiełowo, this bunker served as a hospital.[27]

An authoritarian and strong leader, Tuvia made all the final decisions. Before he did, however, he would consult with his two brothers, Asael and Zus, and his Chief of Staff Malbin. He felt that each of the three was qualified to give different advice. When it came to military strategy and organization, he valued Malbin's opinion. Asael, a hard-working and courageous fighter, was devoted to the otriad. Tuvia would consult him about different threats and send him on dangerous missions. Zus was in charge of intelligence. With a group of riders he would travel all over the area collecting important information. Frequently, the very survival of their group depended on the material gathered by Zus and his men.

One of the central activities was watching the camp. Whoever was assigned to watch duty was supplied with a gun. Men and women both took turns at guard duty. "Sentries were organized in circles . . . one was inside the camp. Then there was a ring around the camp some hundred meters on the outskirts. There was another ring further away, a few kilometers outside the camp. The camp was subdivided into groups. Each group had a leader who was a fighter. It was his responsibility to select people to watch."[28]

Anyone who came close to a guard would give a proper password. Frequently Jewish fugitives were ignorant about passwords and would use common Yiddish and Hebrew expressions. Guards were supposed to shoot at an enemy's approach. Such shots would retard the enemy's attack and alert the people to danger. Any shooting automatically exposed the sentry to peril; some watchmen had, indeed, paid with their lives.

In February 1943, a group of young men went on a food expedition into the Nowogródek region. As sometimes happened, they each had had a few drinks, which made them less vigilant. Part of their captured loot included some freshly killed chickens. On the way to the base, the blood from the chickens dripped, leaving spots on the snow-covered road. In the morning these red dots brought Belorussian and German policemen to the forest.[29]

Oppenheim, an "old" man of sixty, was keeping watch. When he heard the sleighs approach, he thought it was Tuvia returning from a meeting. Once he realized his mistake it was too late. The policemen did the shooting and a bullet hit Oppenheim's nose. He fell. His attackers kicked him to see if he was alive, but he made no move.

Some say he fainted, others that he pretended to be dead. When the policemen continued on their way, Oppenheim managed to crawl into the ziemlanka that served as a hospital. His wound proved superficial.[30]

The shot that hit Oppenheim announced the assault, one kilometer from camp. When the enemy moved closer, ready to attack, the second guard began to shoot. This sentry, Ephraim Kopelman, tried to keep back the aggressors. Indeed, by the time a fatal bullet ended his life, all the people of the otriad had dispersed into the forest.[31]

The police arrived at an empty camp. They destroyed the bunkers and took away cows, horses, and wagons filled with provisions.

Tuvia reached the camp after the attack. He says, "On this sad day there was one note of cheer, the police had not discovered our hospital. When I visited the patients that evening they welcomed me joyously. I inquired if all was well, they said, 'Yes, friend commander, we are all well.'"[32]

Because the assailants had killed only one guard and injured another, they were bound to come back. Therefore, on the next day the otriad moved some fifteen kilometers from Zabiełowo to Boczkowicze. There, unable to dig the frozen soil, they improvised primitive tents. But soon this place proved dangerous. The same happened with another and yet another site. Every few days they had to change places. Even at this difficult stage, with three hundred people on the move, the otriad sent guides to the ghettos, urging the people to come. These letters of invitation argued that one was not likely to die from cold.[33]

At forty-nine, Alter Titkin, Lilka's father, was considered too old to fight or go on food expeditions. Alter disagreed with this assessment. Feeling vigorous, he resented being a burden and insisted on going with the young for food collections.

Tuvia discouraged him. "Think of Lilka, how would she feel if something happened to you?"

"Nothing will happen," was Alter's stubborn answer.

Alter maintained this position and this exchange was constantly repeated.

In March 1943, Abram Polonski was scheduled to lead a food mission in the Nowogródek region and Alter was eager to join it. This time, reluctantly, Tuvia consented to his request. Nine young men, Abram and Ruven Polonski, the Szumanskis and others, all good fighters, were going. Alter Titkin was the tenth.

News about the trip made Lilka unhappy. She begged her father to give up the idea, but he would not hear of it. The father-daughter confrontation ended on an unfriendly note. In a huff, offended, Alter Titkin left to join the nine young men.

That evening, as Lilka sat close to the fire, she saw her father come toward her. Excited that he had listened to reason, she said, "I am happy that you decided not to go. You realized that I was right."

"I just came back because I thought that I did not say goodbye to you the right way."

Lilka remembers that "He hugged me, he kissed me, he waved to me as he was leaving. Then he and the whole group disappeared from my eyes."[34]

As planned, the Jewish partisans filled their wagons with all kinds of provisions. When they were ready to turn back they realized that darkness might give way to light earlier than expected and daylight meant danger. They were a few miles from the town of Nowogródek, close to the police station. To play it safe, they decided to postpone their return.

In a nearby hutor the Polonski brothers had a Belorussian friend, Belorus. They went to his place and asked if they could stay over. "Belorus welcomed them, served them food and drinks, and arranged accommodations for a good rest. They, of course, posted sentries. . . . The hamlet was attacked. Belorus had informed the police about the presence of partisans in his house. He had sent one of his sons to Nowogródek to inform. When our friends awoke in a panic, it was too late to fight. They were all shot. The only one to remain alive was Abram who had hidden in the chicken coop under the oven. When he heard the police and their commander leave, and the family celebrating, he came out of hiding and announced: 'You will all pay dearly for this, the partisans will settle accounts with you!' At once an axe wielded by one of the Belorus sons came down on his head."[35]

Tuvia states, "We could neither forgive nor forget. If we do not take revenge upon such collaborators, what would become of us? Every farmer would feel free to inform! The scouts that we sent out to investigate came upon a group of Jews hiding out in a remote farmer's hut. One of them told the scouts that Belorus' godmother had been an eye witness. The Bielski partisans questioned her. At first she denied that she had seen anything, but after a good beating she told all."[36]

In two weeks they had assembled all the evidence. Asael was in charge of the next step. He collected twenty-four fighters, among them Pesach Friedberg, Israel Yankielewicz, Michal Leibowicz, Ben Zion Gulkowitz, and others.

When Dr. Isler asked to be included, Zus warned him, "You should not be coming. We are going to finish them off!"

"I want to examine their bodies to see if they are really dead," Isler answered. He was allowed to go.[37]

At midnight Asael's men surrounded Belorus' house. Four went inside with Asael at the head. As soon as they entered, they announced that this was a visit to avenge the death of their comrades. Four rifles pointed in the direction of the peasant.

Belorus wrestled with the rifles, but was quickly overwhelmed. The rest of the family had no more luck—in a few minutes all were dead. One of the fighters, Michal Leibowicz, with Asael's permission, exchanged his jacket for a coat that hung on the wall.

They freed the animals, set the entire farm on fire, and left a note explaining the reason for the attack.

They followed Tuvia's orders not to appropriate anything. He wanted everyone to know that this was an act of revenge, not a robbery. Except for the coat, nothing else was taken.

Later, in the coat's pocket, they found a letter from the Nazi commander of Nowogródek. In it the German thanked Belorus for helping to eliminate Jewish "bandits." He also expressed the hope that Belorus' neighbors would follow his good example. From the letter it was clear that in appreciation of the deed Belorus was rewarded with 50 marks.[38]

People continued to come to the Bielski otriad. Since most of these newcomers were unarmed, this created a greater need for weapons. Some guns were bought with money collected from the new arrivals. Luba Rudnicki, for example, gave Tuvia a watch and jewelry with which to buy two guns, one for her husband and one for her.[39]

Some weapons were confiscated during anti-Nazi moves. Some were given as presents by friendly peasants. Many were taken by force. From time to time peasants would tell about each other's possession of arms. The Jewish partisans would follow these leads and make the owners give up their treasures. Occasionally the information was false and the partisans demanded guns from people who did not have them.

In general the local population was caught in the middle. At night the partisans would come with their demands; during the day, the Germans. The peasants were afraid of both. With time the partisans became more powerful and took over entire areas, and the Germans stayed away, making it easier for the local population.[40]

Contact between the Bielski otriad and the ghettos involved more than bringing out inmates. Some entailed important exchanges of goods. In Lida, inmates who were employed in the military hospital would smuggle out all kinds of medication and give them to Tuvia's guides who would bring them to the forest. Stolen radios and batteries were occasionally included in these transfers.[41]

By April 1943, the unpredictable, nomadic existence of the Bielski group was temporarily halted at a place called Stara Huta. After several weeks this respite came to an abrupt end when guards reported sighting a truckload of German soldiers moving toward the camp. This was followed by an order to collect all who could not fight and move them deeper into the forest. The fighters, rifles in hand, were covering their retreat. They waited for the enemy; their aim was to delay the Germans' entry into the forest.

Then, the truckload of soldiers turned around and left. This was strange. Did the Germans see the partisans? Did they go for reinforcements?

With no answers, the camp had to relocate and transferred to the Jasinowo forest. Later the scouts learned that the scare had been a false alarm—the Germans had lost their way. They had not even seen the Jewish partisans.

With time, bigger numbers and better organization enabled the Russian partisans to inflict more serious damage. Derailment of trains and destruction of bridges and telephone lines weakened Germany's supply routes and curtailed the army's ability to fight. The Germans began to concentrate more vigorously on destroying the partisan movement.

The spring of 1943 brought more frequent enemy attacks. The Bielski partisans continued to move. And because the otriad was subdivided into groups, special scouts would ride over the area to maintain contact between the different subunits. Even at this stage the Bielski detachment kept growing.

Among the new arrivals was the partisan Grisha, someone Tuvia had met in one of the villages. A non-Jew, a Russian Cossack, Grisha asked to be admitted into the Bielski otriad with his Jewish wife Malka Zilberman, her father, and two brothers. They were members of the Lenin otriad where Grisha was the politruk, the person in charge of propaganda and political education. At that time Grisha suffered from a hand injury. He felt that the doctors in the Bielski otriad could do more for him than those in his Lenin otriad.

This transfer required official permission from the commander of the

Lenin otriad. When this was granted. Tuvia appointed Grisha as head of a subgroup.[42]

Soon word reached the partisan headquarters that the Germans were preparing a major assault on the Lipiczańska forest. At approximately the same time, four partisans from that forest came to Tuvia. They were the spokesmen for twenty-two Jewish fighters, all members of the Orlanski otriad. They asked to be accepted into the Bielski otriad. Although these Jewish partisans had distinguished themselves as fighters, they were afraid that during the expected "big hunt" their Russian "friends" would take away their weapons. By joining the Bielski group they wanted also to escape from constant exposure to anti-Semitic abuses. They asked Tuvia to arrange the official approval for their transfer.[43]

Eager to help, eager to acquire young armed partisans, Tuvia went to talk to his superior, General Platon. By the time he reached the headquarters Lipiczańska forest was under attack. Platon himself was nervous.

Tuvia mentioned but did not emphasize anti-Semitism or the discrimination that came with it. Instead, he spoke about the advantages that an acceptance of these partisans would have for the struggle against the enemy. He argued that with the approaching German raid these fighters could be exploited for the good of the country. Later, there would be enough time to investigate the situation properly. They could be admitted temporarily. Platon liked that.

"You've done well," he said. "You're a good Bolshevik and a loyal fighter. We must conserve our fighting powers in order to carry out Stalin's orders. Take them! We'll clarify matters later."

With a written order from Platon to temporarily accept this group, Tuvia returned to his base.[44]

It was hard to find provisions, but as one partisan explains, "Russian partisans resented the Bielskis because they collected so many unarmed people whom they saw only as a burden to be fed. . . . I had occasion to go twice on a food expedition and realized how much hostility the Russian and Belorussian partisans showed towards the Bielskis."[45]

Some Russian partisans felt that the local population was becoming more hostile toward all guerilla fighters only because the Jews had been confiscating too many goods. Jews were accused of robbing the local people of forbidden items.

There was a certain amount of truth to these accusations. Some Jewish partisans would take honey, eggs, and meats from villages that were friendly toward the partisans. This was forbidden. At Russian headquarters it was assumed that these luxury items could be confiscated only from pro-Nazi villages.[46]

Unable to fully control his people, Tuvia tried to keep the hostility in check by bribing some of the Russian commanders with an occasional leather coat that he succeeded in smuggling out of a ghetto or received from a member of the otriad.[47]

The Bielski otriad also tried to buy goodwill by extending its hospitality to Russian commanders. Sometimes the arrival of a partisan officer was connected to official or semi-official business. One day an official visit was paid by an otriad commander of who happened to be a stutterer. The Chief of Staff, Malbin, also a stutterer, and Tuvia were entertaining him. As customary, each man had a few drinks and, although each knew how to hold his liquor, their talk was gaining in volume and animation. Then, flushed, the Russian tried to say something. He found it difficult to bring out the words—his stuttering was in the way. Malbin, eager to rescue him, tried to say something. He too began to stutter. For the partly intoxicated guest this was too much. Furious, he jumped up and drew his gun. He was not going to tolerate the insult. Pointing at Malbin, the weapon was ready to defend the man's honor.

In a split second Tuvia placed himself between the two men. Shaken, Malbin listened as his commander explained that his Chief of Staff was a stutterer and that an unfortunate misunderstanding had happened.

Perhaps not quite convinced, yet embarrassed, the Russian returned his gun to its place. Avoiding each other's eyes, the men exchanged a few words that resembled an apology. With many more unspoken words hanging in the air, this uneasy gathering came to an end. Alone, Tuvia and Malbin still said nothing as each went in a different direction.[48]

In mid-1943, feeling pressured and aware that a German attack would lead to a loss of lives, Tuvia, after extensive consultations, decided to bring his people to the Nalibocka forest. Unlike the Lipiczańska woods, the Nalibocka forest was not under attack.

There were some similarities between the two places. A huge jungle-like forest, the Nalibocka also had many marshes and swamps. Here, too, no humans had ever ventured into some of its more remote parts. Whatever roads crisscrossed the area were poorly built and hard to travel. At the edge of the forest and scattered within it were small towns and villages.

Huge size together with its other special features made the Nalibocka woods partly inaccessible.[49] This very inaccessibility was particularly attractive to anyone who wanted to hide from the enemy. The Bielski otriad now numbered about 700 members. To this large group the Nalibocka forest promised immunity from persecution.[50]

As usual, for this journey Tuvia divided his otriad into smaller parts. Each included wagons with a few basic belongings, cooking utensils, shovels, covers, and foodstuffs. The majority had to walk. They traveled at night and rested during the day. On the way more Jewish fugitives joined them.

They also lost some people. On the way, Russian partisans asked for fighters and Tuvia tried to accommodate them. He asked for volunteers and in no time fifty young people stepped forward, among them Dr. Tamara Zyskind. Some, particularly the young, were frustrated. In the Bielski otriad most of their energy had to be devoted to getting food for the people, and those who wanted to fight were glad to take advantage of this new opportunity. Although sorry to see them leave, Tuvia did not interfere with their decision.[51]

Reduced by fifty, the otriad continued its journey. Whenever a group stopped to rest, guards were posted around it. People would get their food from the leader of the subgroup, and cooking was done in groups. Soon food began to dwindle. They passed villages that could have supplied food but they had to be careful about taking anything from them. They had to balance the need for provisions with the need for safety. Food rations were becoming smaller and people were hungry.

It was not easy to keep the otriad moving. Some were discouraged, ready to give up. One of the Bielski partisans, Moshe Bairach, recalls: "Once we were caught in a terrible rain, even the wagons were stuck. It was awful. It was pouring all night. We were soaked through. Tuvia moved around us and tried to keep up our spirits."[52]

They had to use side roads, making the journey much longer. The rivers Niemen and Bieroza crossed their path and many did not know how to swim. As a precaution people would tie themselves together before negotiating a water crossing.[53]

One of the Bielski partisans, Raja, is convinced that no one else except Tuvia could have succeeded in bringing the people to their destination. Seeing the move as a miracle, she paints a picture of what occurred.

"It took a few weeks of walking. Of course, we were searching, groping. I thought that it was a job for Moses and for no one else. We were about 700 people. This huge group moved along. Not all of them listened to Tuvia. Some got lost. We had to search for them. Some could not keep pace. It was like a herd that did not listen to its shepherd. Tuvia had courage, even 'nerve' (chutzpah). . . . He had so many problems, so many difficulties. . . . It was a miracle. One could write books about that move alone.

"I remember because I stayed next to him. He was exhausted, depleted of all strength. It almost killed him. To have the responsibility over a camp of that size, where everyone wants something else, and where no one listens. . . . Each one did what he wanted. It was a heroic deed. To this day I cannot understand how he accomplished it."[54]

Nothing came easy. As Bairach remembers, "When we tried to come into the forest, Russian guards . . . would not let us in. They claimed we had no permit. Tuvia negotiated. After half a day we were allowed to enter."[55]

As soon as the otriad reached its destination, Tuvia began to organize guides to visit Lida ghetto. They were to bring more people and medications.

Food collections were even more problematic in this new base because of distance and bad roads, but food missions were planned immediately. In this new and strange environment, exploring, learning, and building went on and on.

9

The "Big Hunt"

When the Bielski otriad reached the Nalibocka forest in summer 1943, the Russian partisan movement had already granted it some legitimacy. Tuvia and his people had come to a place more fully controlled by the Soviets. Here many guerilla detachments had already established their bases, including the headquarters of General Platon, head of the Russian partisans for the Baranowicze region.[1]

At that time the Soviet partisan movement had also adopted a more liberal recruitment policy. From the Jewish perspective this was a mixed blessing. As an open invitation to all able-bodied men this policy was a most welcome development. On the other hand, because this invitation included Nazi supporters, it led to stronger anti-Semitism.

From that point anti-Semitism among Russian partisans could be traced to two immediate sources: the acceptance of men with a pro-Nazi past, and a reaction to the stepped up German anti-partisan raids. Inevitably, such attacks led to hardships and losses. Relying on an old tradition, instead of blaming the enemy, the partisans blamed Jews for the problems created by these attacks.[2]

Jews then were caught between these anti-Semitic partisan practices and the official Soviet policy of acceptance that gave them a right to exist. Not surprisingly, they emphasized and relied on the official policy. And as this policy became more firmly established, despite some flareups, it improved the conditions of the Jews in the forest.[3]

All along Tuvia was sensitive to official and unofficial undercurrents and tried to use them to his and his people's advantage. In his contacts with the Russians he would lean on the official policy of non-discrimination. With high-ranking partisans, in particular, he would argue that Jews were Soviet citizens, Soviet patriots, who had come to the forest to defend the USSR. Tuvia had behaved as if, from the communist perspective, his otriad was a most welcome development.

Actually he had never asked for permission to collect Jewish fugitives. Had he come to the partisan authorities with such a request, they most probably would have refused. Especially at the early stages, the Soviets felt that civilians in the forest were a burden, an obstacle to their struggle with the Germans.

But the Bielski otriad was simply there. The Russians "were faced with an established fact and did not know how to react. Tuvia knew how to deal with them. He would say to them, 'Remember when our people come back they will ask what you did with the Jews.' They had to tolerate it."[4]

Like most charismatic leaders, Tuvia was spontaneous.[5] He felt free to pursue an agenda that required him to save as many of the doomed Jews as possible. In addition to using his personal ties and diplomacy, Tuvia tried to counteract the negative effects of anti-Semitism by making his group useful to the partisan movement. In the Nalibocka forest in particular, because the group was stationary, he was able to offer different services to neighboring otriads. For these services, the Bielski otriad received a valuable commodity: food.

Bread continued to be in short supply, a sought-after luxury. In the Nalibocka camp the Bielski partisans transformed one ziemlanka into a bakery, producing large quantities of bread. Russian partisans would buy bread and pay for it with grain, flour, or other goods. Profits from these transactions belonged to the entire otriad.

One Bielski partisan, Oppenheim, a man in his sixties, was a talented locksmith. He came to the forest with his son from Lida ghetto. When the otriad was not on the move, Oppenheim was busy reconstructing weapons, specializing in the creation of gun parts. Weapons that had missing parts or that "misbehaved" found their way into Oppenheim's magical hands. His success earned him the reputation of a wizard gun fixer. For his inventions and solutions he was paid with food. These profits also went to the otriad's general food storage.

Another partisan, Mordechai Berkowitz, a blacksmith, established a workshop that offered all kinds of services, especially fixing horseshoes, and partisan riders would seek his help. For this work too the Bielski otriad was paid with groceries.

In the forest a chronic shortage of physicians always existed and the Bielski detachment was glad to share its medical staff. Russian commanders and their mistresses would seek out the services of these doctors and would bring gifts, usually in the form of food. These too helped feed the rest of the people.[6] Although a welcome addition, such earnings could not eliminate the need for food missions.[7]

Surrounded by jungle-like growths and swamps, the new base was less accessible to the enemy so there was no need to spread the ziemlankas over a wide area. People were assigned to dig rectangular spaces for ziemlankas, each to accommodate twenty to forty people. As usual, the part of the ziemlanka above the ground was covered with sticks and tree branches.

Because these living quarters were in close proximity, the camp had one communal kitchen, with a specially appointed cook. Each person was entitled to three meals a day, although the type and amount of food depended on availability.[8]

Here, feeding a group of about 700 was more challenging. The partisans did not want to antagonize the local people who lived in or close to the woods so food had to be confiscated from faraway farms and each partisan detachment was assigned a special area.

For the Bielski otriad this meant traveling about one hundred kilometers from the base. Depending on road conditions and dangers, such a mission might take up to three weeks. Only the young and vigorous could undertake these journeys. At any time not more than twenty percent of the Bielski people could participate in food expeditions. This meant that relatively few individuals had to feed large numbers; it also meant more frequent food missions. Because of great distances, each group tried to collect as many provisions as possible at one time. Larger quantities of food required more people. Sometimes a food mission included as many as twenty-five men.

In the Nalibocka forest the Bielski otriad was in continual flux. Some of these changes had to do with the building up of the base, others with the arrival of fugitives.

Once, on the way from an official partisan assembly, Tuvia and Malbin met a group of fifteen Jewish men. Led by the attorney Volkowyski from Baranowicze, the group had a letter from the head of a Russian brigade asking Tuvia to accept these men into the Bielski otriad. Although not openly stated, this Russian commander implied that these men were burdensome and he had to be rid of them.

Tuvia's assessment was different. "The group had five rifles which, of course, was the best possible dowry. I was captivated by Volkowyski, he made a wonderful impression. He asked me if I would accept his group, to which I answered that we take old people, why not take in fighters, especially when led by him. He was happy to hear my reply and I was very pleased with this meeting. He subsequently became integrated into our group, carried out important missions and was very helpful."[9]

This Russian commander, and some others, used the Bielski otriad as a dumping ground for Jews they did not want. Anti-Semitism rather than realistic evaluations was often behind these decisions. At times anti-Semitism would lead to more serious consequences—forcible disarmament before dismissal or death.[10]

Some Jews, although themselves accepted into Russian detachments, were not allowed to bring their wives or parents with them. A few sent their parents to the Bielski otriad.[11]

Starting in the summer of 1943, devilish rumors began to circulate about Jews. It was said that the Germans were sending Jews into the woods for the specific purpose of poisoning Russian partisans. Another rumor was that Germans were dispatching Jewish women infected with venereal diseases into the forests.[12]

Behind these accusations was the assumption that the Jews let themselves be used for the purpose of destroying the Soviet partisan movement from within. For some anti-Semitic guerillas, such widely circulated charges justified the mistreatment and even murder of Jews. In addition to seriously affecting Jews, these charges also affected the lives of some non-Jewish partisans who were attached to Jewish women.

Josef Marchwinski was a Polish communist, second in command of a Russian otriad. He was also married to a Jewish woman. Marchwinski received a letter, signed by the head of his brigade and co-signed by General Platon, stating that his wife and one other Jewish woman, married to another partisan officer, would be transferred to the Bielski otriad. When Marchwinski's strong objections failed to change the order, he, his wife, and the other couple joined the Bielski group.[13]

As an old-time communist who had spent years in Polish prisons, Marchwinski was a prominent figure in the Soviet partisan movement. His communist past, by itself, challenged Tuvia's authority.

In the Bielski otriad he had no official position, yet people expected him to wield power. Tuvia and his staff were cautious. It would have been imprudent, even dangerous, to oppose an old-time communist. Tuvia pretended not to mind, not to see, when Marchwinski took over some important duties.

From the start this newcomer objected to what he regarded as blatant lifestyle inequality. He was convinced that in the Bielski otriad the majority suffered from hunger while an elite minority ate more and better foods.

To improve the lot of the underprivileged majority he took over the economic dealings that the Bielski otriad had with other partisan detachments. He claimed to have made the Russians pay more for services rendered and that this, in turn, had improved the economic situation of the entire otriad.[14]

Marchwinski also gathered around him a group of teenagers whom he indoctrinated with communist ideology and taught to fight. Later some of these youngsters participated in anti-German moves.

Most adults in the otriad were suspicious of Marchwinski's motives and felt uncomfortable about his presence.[15] However, a minority not only welcomed him personally but also his pro-communist propaganda and activities.

Grisha, the Russian partisan who had joined the Bielski detachment before it came to the Nalibocka forest, was part of this politically conscious minority. Another was Goldwasser, a recent arrival with attorney Volkowyski's group. The third openly to join the others was Arkie Lubczanski who had given Tuvia trouble before, when the group was settling for the 1942–1943 winter.[16]

Supposedly, during the Soviet occupation Lubczanski had cooperated with the authorities by denouncing rich Jews. In the forest, dissatisfied with Tuvia's apolitical posture, he talked against him to any Russian partisan who would listen.[17]

While Marchwinski, Grisha, Lubczanski, and Goldwasser were politically active, there were others who might have been tempted to join them but did not. One potential candidate was Israel Kesler.

Earlier, the Russian commander Panchenko had threatened Kesler and his followers with death unless they joined forces with the Bielskis. Reluctantly they did. A common, uneducated man, Kesler was an illegitimate child, born to a lower-class woman. As a professional thief, he had spent many years in prison.

Without denying Kesler's shady past, some people admired his leadership qualities, his independence, courage, and love of the Jewish people. Like Tuvia, he refused to go to a ghetto and had collected a group of Jews whom he succeeded in arming.[18]

Those who came with him to the Bielski otriad, including his new wife, Rachel, looked up to him as their leader and he seemed to care about their welfare.[19]

Tuvia gave Kesler a great deal of autonomy. Some say that Kesler admired Tuvia; others are convinced he was envious of the commander's authority. Perhaps there is truth in both assessments.

All agree that Kesler was smart. He flirted with the new politically active group but did not join it. He watched and waited. Tuvia also watched and waited as Grisha became the spokesman for the group. First, Grisha talked about the need to have political reform. Then he petitioned his commander to appoint him the otriad's commissar, a political position tied to the Communist Party and to the Secret Service, the NKVD, later known as the KGB.[20] When Tuvia refused, Grisha attacked him in the presence of two high officers, Panchenko and Sinichkin.

Tuvia reports: "Grisha claimed that I was interfering in government procedures, opposing party activities, preventing the establishment of a party branch. Sinichkin heard him out. He then asked me for my reaction to these accusations. I felt that self-defense would not be advisable under the circumstances. It would be much wiser to attack. This I proceeded to do.

"'It does not behoove me to defend myself,' I said. 'I am here to carry out Stalin's command. At a time when people like Grisha were prisoners of war, waiting for the German 'hand out' of 100 grams of bread a day, my brothers and I united to fight the enemy, to save the lives of Soviet citizens, and to destroy all those who oppressed us. Even before receiving orders from our homeland, we formed an alliance with Commander Victor Panchenko. Together we executed attacks upon the enemy, killed and gathered booty. Together we destroyed a train station. There are hundreds of members in my company. We've never been a burden to anyone. Grisha came to us to heal his injured hand, and I will not permit him to become master over me.'"[21]

This confrontation led to two victories. Tuvia retained his position of commander and Grisha became the commissar of the otriad. Since a commissar was under the authority of the commander, it was not at all certain whether Grisha would be satisfied with his promotion. He wasn't.

Once more Tuvia was asked to appear before a Russian commander, Platon's deputy. This time the plotters, headed by Grisha, demanded the removal of the Bielski brothers from the otriad. Tuvia was accused of not caring that at one time his Chief of Staff, Malbin, belonged to a Zionist organization to the right of the political spectrum. Malbin, they claimed, was a fascist. Tuvia argued that Malbin "was a good friend, an excellent fighter, knowledgeable and experienced in military matters, what else could I ask for?"[22] Placated by this answer, the interrogators continued by demanding an accounting for a large sum of money, 40,000 rubles, money that they felt belonged to the Soviet Union. Although surprised that his accusers knew about these financial dealings, Tuvia was ready to answer to the charges. He explained that this money had been obtained by confiscating German goods that were being transported by a Belorussian farmer. In fact, the Bielski partisans protected the man by giving him a letter stating that they had taken away the provisions. Such a letter they had hoped would shield him from German reprisals. After they sold the goods, they gave this farmer a small share of the money.[23]

When this transaction took place the partisan headquarters sent an order to the commanders of the otriads to try and supply motorized partisan units with gas. Tuvia delivered gas worth 20,000 rubles to them. Additional funds went to the procurement of arms and other necessities. He was fortunate that Panchenko was there to verify all these dealings.

This hearing ended with the Russian investigators saying, "Make them work hard and they won't be occupied with nonsense."[24]

Tuvia had no illusions. This was a fragile truce. Many powerful people, for a variety of reasons, resented the very existence of the Bielski otriad.[25] Around him he could hear grumblings—"Why should Jews amass so much power?"

One wrong move and the Bielski brothers could lose their positions. And if that happened, who would take care of the people?

But then changes that went beyond the powers of the adversaries took place. Marchwinski, the man with the real political muscle, had to leave. He was transferred to the Komsomol brigade, which consisted of Russians, Belorussians, and Poles. He was assigned to important duties. As a Pole, loyal to the USSR, Marchwinski was a political asset. Soon he was offered an even higher position. He helped establish a special Polish division that was subordinate to Moscow.[26]

Marchwinski's promotions reflected the competition for the control of Poland that existed between Moscow and the Polish Government in Exile in London.[27] But it took a while for the effects of this conflict to reach Western Belorussia. By mid 1943, the Soviets had succeeded in establishing a firmer grip over the Russian partisan movement, without, however, exercising full control. Unwilling to lose the power they had, they compromised by bowing to some local demands. They continued to tolerate a variety of subgroups with distinct, at times even hostile, aims.

The Soviets were numerically and politically dominant. But most Russian partisan units remained an ethnic mixture of Russians, Belorussians, Ukrainians, Jews, Poles, and Lithuanians. Each ethnic group was, in turn, politically and socially heterogeneous.

Predictably, most Poles resented the Soviet domination over the area's partisans. Still, this did not prevent some of them from joining Russian ranks. Poles who became part of the Russian units were under direct Soviet domination. Some were communists, others found themselves there by chance. The majority of Poles, however, had refused to be under direct Soviet command and insisted on forming their own fighting detachments. Quite possibly, if given a chance, most Poles would have preferred to be under the authority of the Polish Government in Exile, in London.

Before the summer of 1943, a local Pole, Lieutenant Miłaszewski, organized a separate fighting group, called Kościuszko. Most members of this Polish otriad identified with the AK (Armia Krajowa, Home Army), the official fighting force of the Polish underground, that was under the command of the Polish Government in Exile, in London. Miłaszewski, although an AK man, cooperated with the Russian headquarters, coordinating his anti-German moves with theirs. Politically, his partisans were heterogeneous, ranging from the left to the right. Some were even affiliated with the Fascist part of the underground, the NSZ (Narodowe Siły Zbrojne, National Armed Forces).

But Miłaszewski and his men never openly admitted their nationalistic inclinations, which would have placed them in a precarious position. For quite some time the Soviets were under the impression that the Kościuszko otriad was made up of local Poles free of political aspirations.

A communist and partisan leader, Hersh Smolar explains, "Only gradually did we realize that Miłaszewski and his men were connected to the Polish Government in London. I guessed it only from the kind of literature they gave me to read. We were quite isolated for some time. We did not know who they were and what they were after."[28]

Miłaszewski's men had a reputation for being courageous fighters. For the Russians, determined to fight the Germans, they were an asset. This Polish detachment was divided into a regular partisan group and a cavalry group. Estimates about the size of the entire group range from 400 to 600.

In the summer of 1943, with the approval of the Soviet partisan authorities, the Kościuszko otriad established a base in the Nalibocka forest. They were separated from the Bielski otriad only by one kilometer.[29]

Miłaszewski and Tuvia took an instant liking to each other. Both had a passion for chess and would meet and play in each other's camp. Without expressing it in words, the two had a certain basic understanding, an appreciation of each other's needs. And yet, despite this affinity, Tuvia admits he did not quite trust Miłaszewski. In fact, he referred to his Polish friend as a "bandit." Then, with a chuckle he added, "I was as much of a bandit as he."[30]

Their friendship continued for as long as both remained in the Nalibocka forest.

As the Bielski partisans were adjusting to the new place, echoes about an approaching storm kept invading their lives. These involved the German determination to destroy the partisan movement and block all future regrouping. Although frequent, these messages contained few details.

A German document dated July 7, 1943, describes how and when the partisans in the Nalibocka forest would be attacked and eliminated. This operation was known as the Hermann undertaking. Several of these official communications show Nazi awareness about the difficult terrain of the Nalibocka forest and the obstacles their men would be encountering.[31] It is not certain if and how many of these documents found their way into the hands of the Russian high command.

Shortly before August 1, 1943, the official starting date of the "big hunt," a food mission headed by Asael returned to the Nalibocka base. With fifteen cows, many horses, and wagons filled with all kinds of provisions, this journey had been a success. The otriad had reason to rejoice.

However, the news accompanying these acquisitions prevented a celebration. The Germans were getting ready for a major assault. For this important undertaking they had committed several divisions from the front and other specially trained units. No place would be safe, not even the Nalibocka forest.[32]

The partisans could not stand up to these superior forces. An order went out to all commanders: do not engage the enemy in open battle.

Still the partisans refused to remain passive. They were eager to reduce the effectiveness of the attackers. In the Nalibocka forest each partisan detachment was asked to send a portion of their men to headquarters. These fighters were to participate in a collective effort. One hundred armed men were sent by the Bielski otriad.[33]

Some of these arrivals were organized into special ambush groups and dispersed at different entrance points to the Nalibocka forest. These units were charged with placing mines on roads that German vehicles were expected to travel on. Then they were to hide on both sides of the mined roads, ready to attack those who survived the explosions.

A few ambushes failed because of spies who warned the enemy.[34] A Jewish partisan from the Ponomarenko otriad tells about a successful ambush. "As a truck with thirty Germans approached, it exploded. They all flew into the air. All were killed. This was on the road leading to the Nalibocka forest. But after two days we realized that we could not fight them openly. They were a well-equipped army and we did not have enough power."[35]

Indeed, while many of these ambushes were a success, they failed to block the enemy's entry into the forest. It appears that the Germans wanted to surround the guerilla forces, trap them in one area, and then destroy them.[36]

In anticipation of this strategy, Soviet headquarters devised a plan for curtailing the enemy's mobility within the forest. Although the swamps and wetlands themselves interfered with the movement of military vehicles, the partisans wanted to further obstruct the mobility of the marchers. Ideas about what happened next diverge.

Some reported that the partisans built a bridge or bridges over the swamps with cut-down trees. One person claimed that a huge bridge was built only as a diversion to make the Germans think the partisans would use it. The enemy, eager to engage the guerillas in battle, would try to follow and use the bridge. In the meantime the partisans would go in the opposite direction.[37]

Others assumed that a huge bridge was built for the partisans. One partisan noted that "The main command decided that they would build a bridge over the wetlands, the marshes, to run for a few kilometers. When no longer able to withstand the pressure of the Germans we would run away on this bridge. Then we would destroy it and the Germans would have no time to build another one."[38]

Still others denied the existence of a bridge. Indeed, the Chief of Staff, Malbin, said he never heard about a bridge. He talked about the huge space and asked, "How could one have built a bridge? It would have been like a pin in the sea. No, we built no bridges."[39]

Perhaps those who mentioned a bridge or bridges are referring to trees cut down by partisans to block the forest roads. Tuvia and others spoke about partisans who were busy putting down trees to slow the enemy's progress.[40]

Beyond the one hundred fighters, headquarters demanded nothing else from the Bielski otriad. With the threat so close, Tuvia turned to his superiors for advice, but they had none to give. Uncertainty was in the air. Tension within the Bielski camp was mounting by the hour.

Tuvia's neighbor, Miłaszewski, came to visit with a few of his men. As the two talked, riders stopped to announce that the Germans were only two kilometers away.

Tuvia remembers that "It was noon. Miłaszewski got up and shook my hand. 'Be brave, sir,' he said, 'we're surrounded.' He was not concerned for himself; his entire company consisted of fighters. He saw the women, children, and elderly in my camp. He turned, and accompanied by his entourage, returned to his camp."[41]

Tuvia watched Miłaszewski leave. A few fighters from a Russian detachment stormed by and they did not even stop to greet him. Some distance away one could hear shots. Eager for news, for suggestions, Tuvia went to see the Russian commander Kovalov. "He told me that the Polish partisans had a skirmish with the Germans and had to retreat. The Germans were coming closer. They would certainly reach us by morning. They would not advance at night.

"'What do you intend to do?' I asked Kovalov.

"'What a silly question! Obviously, since we cannot stand up to the Germans in open battle, we have to get away.'"

"'But where to?'"

"'The road is open.'" He said no more, he was very nervous.

"I tried again: 'What do you think I should do?'"

"'Whatever you think best.'"

"I went back to our camp. Everyone was tense, waiting to hear what I had to say. It was essential to encourage the people, to strengthen and warn them at the same time. I addressed them and said there was no alternative, we would have to leave the camp."[42]

More shots were heard. This time they seemed close by. Some people panicked and started running haphazardly in circles. Pesia Bairach was among those excited runners. Her husband Moshe thinks that Pesia wanted to reach the Moscovici otriad. A good friend of theirs, a courageous fighter, Baruch Lewin, was in this Russian detachment and she probably expected him to protect her. But she had no way of knowing that the Moscovici otriad had already abandoned their base.

Bairach tried to stop Pesia from circling around. The chaos continued. People screamed at each other, and at no one in particular; nobody listened.

Bairach recalls that "It was as if Pesia had lost her senses. I ran after her but could not stop her. Suddenly a rider, Tuvia, appeared from one direction, and Zus from another. With gun in hand and curses, Tuvia screamed, 'All of you return to one place! I will shoot you like dogs if you don't!'"

"Tuvia came closer to me and saw how I begged Pesia to stop. He shouted at me: 'Bairach, take her back right away. If not, I will kill both of you!'

"Still on horseback, he collected the people and threatened that he would shoot anybody who tried to run away.

"Then Tuvia made them stand in rows. And he spoke to them in a most painful voice. 'Jews! Have mercy, don't lose your minds! If we don't keep together and calm we will all be lost. We have hope that we will get out of this. You must only listen to me.'"[43]

As Tuvia tried to establish order, he was thinking, weighing the next steps. He made a supreme effort to appear self-confident and continued to address the crowd before him. At his side, he noticed men from the Pervomaiski company.

He remembers turning to these Russian partisans. "I asked where they were going, but got no reply. I guessed at their plan. I then saw the Jewish Zorin company following them. This was an otriad made up of people who were unable to fight. These were Jews from Minsk and its surroundings. I again warned my company about discipline and repeated that I would shoot all rebels. In truth I was quite confused myself and could see no way out. Some unknown instinct told me not to follow the Russian partisans. I stopped Zorin and warned him not to follow them either. But he would not listen to me."[44]

They continued to walk over the trees that had been placed there by the partisans not so long ago. The tree cutting had been a collective effort but now, with the enemy advancing closer, the cooperative spirit seemed to have evaporated. The Russian detachments with their able-bodied young did not want to bother with those who could easily become a burden.

While Tuvia was agonizing about what to do, two of his partisans, Michal Mechlis and Akiva Szymonowicz, came up with an answer.

Before the war Mechlis had been a surveyor of these woods. Akiva had also worked in and around this forest, selling and buying second-hand goods. Both men were trustworthy and both felt at home in the Nalibocka forest. Mechlis approached Tuvia. "Let's go to the big swamp. Twelve kilometers within this swamp there is an island, Krasnaya Gorka. Once we reach this island we can hide there and wait for the Germans to leave."

Tuvia asked, "How can you be sure that the Germans would not come to Krasnaya Gorka?"

"One cannot be certain, but should they come, the forest is always before us!"

Akiva agreed that this was their best chance. The Germans were unlikely to enter treacherous, unfamiliar wetlands.[45]

Another possible alternative would have been to follow one of the Russian detachments that kept passing over the cut-down trees. But Tuvia had a strong feeling that the road these partisans were taking was unsafe, so he decided to follow Mechlis' and Akiva's suggestion. He then had to make sure that their departure would be properly executed.

To calm the crowd, he announced that they would be leaving soon for a safer place. They would have to travel light; except for weapons and food everything else would stay behind.

Tuvia gave the order to untie all the animals and let them free. He had the food storage place opened and invited his people to fill their pockets with non-perishable items. When all pockets were bulging with bread, grains, beans, and all kinds of groceries, the people once more were asked to arrange themselves in rows. The crowd, anxious, but silent, awaited instructions.[46]

To the side of their camp, covering a huge part of the ground, were the cut-down trees. Several partisan detachments had already used these trees as a pathway and the Bielski partisans were convinced they would go this way too.

Calling these fallen trees a bridge, one partisan comments, "We see that all the other groups move onto the bridge, but Bielski keeps us back. Bullets were practically over our heads. It is night. Suddenly Tuvia gives an order that we are to move in a different direction. Not to the bridge. . . . He had decided that the bridge was very dangerous. Later on we discovered that he was right."[47]

Tuvia told his people that they were going toward the big swamp, not far from their camp. No one objected. He told them to be patient, to wait their turn, and all would be well.

Mechlis and a few armed young men went first. Tuvia stayed. He wanted to make sure that no one would be left behind.[48] Tuvia describes the move. "The children were carried on our shoulders. The people with the children went up front. The swamp began several hundred meters from our camp. It deepened as we progressed. The mud thickened and stuck to us. However, there was no fear of meeting anyone in the swamp. The stronger ones and the ones carrying arms walked last. In some spots we sank up to our navels, though this was not for long distances. We finally crossed, but by supreme effort. The water became more shallow. After each deep spot we had to stop and check, to make sure that everyone had crossed safely. We made very slow progress in the water and mud, it took three hours to cover a distance of three kilometers. We found a dry spot and sat down to rest. People fell asleep in spite of being wet and weighted down by mud. It was not even cool; the swamp was stuffy. We did wake the snorers to maintain quiet. I made it clear that we had to go back to whispering just as we did in the first half year when we were unarmed. At dawn we talked it over and decided that it might not be a bad idea to stay there all day."[49]

They rested. All was quiet. Encouraged by the calm, Tuvia and Mechlis went back to investigate in the direction of their camp. Just as they were beginning to feel reassured by the surrounding silence, they were startled by a barrage of shots. Bullets flew past their heads. Tuvia and Mechlis fell to the ground. Rough, excited voices speaking German kept reaching them: "Hurrah, catch the Jews, catch! Catch! Catch!" The attackers had probably hoped that hidden Jews would start running and thus be hit by bullets.[50]

When these voices began to fade, the two partisans, half crawling, half running, rejoined their people who must have heard the shots.

Then Tuvia saw them "lie in a living convulsive heap, on the ground, petrified. It was a terrible sight, men, women and children lying there in utter confusion and helplessness. Mechlis and I had decided not to tell them of the seriousness of the situation but rather to encourage them, for things really looked bad. From our deserted camp we could still hear the shouting and shooting. We got everyone up and continued marching forward."[51]

Soon the Germans were close by. They were talking, "Sigi die Kuhe, halt!" (Sigi the cow, stop!).[52] In silence the Bielski otriad moved on. For a while the enemy seemed to be chasing them. With the mud getting deeper the Germans gave up. Although these pursuers were gone, they continued to shoot. Shells passed over people's heads, exploding some distance away.[53]

Chaja's brother carried his six-year-old daughter on his shoulders. Another man, Kolodnicki, had two children, aged two and four. He placed each on one of his shoulders.

Chaja recalls that "The more we walked the deeper the mud was. My mother became weak, she could not move her legs. The mud reached up to her hips. My brother tied her with a string to himself. He dragged her after him. We walked and walked, but were still close to our camp. The mud covered a wide area. The reeds in it were high. . . . When you step into the marsh with one leg it is hard to pull out the second one. But if hundreds of people enter at once then everyone sinks more deeply."[54]

To avoid drowning or getting lost, others also attached themselves to each other with whatever they had, belts or ropes. Bairach says that "there were trees in this mud. We tied ourselves to the trees when we were resting. We were sinking in the water. We rested while we were tied to the trees. We were holding onto the bushes, trees, whatever we could, in order not to drown."[55]

They would stop walking during the day and resume marching at night. Tuvia urged his people to eat in moderation, to save whatever they could. They ate what they had put into their pockets: dry beans, grains, potatoes, and other foods. All their provisions were covered with mud and it was hard to recognize them, but they ate despite that.

The children were very quiet and ate the little food that was given to them. They were hungry but did not ask for more. Somehow they must have understood; all of them were silent and undemanding.[56]

A few people from the otriad were not with them. They had been sent on special missions that were already underway when the main group left the camp. One of them, Dworecki, went to the Lida ghetto to obtain special goods unavailable anywhere else.

Tall and blond, with blue eyes and an upturned moustache, Dworecki looked like a typical peasant. On the way back from the ghetto he heard about the big hunt. Eager to join his people, especially his wife and two teenage daughters, he caught up with them after they had been on the move for almost a week.

The Dworecki sisters recall that "Father came with a big loaf of bread, dark peasant bread. He said that we must give the bread to Tuvia Bielski to eat because he, as a leader, must be well fed. But Tuvia refused to take it. Our father begged him, but he would not. 'You must give it to your children. I cannot take it from you.'

"Tuvia cried like a child, but he absolutely refused to touch the bread. I will never forget the picture, how these two men argued about the bread."[57] In the end the Dworecki children shared this treasure with other children and made it last a while. Even though the situation was deteriorating further, with food becoming more scarce the children did not demand a thing.

People suffered, yet they cared for each other. All through the march no one tried to run away. Only size and circumstances made them break up into smaller groups. Occasionally a few would stop and rest and then try to catch up. Tuvia continuously moved back and forth and checked. He also made others go and see if anyone was missing. He would try to encourage his people. At the same time he warned them about the importance of discipline and order.[58]

After more than seven nights in the swamp, sinking beside the tall grass, not knowing where they were, hungry and tired, they found it impossible to move.

Tuvia recalls: "I loosened the strap of my submachine gun, tied it to my belt and wound the two around myself and the tree. I then sat down. I drowsed. Almost everyone followed my example. Some actually climbed the trees and slept up there. We spent the night just sitting there. The morning broke sunny and bright. We came to very deep water. It was almost impossible to check the end of the line to see if everyone crossed safely. We were certain that here we might really lose some of our people. After several hundred meters we suddenly came upon dry land, a lovely hill, 'Krasnaya Gorka.' . . . We felt removed a fair distance from the enemy. They (Germans) would not cross the swamp and alternate routes were distant. . . . We began drying out our clothing and underwear. The sun shone, we were warm. We sat and relaxed on the grass. We were very hungry, but the joy of survival strengthened our spirits. The hunt in the forest was still on, we had to wait here."[59]

On the first day at Krasnaya Gorka the people were counted. Six were missing: Kolodnicki and his two children, old Sioma Pupko, Monik Przepiórka, and another man from Lida. Three men volunteered to go look for them. After a day they found them and brought them back. The missing group had simply been too exhausted and too weak to keep up.[60]

Now the entire group settled down in the hope of not being disturbed, but it was not to be. As one of the first to keep watch at Krasnaya Gorka, Bairach describes what happened. "My job was to stand guard. Suddenly in the middle of the night my partner and I heard some noise. We stopped the person. It turned out to be a woman, a Belorussian. She claimed that she lost her way. This was suspicious. I left her with my partner and went to the camp to see what they suggested we do with her. Tuvia and someone else came and began to ask her questions. She mumbled. They hit her. She admitted that the Germans asked her to go and check if there were any partisans in this area. She was a spy."[61]

The unwritten partisan law demanded the execution of spies. The woman was shot without delay.[62] This spy was the only person to pay them a visit. Other outside intrusions were limited to shooting and sounds of explosions.

Undisturbed by outsiders, people collected mushrooms and berries. These they mixed with the remnants of food still left in their pockets. Only very few had flour or beans. As a last resort, they made flour out of tree bark, added to it water and whatever little else there was to eat.

Tuvia reports: "Ten days had passed since we had left our camp. I noticed that one of our women had blisters around her eyes. Our doctors told me that those were symptoms of starvation. Our doctors collected the bits of food that some still had and divided it up among the sick and the children.

"One night I sent Akiva and a few others to the village of Klaszczyce to see if he could get food. He came back empty-handed but the information he brought was important. He found the village lit up by the headlights of many German vehicles full of German soldiers.

"Finally the shooting ceased. We heard later that the Germans had moved out all the residents of the village Kleszczyce in vehicles and torched the village itself, burning everything in sight. The farmers were taken to Germany except for the few who had escaped. Some animals the Germans took, some they shot, others perished in the flames. During that period the Germans put to the torch seventeen villages and hundreds of homes on estates. The town of Naliboki was in ruins. The Germans wanted to eliminate the partisans' sources of food and hiding places."[63]

Hunger and the expectation that the Germans were winding up the big hunt made Tuvia dispatch a few scouts in the direction of their abandoned base to determine whether it was safe to return.

On the way, without being seen themselves, these scouts encountered a large group of German soldiers.

No, they could not return to their camp. But neither could they continue to stay on this island. If they stayed they would starve to death. Zus, who was particularly sensitive to the gnawing hunger pains, wanted to investigate the possibility of breaking through the enemy's ring and leaving the Nalibocka forest.[64] It was decided that if after two days Zus and his men would not come back, this would be a sign that there were openings and that the enemy ring around the forest was not tight.

Eighty armed young fighters left with Zus. For a few kilometers Tuvia, with twenty men, accompanied Zus' group. Tuvia reports that "I met a Jewish woman carrying a child in her arms. Both were barely alive. She was a member of the Zorin company. She told me that the Zorin detachment had been destroyed, many had been killed, and the rest were scattered about. Among the dead were many children. I sent her to our base. After a while the Zorin company regrouped and she rejoined them."[65]

On the road, Zus split his men into several units. Separate yet close to each other, they came upon fires only recently abandoned by retreating German soldiers. Scattered around these fires was leftover food, a most welcome reward. At night they reached the banks of the river Niemen and found a boat. They crossed the river and still no Germans. It seemed that the big hunt and the Nalibocka forest were behind them.[66]

Those who remained at Krasnaya Gorka were getting ready to leave. Some were very weak, some were resigned and unwilling to move. Tuvia encouraged them. The war was coming to an end; they must try.

Monik Przepiórka suffered from a lung ailment. He was one of the six who had been lost and brought back by the volunteers. The stay on

Krasnaya Gorka had depleted the little energy he had left. During the ongoing preparations, feeling sickly, Monik came over to Tuvia. His feverish eyes were glued to the commander's gun. In a resigned voice he said, "Please give me your gun, just for a minute."

Tuvia was startled. Tears filled his eyes. Gently he patted Monik on the back, and said, "I will not abandon you, nor anyone else. Either we will survive together, or die together."[67] A basic rule of the Bielski otriad was not to leave anyone behind. Yet there was one exception, a youth who suffered from a leg injury and was totally incapacitated. Without medication the doctors had no way of treating him. The injured boy begged Tuvia to shoot him, but Tuvia refused. He explains, "Before our group left, we provided the injured youth with some water, matches, a bit of flour, anything to help him survive. He could see no chance of survival. I tried to encourage him and said, 'There are no dangerous animals about, you'll be able to survive for a while on the food we're leaving you. Your leg will get better. Other partisans will be passing this way and will surely share their food with you; by that time you'll be able to join another group of partisans.' I later heard that the Zorin partisans found him and took care of him. When we returned he rejoined us, well and happy."[68]

There was still another exception—Israel Kesler had an unusual request. He and about fifty of his followers were asking permission to remain in the Nalibocka forest. They argued that as natives of the nearby small town of Naliboki and other close-by communities they were familiar with the area. They could easily elude the Germans, especially now when the Germans seemed to be retreating. Besides, the otriad could benefit from their staying—they would prepare the place for everyone's return. Tuvia agreed.[69]

Before they left, Tuvia announced that since they would be splitting into smaller groups they were bound to lose sight of each other. They should all assemble in the Jasinowo forest close to one of their former camps. He had expressed the hope that everybody would be there within two weeks.

"We staggered the exodus so that a group left every hour or two, so as not to march together. It was impossible to keep an equal number of people in each group. . . . There were undisciplined members who could not stand certain leaders. They would quarrel with them, stay behind, or return to the camp. Thus there were smaller and larger groups. Some people stayed behind and waited for me. I had planned to be the last to leave, to keep track of the move. Asael and the Chief of Staff Malbin were with me."[70]

Not all the young leaders were equally willing to take the weak and the old. Sensitive to this possibility, Tuvia repeated again and again that no one will be abandoned and that they had to protect each other. And because he watched over each departure no one dared disobey.

Pesia, who was close by, says, "Tuvia waited until everyone left, only then did he go. I could have left earlier with Moshe but I refused. I was convinced that with Tuvia I would be safest so we waited. At the end there were fifty of us and he went with us and took good care of us."[71]

On the road even the smaller groups were often forced to split into still smaller units. For a variety of reasons people would continuously lose sight of each other.

The journey was rough for Tuvia's own group. Food was scarce. All kinds of partisans and unsavory characters were swarming the region—and all were in need of food. The local people who remained in their homes had little to give. More than ever, their meager offerings were mixed with fear, hostility, and anger.

When Tuvia and his people came close to the river Niemen, a farmer warned them that the German police were on the way. He advised those who could swim to go as quickly as possible to the other side. One young man, Gornofelski, although he knew how to swim still drowned. There was one boat for non-swimmers. Tuvia waited until all of them crossed; this other side of the river promised some safety.[72]

After the crossing, two families, the Dworeckis and the Taubs, told Tuvia that they would like to remain in the area, in the homes of Christian friends. The Dworecki sisters explained that "We were in a terrible condition. We had wounds, lice, we were filthy, exhausted. My father felt that maybe we should stay for a while with my father's Polish friend, G. Filipowicz. When father asked Tuvia what to do, Tuvia said: 'Do as you like, but if you want to return to us, do.' We indeed went to the Pole. We were there for the winter. They helped us build a ziemlanka in a nearby forest. But then we decided to leave. It was hard. We were afraid. Also, Polish partisans would come there. It was getting dangerous. We went back to Bielski."[73]

A part of Chaja's family, including her old parents, also made arrangements to stay in the home of Christian friends.[74]

The big hunt resulted in tremendous loss of life and the breakup of partisan units. Discipline eroded and robberies became the order of the day. Often their targets were Jewish fugitives, less frequently armed men.

One Bielski partisan told about his experiences while separated from the large group. "Because we were split into many small groups some Russian fighters took advantage and attacked us. A friend of mine and I met a group of these Russians. I had old, torn shoes and they had nothing to take from me. They forced my friend to take off his boots and made him give up his shotgun. Then we met our people and we reassembled. Those I met were brave young men. Once I joined them I had no problems. We were in a small forest. At night we would go to the peasants for food. Finally the happy day arrived when I came to the forest of Jasinowo where we met our commander and hundreds of people. Others continued to appear."[75]

Undisciplined rough behavior was not limited to Russian partisans. When most of the Bielski people reassembled, complaints about one of their own group leader, Kaplan, were also heard. Local farmers on whose goodwill they depended accused Kaplan and a few of his men of robbery. These forcible confiscations included money and valuables.

Before the big hunt, Blacher, Tuvia's aide-de-camp, went to Lida ghetto on official business. He rejoined the Bielski otriad only after it had left the Nalibocka forest. In her husband's absence, Blacher's wife and her little child were assigned to Kaplan's unit. After they reassembled in Jasinowo forest, Blacher's wife accused Kaplan of trying to chase her away and of screaming, "Go to Tuvia. He took the gold, let him also take the people!" He was accusing Tuvia of collecting money for his personal use.

When Zus confronted Kaplan with this incident he got a similar reaction. "Those who took gold from the people can have the people." Zus got angry and reached for his gun: "I will shoot you like a dog." Kaplan started to run. Asael warned his brother, "What are you doing, don't shoot!"

Too late. Zus touched the trigger of his gun. A shot. A body hit the ground—Kaplan was dead. Right away Zus regretted what he had done. Over the years the regret turned into guilt and refused to go away.[76]

10

Building a Forest Community

With the manhunt behind them, the Soviet partisan leaders were eager to reestablish order. Through this reorganization they had intended to get a firmer grip on the different partisan groups in the area. The Bielski otriad, caught in this effort to reorganize, became officially attached to the Kirov brigade. Vasiliev, its head, had special plans for the Bielski partisans.

Tuvia was told that his unit would get a Russian deputy and a Russian Chief of Staff. He recalls that "In the guise of 'advice' we were asked to send all of our 'non-fighters' to the Nalibocka forest. The fighters would remain here and would from time to time send food to the families in the forest. I realized that our independence was being seriously threatened. All kinds of thoughts occurred to me. Perhaps I should detach myself from the brigade and leave the area?"[1] But Tuvia knew that headquarters would not tolerate this kind of disobedience—they would probably kill him. And if this happened his people would be unprotected.

When he consulted Malbin and Volkowyski, both thought he had no choice but to agree in principle to the Russian proposal. During the negotiations that followed, the commissar of the brigade told Tuvia that the area they were in had to be cleared of all civilians within five days.[2]

Tuvia's efforts to negotiate were only partially effective. In the end the Bielski otriad was split in two. The family group became officially the Kalinin otriad. In reality it continued to be called the Bielski otriad. The second group, with one hundred eighty Bielski fighters, was named Ordzonikidze. This newly formed unit had a Russian commander and a Russian commissar. Except for Zus, the head of reconnaissance, a position he held in the Bielski unit, all other high posts went to Russians.

One positive note in this new development was the permission to include in Ordzonikidze the fighters' wives. Among these women was Sonia, Zus' wife, Dr. Isler's wife, a nurse, and Luba Rudnicki, the wife of Janek Rudnicki. Ordzonikidze never succeeded in becoming an integrated community; there was always a wedge between its Russian leadership and its Jewish majority.[3]

The Soviet command was eager to see more fighters leave the Bielski detachment. Because of Asael's reputation as an experienced and excellent fighter they ordered his transfer from the family group. Not to cause trouble, Asael joined the headquarters of the Kirov brigade and became head of intelligence. But, determined to save Jews, Asael felt that his place was with his people. Although it was a risky step, after a few weeks, he returned to the Bielski otriad.

After Asael's arrival, he and Tuvia went to General Platon to legitimate the transfer. The General agreed on condition that Asael would participate in attacks against the Germans. From then on whenever Asael organized moves against the Germans he had to report about them directly to General Platon. At that time, Tuvia Bielski, Lazar Malbin, Herzl Nachumowski, and several other Bielski partisan fighters successfully declined invitations to join Russian detachments.

Shortly after the big hunt the Germans left the Belorussian woods, including the Nalibocka forest. They also relinquished control over some surrounding territories. Once more parts of Western Belorussia were identified as "partisan lands." And again Nalibocka forest became home to the partisan headquarters, with General Platon as its head. Similarly, many of the recently regrouped Russian detachments began to reconstruct bases in and around it. Some camps were used for specialized needs. For example, the Iskra otriad built a hospital at their forest base, while the rest of its partisan unit made their home in the surrounding villages.[4]

In addition to the Bielski otriad, Soviet headquarters ordered all other unprotected civilians who roamed the countryside into the forest. In the past practically all homeless civilians had been Jewish ghetto runaways. However, by the fall of 1943 there were also Belorussian, Polish, and even Gypsy refugees.

Most of these non-Jewish fugitives had been left homeless as a direct result of the manhunt. While destroying the communities in and around the Nalibocka forest, the Germans divided the local people into two categories: the fit and unfit for work, the useful and the useless. Those defined as useful were deported to Germany for slave labor. Of the remainder, the majority were murdered and a minority succeeded in escaping. Some of these runaways joined partisan units. The rest were ordered by the Soviets into the Nalibocka forest where they became a part of special family camps.[5]

As part of the new Soviet directives the Bielski otriad, reduced in size, headed by its commander, Tuvia Bielski, was soon passing through ghost-like communities, emptied of all human life. Here and there they were joined by some of their former members, by recent ghetto runaways, and by fugitives directed to them by Soviet partisans. Not only did Tuvia accept all of them, but he left behind special scouts with instructions to help Jewish fugitives to catch up with the Bielski people.

Although the Germans had wrecked many communities and many more lives, they left much that could sustain life. The fields were filled with potatoes, wheat, and all kinds of vegetation. Trees were laden with a variety of ripe fruits and beehives were filled with honey.

Burnt skeletons of animals covered much of the ground, as live animals continued to roam the area, no doubt in search of food. These homeless creatures seemed to be welcoming the humans who might have reminded them of their former masters.

And while no buildings were left intact, doors, windows, and other items were still in usable condition. Inside these partially demolished structures were many undisturbed household goods and building materials. Winds and rains must have conspired, preventing the fires from completing their jobs.

A portion of the Bielski otriad, Kesler with his fifty partisans, had never left the forest. When German hostilities stopped, this group set to work. They began by building a new base, some seven kilometers from the ghost town of Naliboki. One part of this group collected the homeless horses, cows, pigs, and fowl. Others assembled the abandoned carts, fixed them, and then used them to gather potatoes, wheat, vegetables, fruits, and items deemed useful for the establishment of a camp. They also built special warehouses for storing these newly acquired goods.[6]

The Kesler group, just as its parent unit, continued to grow. In their case too some of the newcomers were people who had once belonged to the Bielski otriad and were only now finding their way back. Others were Jewish fugitives who had recently escaped from ghettos and were rejected by other partisan detachments. By the time Tuvia and his people reached the new base in the Nalibocka forest the Kesler unit had grown to about one hundred and fifty.[7]

Perhaps in recognition of Kesler's diligent work, at this point his life-long career as a professional thief was "only whispered about."[8] Nor were people likely to mention that he had learned how to write his name only in the forest and only with great difficulty. Instead, Kesler, the man, was showered with adjectives like ambitious, energetic, enterprising, brave, and capable.[9] Those who were close to him described him as someone who would "share his last slice of bread with the needy."[10]

Inevitably, Tuvia's return cast a shadow on Kesler's authority. As commander of the otriad, Tuvia had control over Kesler's unit. But instead of challenging the man's independence, Tuvia seemed to be giving him special freedom. This he did by announcing that the building of the main camp would take place more than a kilometer away from Kesler's camp.[11]

Kesler and his people remained in their recently established base, which called for a separate kitchen and thus control over the distribution of food. In addition, Kesler was entrusted with a new assignment. On his premises he was to build a ziemlanka that would be used as a hospital.

While Kesler had been busy supervising the building of the hospital, most of his people were incorporated into working teams involving all members of the otriad. These cooperative efforts helped erase the differences between those who belonged to the Kesler group and members of the larger unit.

Tuvia improvised, continuously adapting to new conditions. He divided his people into teams and subteams, each assigned to different tasks—to help build a new community, to collect food and materials to sustain such a community, and to protect the community within a hostile and unstable environment.[12]

The building teams would search through the deserted villages and small towns and bring back window frames that sometimes had intact glass, along with ovens, boilers, barrels, and all kinds of kitchen utensils. Among the rarer finds were items that were important for setting up services such as a flour mill, a communal bath, and a variety of workshops.

A less specialized group would go and plough the fields, dig for potatoes, search for other foods, and then store them.

A separate team stayed at the campsite and concentrated on building living quarters. Initially, these were temporary bunkers made out of woven tree branches that were covered with soil, dry leaves, blankets, and whatever else people thought might protect them from poor weather.[13]

Sometimes even these accommodations were in short supply and members of the otriad had to sleep in primitive tents. Luba Garfunk, who joined the otriad toward the end of September 1943 with her husband and four-year-old son, remembered how her boy used to ask in the evening, "Mommy, under which tree will be our house today?"[14]

Eventually the rush to meet the approaching winter led to the building of more protective quarters with engineer Ribinski in charge. First, rectangular foundations were dug, eighty to one hundred centimeters in depth. These were lined with tree trunks wired together. The spaces between were filled with moss, covered with unplaned boards, and then a layer of earth. As usual, the roof was camouflaged with tree branches.

The age and size of the Nalibocki trees created special building opportunities. For example, already in August, the bark of the trees would come off in a single piece and would be used as roof coverings. Built in a tent-like shape, two-thirds of the structure was buried in the ground. The rest, mainly the roof, rested on top of the soil. The upper part consisted of wooden sticks that touched each other, forming a triangle. These sections of wood were in turn covered by the bark sheets.

At the entrance of each ziemlanka there was a door with a few steps leading into the hut. On entering, one was enveloped by darkness. Soon the eye would become accustomed to the little light that seeped in through the tiny window and the opposite door. One could discern wooden bunks along the walls on both sides. The bunks, "upholstered" in straw, served as beds for up to forty people who slept lying next to each other. As in other ziemlankas, in the middle was a stove, salvaged from one of the demolished homes.

These living quarters were built in two rows, on both sides of the camp's "Main Street." Each ziemlanka had a number, sometimes a name descriptive of its inhabitants. For example, hut number 11 was known as "Intelligentsia" because the residents were Dr. Hirsh, a dentist from Minsk, the attorney Salomon Volkowyski from Baranowicze, and others who were better educated or from a more privileged background.

Going back down Main Street was a large square, the "nerve center" of the camp, with the headquarters of the camp on one side, and different workshops on the other. The collective camp kitchen was located at the end of the street, some distance from the huts and practically across the street from headquarters.[15]

Outside Main Street was a fenced-in area for the camp's cows and not far away was the grazing area for horses. Carts were also kept there. Cows were milked early in the morning.

A hut serving as a clinic stood close to the commander's ziemlanka. Here Dr. Hirsh, the chief physician, the dentist, and nurses would attend to patients. In the morning one could already see people lined up in front of the clinic.[16]

Construction of new structures continued almost up to the time of liberation. Also, until the end, special scouts circulated within the region whose main assignment was to gather intelligence. In addition, they would search for Jewish refugees and bring them to the camp. Whether with scouts or on their own, new people constantly arrived at the otriad.

The Jewish partisan from the Russian detachment Iskra, Jashke Mazowi, recalls a scene. "I came close to Ordzonikidze, there I saw Zus stand at the edge of a small bridge. Around him was a group of Jewish refugees. They looked disheveled, exhausted. Rejection and disappointment were written over their faces. But then I heard Zus say, in a warm voice, 'Come to us brothers, don't worry. Later I will bring you to our family camp. . . .' This scene made a deep impression on me. I will never forget it. Zus took them to the Bielski otriad.

"In my detachment Jewish fighters felt okay. But we knew that if something happened we could always go to Bielski. Bielski was for us what Israel is for the Jews now, an insurance."[17]

Indeed, Jashke Mazowi told about his stay in the Bielski camp. "In our otriad we had one machine gun, a heavy one. One man carried it. It was our best gun. I was usually the one who carried it. We went through a forest, this time the one below me in rank carried the machine gun, two other youths had carried the boxes with the bullets. Suddenly, Germans began to shoot at us. We dispersed. I could not find my men. I was responsible for the group.

"To come back without the gun would have meant death. If some one lost a gun, he would be shot by the partisans. I walked around, looking for the gun, and it was not there. I inquired, but no one knew. I came to the otriad empty-handed. I told the story, and they believed me, but they asked me to go and find the gun."[18]

He searched in vain. On the way back he stopped over in a village where he met Tuvia and told him about his predicament. "Tuvia said, 'What do you care? Come to us. I will give you another machine gun.'

"I stayed with him a week. They suspected that I was at Bielskis, so a 'razwiedka' patrol from Iskra came. Tuvia said to me: 'Run!'

"I was so sure that I was right that I stayed.

"The Russian said 'We found the youth with the machine gun.' He fell on my neck and embraced me.

"Tuvia turned around and spoke to me: 'Jashke, don't believe them. They will shoot you.'

"But I could not desert. I went back to my otriad, all was well."[19]

Mazowi was only one of many Jewish partisans in Russian detachments to whom Tuvia offered protection.[20] But even for those who had no direct contact with the Bielski otriad, just the promise of unqualified acceptance gave them courage and hope.

When runaways from ghetto Lida continued to pour into the Bielski otriad in the fall of 1943, Zorach Arluk recalls one expression of their acceptance. "I returned to Iskra, Tuvia was there visiting us. I told him, 'I just met twenty people, they were going to join you. How will you take them in? How will you manage?'

"'Zorach, you know what a Belorussian bread looks like (it weighed ten pounds and was round). We have a man in the kitchen who cuts the bread. He has an order from the cook to cut it in twenty to thirty pieces. After the new arrivals his order will be to cut the bread into forty pieces, so each of the slices will not be as big. To the bucket of soup we will pour another pail of water and we will add some more potatoes and wheat. Rest assured that from hunger no one will die. If more and more will come that is good!'"[21]

Not only was Zorach impressed with Tuvia's attitude, but he also had reason to be personally grateful. In his Russian otriad, Iskra, Zorach had become a distinguished fighter. But the commander of Iskra had refused to accept Zorach's relatives. They were civilians, unfit for combat. Tuvia was happy to take them.[22]

On Asael's return from the Kirov brigade to the Bielski otriad, he was officially reappointed as an assistant to the commander and commander of the fighting forces. Responsible for the safety of the otriad, Asael and his men also participated in joint military moves with Russian partisans.[23]

Some remember Asael as a young man who "seemed to be constantly everywhere, in camp riding his horse, inspecting every corner. He was brave, dedicated to his missions, and carried them out faithfully. His simplicity and honesty were mirrored in his blue, straightforward-looking eyes. His handsome, smiling face was open and friendly. He was liked by his friends as well as by his superiors."[24]

Asael's good nature was expressed in some specific gestures. When, for example, he noticed his cousin shivering from cold he took off his warm jacket, saying, "Take it, I will get another one."[25] Nor was his generosity limited to family. Asael's wife, Chaja, recalls that "One day in the winter outdoors I saw a tall man in a black coat, stooped shoulders, long legs. On the feet, he has rags instead of shoes.

"I went back to our ziemlanka and said, 'Asael, I saw an old man with rags on his feet. It is so hard for me to see it. We must do something about this.'

"He took off his boots and said, 'Chaja, take them and give them to him right away!'

"I did not know who the man was. I inquired where he lived and brought him the boots. I told his men that Asael had no boots. He was so devoted to them and they loved him. Immediately they found boots for him."[26]

This old man was Dr. Amarant, who became the official historian of the otriad.

When comparing the two brothers, Asael and Tuvia, some say that "Tuvia was harder, more domineering, more rigid. Asael was more open, more easygoing. Asael would come to the different ziemlankas, talk to the people. Tuvia kept a distance between himself and those that were below him, his subordinates."[27]

More accessible and more visible than his older brother, Asael would smooth over internal conflicts that came to his attention. However, "with more serious problems people would go to Tuvia. Tuvia would in a minute cut through the difficulty and impose his will. He had a 'no nonsense' attitude. They listened to him."[28]

The difference in the accessibility of the two brothers may be an expression of differences in personality and the positions each held within the otriad. Tuvia comes close to the ideal of a charismatic leader. He had a special intangible gift that made people want to follow his orders. Asael lacked charisma. Tuvia was a gifted diplomat while Asael was direct, to the point of being blunt. Asael was devoted to his fighters and they had a great deal of affection for him. Tuvia was admired and listened to because of his special charisma, not because of personal attachments.

Tuvia's trait of distancing himself from his people came together with other traits expected in a charismatic leader. Thus he left the management of everyday routine activities to his subordinates.[29]

Also, as with most charismatic leaders, Tuvia was surrounded by a special entourage. Some feel that those around him created a wall between him and the rest of the otriad. In a negative, yet tentative way, it has been suggested that "He listened too much to his inner circle. Perhaps he should have paid more attention to those who were removed from the center of power."[30]

Others note that, particularly in the last year, from the fall of 1943 through the summer of 1944, people had little access to him. Some even allow that the power he wielded might have gone to his head.

And yet, even those who object to Tuvia's distance concede that under the circumstances this was inevitable. After all, Tuvia was in control of a large community that eventually grew to more than 1200. This was a socially heterogeneous, undisciplined group of people with many contradictory demands and expectations. Tuvia had to supervise the running of the otriad as well as its relationships to a hostile and unstable environment. He commanded wide respect. Free access to his person might have reduced his prestige and interfered with his effectiveness.[31]

With all these unusual traits came an ability to compromise. And in this spirit of compromise Tuvia had accepted into the Otriad a Soviet commissar sent by General Platon. Although officially under a commander's authority, a commissar wielded much political power. Unofficially, a commissar's main loyalty was to the Communist Party and to the secret service. For a head of an otriad to have a commissar was a little like having a political spy.[32]

But the new commissar, Shematovietz, was different and special. A loyal communist, in his late fifties, he was a heavy man, with bear-like movements. His slow, heavy walk contrasted with the alert expression of his deep-set eyes.

When the German-Russian war began he had lived in Minsk with his Jewish wife and daughter. Soon after the German takeover his wife and child were murdered. For a while Shematovietz continued working as a railroad official, a job he held before the war.

Shematovietz is remembered as being free of prejudices, as someone devoted to helping the persecuted, particularly the Jews. Through his railroad connections he made it possible for Jews to travel on trains, an illegal activity at the time. After the establishment of the ghetto in Minsk he cooperated with the Jewish underground. Warned about his impending arrest, he escaped to the forest and became a partisan commander.

In the forest Shematovietz met an eighteen-year-old Jewish girl and her father, took care of them, and became the girl's lover, "husband." The two men were approximately the same age. Shematovietz was a communist, his "father-in-law" an orthodox Jew, preoccupied with prayers. Although the Russian had no illusions about their future life together, he was devoted to his mistress and her father. When he was ordered to send them to the Bielski otriad he refused to separate from them. Instead, he accepted a demotion when he was appointed commissar in the Bielski otriad.

Tuvia welcomed these new arrivals and was soon glad to discover that the new commissar was interested in two things: how to ease Jewish suffering and how to find vodka. Shematovietz identified with the Jewish plight and spoke on behalf of Jews at headquarters. He participated in running the otriad only when he could be of help. Unobtrusive and cooperative, he became the camp's valuable addition.[33]

With the exception of Shematovietz and Asael, all those who worked at the headquarters were Tuvia's appointees.

Charismatic leadership, unless backed up by an organization, by a bureaucracy, is unstable. There is, however, a certain inconsistency between bureaucratic and charismatic authority.[34] Like all large, expanding units, the Bielski otriad had to rely on bureaucratic procedures. Soon, the group established a balance: the head, Tuvia, was a charismatic leader, those below and around him were not. Tuvia's power was absolute, except for instances where he himself chose to share it. He was surrounded by people he knew well; most were relatives and friends. They were loyal, but not necessarily capable. Tuvia must have recognized their limitations, and because he was not particularly bound by rules he made exceptions.

Lazar Malbin, not a friend, not a relative was one exception. On meeting Malbin, in the summer of 1942, Tuvia offered him the important post of Chief of Staff, a position he retained until the summer of 1944. Malbin was responsible for most bureaucratic innovations and was known as a stickler for order, precision, and efficiency.[35]

At the headquarters, in addition to Tuvia, Asael, Shematovietz, and Malbin, there was a Chief of Special Operations, Salomon Volkowyski, the attorney from Baranowicze. He dealt with disciplinary and investigative problems. His job gave him the opportunity to suggest punishments. Transgressors would often be sent to prison, which was a special hut located behind the food warehouse. As a rule prisoners were guilty of transgressions like falling asleep while keeping watch, stealing, or refusing to do a job.

Gordon, another member of headquarters, was the official assistant to the commissar. Pesach Friedberg retained the position of quartermaster in charge of food distribution.[36] Similarly, Raja Kaplinski continued as the secretary of the otriad. The two clerks were Viner and Bedzowski.

The otriad's headquarters consisted of two rooms with maps hanging on its walls. Later a typewriter was retrieved from the ruins of one of the surrounding communities.

Work began at eight in the morning. Twice a week headquarter employees met to discuss special problems. During these meetings Raja took notes. Official reports that must be sent to general headquarters or to different Soviet officials were written here.

Here also people were assigned to different jobs. Most had the job of guarding the camp. Information about watch duties was posted outside headquarters and everyone was required to check the lists.

Those who did not have any special skills were likely to move from one job to another as were those who simply liked changes. Women more often than men were moved from job to job. Of the women sent for guard duty some were afraid, particularly of the wolves that would come close to the camp at night.[37] Skilled workers were assigned to more permanent positions.

In general, there was a wide range of reactions to work. A small minority was eager to keep busy and would accept any job. An equally small minority was reluctant to work at all. Some even objected to doing guard duty.

One partisan observes, "Occasionally, Tuvia himself had to step in with a 'You must go and guard!'"

"'Why I, commander, why not someone else?' With the Russians, they would not dare to argue."[38]

Among those who wanted to contribute to the otriad was young Herzl Nachumowski from Iwje. He was very interested in defending the place. At first, when he arrived, he had to cut trees and carry water. Nachumowski recalls, "But I got tired of it and asked Boldo: 'How long do I have to carry water and cut trees?'

"'When the commander decides on something else he will let you know,' came the answer.

"I went to Tuvia with the same question.

"Tuvia said, 'You don't have to do that.'

"'Since when?' I asked.

"'Since yesterday.'

"With this I ended my career as a water carrier and wood cutter. Instead I would join all kinds of groups and for different expeditions."[39] Soon Nachumowski distinguished himself as a hard worker and brave partisan.

Despite Tuvia's organizational talents and the establishment of bureaucratic procedures, certain underlying currents interfered with the group's order. Many members of the otriad were connected to each other by familial and friendship ties. These attachments led to nepotism and a refusal to follow rules. In addition, the open and unrestricted admission policy of all Jews brought in a diversity of people. Diversity sometimes translated into contradictory demands and disobedience. Another closely connected, yet basic, reason for nonconformity can be traced to excessive Jewish individuality.

The Bielski otriad was an extraordinary community that emerged in extraordinary times. It went through several phases of development. Despite these changes, certain basic social realities and arrangements continued throughout its existence. This was true for the community's social composition, general lifestyles, and its basis for prestige rankings.

The history of the otriad shows that for most of the time about three-quarters of its members were "older people," women, and children. Similarly, the group of its young armed men and those capable of using weapons fluctuated between twenty and thirty percent. The overwhelming majority was poorly educated. Not unexpectedly, a small minority had an upper- or middle-class background. Of this minority, most were women. The majority of the fighting men, the skilled and unskilled workers, were poorly educated and before the war belonged to the working class.[40]

As special targets of the Nazi policies of annihilation from the beginning, the Jewish elite, particularly many of its men, were murdered.[41] A few Jewish leaders succeeded in escaping. Those who stayed and survived the initial waves of persecutions were ill equipped for life in the forest. They were city dwellers and only a few moved into the woodland.[42]

In contrast, working-class men were more resourceful and better prepared for life in the countryside. Basic activities, like riding a horse or milking a cow, came easily to them. Not surprisingly, in the forests they were more likely to fight, more apt to go on food expeditions than former city dwellers.[43] Because their jobs were fundamental to survival they became important. They switched social ranks with the prewar elite and became the new elite.

The continuous struggle for survival and the accompanying social upheavals made politeness and good manners irrelevant, almost inappropriate. In the forest life was reduced to its basics. Physical strength, an ability to adjust to the outdoors, and fearlessness were qualities that mattered. A man's prestige depended on the extent to which he exhibited these qualities. Women were usually not included in these calculations.[44]

Because the working-class people were better suited to life in the forest, they looked down on those who had a hard time adjusting to the rough conditions and to manual work.[45] In fact, they resented behavior that signaled a higher class background.

Tamara Rabinowicz recalls that "Once my husband and I spoke Hebrew with each other and one of them told us, 'Do not speak Hebrew, this is Russian territory and you are not supposed to speak Hebrew.' We did not. After that, when we did want to speak Hebrew, we did it so that no one could hear us."

At another time she was approached with the following questions: "Did you study, in high school?"

"I did."

"Did you study at the university?"

"I did."

"So what does it give you? Can you make a thicker soup out of it?"

"No."[46]

This anti-intellectual posture was also expressed through the use of coarse language. For the few who came from an upper- and middle-class background this could present problems. When Cila Sawicki met an old friend in the forest who, already well adjusted, greeted her with a string of vulgar words, she was shocked. "My ears stood up and I said to my husband what will be? He said, 'We will get used to it.' I never got used to it. I could never use this kind of language."[47]

Similarly, right after the well-educated and refined Sulia came to the forest, a few of the common women called her a whore, for them a frequently used expression. In her innocence, Sulia argued that she was sexually inexperienced and therefore not a whore.[48]

Among the Russian partisans, cursing was widely accepted as was heavy drinking. Lower-class Jewish youths had no trouble adjusting to these patterns.

The disapproving voice of the intellectual Geler says that "People did drink before an expedition and after an expedition. In the winter people drank to feel warmer and in the summer not to feel the heat. Everyone did drink, even women. It was seen as proper behavior. Some partisans paid with their lives for this weakness. Sometimes they were attacked when they slept after a drinking spree.

"People in the Bielski otriad did drink. Tuvia, his brothers, and the rest of the members of the headquarters gave a poor example to the rest . . . when bread was not plentiful, they would take bags of flour and make out of it vodka for the use of the commander and his relatives. People saw these abuses, but did not talk much about them."[49]

While Geler's assessment might be too severe, no one denies that drinking made life easier. It helped them to forget.[50] Tuvia's wife, Lilka, at the time not yet twenty, admits to drinking "a cup, a glass, but not more. All of us used to drink. I was never drunk. The doctor told me to drink, it kept away illnesses."[51]

In fact, some feel that Tuvia's success as a leader was enhanced because in his continuous dealings with Russian partisans "he could outdrink them and he could curse like they did."[52]

As in all groups, here too people were ranked in terms of their importance. The elite were the Bielski brothers, their wives, relatives, and friends. A part of this top echelon included all those who worked at headquarters.

Next in importance were young men with guns. Below them were craftsmen and artisans. Particularly after the group had come the second time into the Nalibocka forest, these people had acquired more privileged positions. The bulk of the people, those with no special skills and no guns, were at the bottom of the social ladder.[53]

In the fall of 1943, the Bielski camp focused on becoming transformed into a less nomadic and more stable community. Indeed, this otriad was soon organized into a "shtetl," a small town, with many different workshops.

11

The Emergence of New Social Arrangements

"When you are in the forest you see individual trees. Only when you leave it, outside from far you see the entire forest in all its glory. The same happened to me. I saw much more as a tourist than the people that were there. I was amazed at what Tuvia did. I saw the workshops, the children (cries) . . . I saw an orthodox Jew pray. People worked, they fixed watches, made shoes, they made leather from cowhide. People from all the surrounding areas came there to have things fixed.

"My position in my otriad, Iskra, was very good till the last day. But if I did not like something, then they could all go to hell. I could always go to Bielski."[1]

The speaker, Zorach Arluk, never transferred, but was a frequent guest in the Bielski camp. Once, when his otriad confiscated twenty-three cows from the Germans, Zorach was assigned to deliver the loot to their base in the forest. On the way he left one of the cows in the main kitchen of the Bielski otriad. This was only one of Zorach's many gifts. Each time he visited the Jewish camp he felt emotionally refreshed.[2]

Hanan Lefkowitz, a member of the Russian otriad Stalin, more isolated than Arluk, had thought that except for a few scattered Jewish partisans all other Jews were dead. While in the hospital being treated for a serious injury, he overheard some men talking about a Jewish detachment, three kilometers from their base. Soon, not even fully recovered, Lefkowitz went to the place.

"I was amazed," he said. "I thought that it was all a dream. I could not get over it . . . there were children, old people, and so many Jews. When the guard stopped me I spoke Yiddish. I met people who knew me. That first time I could stay only an hour. After a few days, I went back and then again and again. . . . Once I saw a roll call, soldiers stood in rows, with guns. I saw two men come out, tall, handsome, leather coats . . . I asked who they were and was told that these were the Bielski brothers. They were giving orders to the fighting men. They were going for an expedition, Tuvia and Asael . . . the two jumped on the horses like acrobats. I imagined Bar Kochva to look like that . . . Maccabee, King David. . . . It gave me hope."[3]

Like Arluk, Lefkowitz wanted to help. His otriad had a storage room with weapons and parts of weapons. Each time he went to the Bielski base he would smuggle guns and parts of guns by hiding them under his coat. He would give those to old friends and people he just met.[4]

Although to these outsiders the Bielski otriad looked like a unified whole, on the inside it was a differentiated community, with different social arrangements and lifestyles, all closely connected to food.

From the very start, Tuvia insisted that all who joined the otriad were entitled to food, regardless of who they were and whether they did or did not contribute to the group. Being fed was a right, not a privilege.

In principle, everybody was supposed to have the same amount and kind of provisions. In practice this did not happen. Even the official rule that entitled everyone to a share of food had some built-in exceptions. Children, the sick, and the weak were given extra nourishment, usually milk and fat.

During the first year of its existence the otriad was frequently on the move. Meals were prepared in groups of ten to twenty, which in itself led to the preparation of different meals.

More important, those who participated in the dangerous food expeditions felt entitled to extra provisions. While this never became an official rule, in practice the food suppliers got their way. In addition, those in control of the otriad and those close to them, the top echelon, ate better than the majority of the otriad.

In the fall of 1943, after the return to Nalibocka forest, some eating differences became formalized. Two separate kitchens were established, one for headquarters and their families, and another, a big one, for the rest of the otriad. For this new arrangement some blamed the flatterers who surrounded the Bielski family. One, the cook Engelstern, was particularly vocal in his support for separate kitchens. Chaja Bielski feels that headquarters should not have allowed this to happen.[5]

Even before the creation of this special kitchen everyone knew that a connection to headquarters meant better food. These people were regularly served cutlets, pancakes, eggs, and all kinds of delicacies. Indeed, more than forty years later, when Hersh Smolar talks about meals he had at Tuvia's place in the forest, a nostalgic smile spreads across his face.[6] Still, these informal preparations of special meals were not as demoralizing as an official separate kitchen.

Because this special kitchen created too much anger and resentment, it was soon eliminated. But removal of a separate kitchen did not guarantee equality in eating.[7] It only meant that those who had extra provisions would be less conspicuous in the preparation of meals.

The elite were reluctant to admit that some people, the malbushim, could be hungry. Even some who were not part of the elite but had acquired a little extra food refused to accept the idea that others did not have enough. One of them, Pesia Bairach, argues that "A big pot of soup was cooked several times a day. Everyone received food three times a day. No one was hungry. If someone told you that they were hungry it is a lie. There was a time that I lived only on what was given to everybody else and I was not hungry. They complained not about lack of food but about the fact that others had more or better things to eat."[8]

Riva Reich, the nurse, defends Tuvia. "Once I went to fetch milk for my sick patients who had typhoid fever. I was told that they had taken the milk to the headquarters.

"I went to Tuvia and told him, 'Comrade commander, I need milk for the sick. I was told that they took it to the headquarters.'"

"He said, 'What do you think, that I take a bath in milk? I don't need milk for myself. Go and take as much as you need.'

"Some of those around him did take extra food for themselves. But this was not because of Tuvia. He was always ready to give, to share. There were people that were flattering him. They knew how. He tried very hard to be correct, to be right. He would listen to everybody. I know that he did not like the flatterers. He liked hardworking, decent people."[9]

A wife of a fighter, someone who had enough to eat, says that "It was very hard to be just. They called Tuvia's family and his friends the Romanoff court. If it were my family, I would have seen to it that they should also be well provided for. He could not be more just. He did not know everybody. There were too many people. They had to fend for themselves."[10]

The Bielski partisans draw contradictory pictures about the character of their commander. While some described Tuvia as being surrounded by flatterers, others were convinced that he was wary of them. Similarly, some felt that he listened to everybody, others regarded him as someone who distanced himself from his people. Quite possibly these different traits existed simultaneously but surfaced at different times.

The young men, the fighters of the Bielski otriad, devoted most of their energies to gathering provisions. Coming back from a food expedition they "would drive up in the wagons to the storage place and leave all the food there. If they had brought luxury items, like cream, eggs, etc., they could take them for themselves. This was their privilege. But the basic goods belonged to the camp. Then it was distributed. There was a special person that gave flour to the bakers, vegetables to the kitchen, etc."[11]

At first Tuvia tried to enforce the rule that everything collected during a food expedition must be delivered to the otriad. But when the food collectors refused to obey, without formally acknowledging defeat, Tuvia pretended not to see. Taking part of the provisions before delivering them to the otriad, became a semi-legal activity. One participant describes how it was done. "When we would come back from a food expedition with seven or eight cows, about three to four kilometers before our base we would select the best cow, kill it, and divide it among the ten to twelve of us who were a part of the expedition. We had people among us who knew how to slaughter a cow, take the skin off, cut up, and clean it. In the winter we would put the meat into the snow and it would be preserved."[12]

As expected, most of those who benefited did not object to these inequities. But members who ate only what was allotted to them by the otriad reacted differently. Some accepted the situation, as they said, "We were not starving, we also did not have an abundance, some had more than others. We had enough. We would collect berries in the forest, mushrooms. This helped too."[13] Others seem to support this view when they report that people were neither hungry nor full, adding that there was enough bread and that the bread was of excellent quality.[14]

But some people were less satisfied. In fact, quite a few insisted they were hungry.[15] One recalled: "When I would come to the kitchen I was dizzy from hunger. I am telling you the truth. I don't accuse anyone. The Bielski family ate better. It was their right, it was coming to them."[16]

Others too, although resigned to their situations, emphasized the differences. "Those in power had to eat better. There is no equality in any place and there was no equality in the forest either."[17] Garfunk, who insists she was not bothered by her limited diet, continued, "I was not a communist, never thought that people should have the same. Those who did not go for expeditions would get soup, sometimes bread and potatoes in peels. There was no salt. If the cook liked you, you would get a piece of lung or some such object that was swimming in the soup. If he did not like you, you would get watery soup."[18]

Riva Reich was convinced that "Tuvia always worried if people had enough to eat. He would come where they were distributing food and look on if people had adequate food."[19]

No matter how people felt about it, food assumed a central place in the forest. For most, the day started with a trip to the kitchen, which in the morning consisted of a huge hole filled with burning wood. Suspended over it was a big kettle hanging from a tree. The kettle contained brewed chicory, a substitute for coffee. With this drink were added two unpeeled boiled potatoes or bread. This was breakfast. After that people went to work.[20]

The midday meal was more substantial. "Long lines of people carrying food containers would form around the kitchen. Those present pushed and shoved to be at the head of the line to receive the soup at its thickest. There was no lack of fighting and quarreling over the place in line. Frequently, the daily bread portion was given out at the same time. On festive occasions, such as May 1, they would distribute portions of sausage."[21]

With time, many "availed themselves of the 'unlawful' opportunities to acquire extra food which they cooked near or in their huts. The camp kitchen gradually deteriorated to serving the 'weaklings' of the camp who had to resign themselves to accept whatever rations they were given."[22]

In addition to a limited diet, being at the bottom of the social ladder had other disadvantages. Cila Sawicki, who before the war had belonged to the educated class, talks about her situation in the forest. "The intelligentsia was down, we were depressed. We were not worth much, they made fun of us that we were malbushim. We were not fit for this kind of life. We had no experience with horses, nothing. The rest, the majority of the people, were uneducated, close to the soil . . . I had little in common with them. I really did not know them. I wanted to be closer to them, but they did not want us. I worked all the time so they would not make fun of me."[23]

A young girl, a teenager, describes her father who used to have a high position in the brewery in Lida. "In the otriad he became a malbush, he did nothing . . . he was intelligent, educated, but not resourceful at all. He was dirty, neglected. He was not counted as a human. No one would have recognized him. There were many disappointed people like him."[24]

Garfunk's husband also had a low social position. She wonders, "My husband came without a gun. He was not a great fighter. He had a high school education and he finished a technical school in Vilna. . . . How the heroes are made is a question one always asks. In general, the intelligentsia was not prepared to fight. We had one, Baran, a common man, limited in many ways, till this day we don't know if he was a hero or an idiot. He would only want to fight and wanted to go on the most dangerous expeditions. He was a great fighter, not afraid of anything."[25]

The malbushim seemed more apathetic, less interested in what was happening. They saw less and found it more difficult to elaborate on different aspects of the otriad. For example, three intelligent women, all former malbushim, could describe practically nothing about the workshops—each was almost unaware of their existence. Placed at the bottom of the social structure, they were uninformed about the workings of their camp.[26]

As in all communities, in the Bielski detachment people could somehow improve their lot. A man, a malbush, could change his fortune by getting a gun. He would then be included in food expeditions which could provide more food and reduce dependence on the otriad's handouts.[27]

Occasionally, the transition from a malbush to a full-fledged partisan fighter was bumpy. Geler reports about his own experiences. "Once I got a rifle, I was sent for food expeditions. First, I did not know how to do it. Therefore, they would have me stand guard while they were collecting food from the peasants. Later on I joined the others. During one of those expeditions I saw a woman's fur coat. My wife could use such a warm coat. I turned to the Polish peasant, 'Will you allow me?' For an answer the Pole cursed me and took away the coat.

"Next to me stood a butcher from Nowogródek. He swore at the peasant, promising him a beating.

"The butcher looked at me with anger and said, 'You miserable intellectual, you don't ask permission from the peasant! Did they ask permission when they were robbing Jews?' Then he added, 'Even with a rifle in hand, a malbush remains a malbush.' The warm fur coat was soon on the butcher's wife. . . . Eventually I learned how not to ask permission."[28]

Other members of the prewar elite also experienced the difficulties of shaking the role of malbush. In this connection Cila Sawicki talks about her husband. "Usually one was permitted to take some of the food for the family or for himself, the extra. But my husband was not lucky at all. If he did get his special share somehow he would lose it. At times his friends would steal it from him. At one point he had a few loaves of bread that he put at the back of the wagon, and the horse that followed the wagon ate it all up so that he had nothing. . . . My husband had no idea how to deal with a horse. (He had studied medicine before the war, in Italy, for two years.) Toward the end he got used to this life."[29]

Being a malbush had other less tangible, although painful implications. At fourteen Riva Kaganowicz-Bernstein joined the Bielski otriad; she had lost all her family. In the forest, she met a woman of forty with her young daughter. The older woman took an interest in Riva and adopted her as her own. In the camp, like most unprotected women, the three automatically were placed into the category of malbushim.

Although Riva admires Tuvia for the protection he gave to so many, she feels hurt when she relates that he never even noticed her. She was not important enough to be visible. About Tuvia's youngest brother, Arczyk, whom she knew quite well and who was her age, she says, "Arczyk did not pay attention to me. Who was I? A poor, poor girl."[30]

As a malbush, Riva was hungry and tried to supplement her meals by extra work. She would volunteer for jobs in the kitchen and was paid with soup and potatoes. What she earned, she brought to her ziemlanka to share with her adopted mother.

But a little more food did not change Riva's low position as a malbush. Occasionally, she would approach the quartermaster, Pesach Friedberg, and ask for things to which she was entitled. These contacts proved painful, often humiliating. She felt that Friedberg played favorites, but she was not one of them.[31] Her negative views of the quartermaster seem to be supported by others.[32]

One way in which a malbush could improve his or her lot was to volunteer for guard duty—all those who watched the camp were fed first. Riva often did extra guard duty. Sawicki also supplemented her diet this way. She recalls that "Frequently during the day I would sleep because for hours at night I would keep watch. When I was keeping watch I would have enough food. There was a special supply of food for those who kept watch."[33]

Before Geler got a gun he too would volunteer to do guard duty. He described the job: "Each post consisted of three people. Those who were keeping watch but did not have their own weapons got a rifle from headquarters. Three to four times a week I kept watch, once at night and once during the day, then again at night, and so forth. After a guard stood for two hours he could rest for two hours. This is how they changed, two hours watch, two hours sleep, in the nearest ziemlanka.

"In the winter, at night, the hours were very long. I was always tense, listening to the quiet, to every noise. My task was never to let a person come close to my place. Whenever one heard steps from far one had to call and ask for the password.

"Whoever came close to you had to answer with a second part of the password. Sometimes people from headquarters would come to check the guards. I used to stand, and my eyes would close themselves automatically, but I resisted. I knew the responsibility and also the punishment that would await me. One waited continuously for the relief. The rest was very welcome."[34]

One night Geler decided to join his wife who had guard duty. "It was quiet. One could only hear dreamy "moos" made by the cows. We had thirty to forty cows. The milk was kept for the children. Each day a glass per child, sick, older people . . . I and my wife had not tasted milk for months. I had an urge to drink some milk. This strong craving overpowered all my other senses. Without much thought I took a small pot and began to milk a cow. I had difficulty. One drop went into the pot and ten around it on the ground. The cows, disturbed in their sleep, tried to push me away or run off. After hard labor I succeeded in filling half a pot with milk.

"My wife and I tried to decide what to do with the two or three glasses of milk. We thought of warming it and drinking it. Then we thought that it was a shame to waste, at once, such a valuable product. Finally, we had decided to cook some milk soup with potatoes. We could even feel its taste.

"It was becoming light, dawn, soon we would be able to leave. But a visit by two supervisors destroyed our beautiful plans. The two noticed the pot and the milk. One of them asked, 'What do you have in the pot?'

"The second one, without waiting for my answer, kicked the pot with his foot. The white liquid spilled all over the green grass.

"At ten in the morning we were ordered to appear at headquarters. I was ashamed to look into the eyes of my commander. Tuvia Bielski until then had always treated me and my wife with consideration. We stood there, beaten, with eyes cast down, I answered Tuvia's questions.

"'Is it true that the supervisors found milk at your guarding place?'

"'Yes, commander.'

"'Did you have permission to milk the cows?'

"'No.'

"My face was burning from shame. There were all kinds of people around us.

Tuvia Bielski, Commander of the Bielski Partisan Unit. Photo taken in the Polish Army, 1927–1929 (Courtesy of Lilka Bielski).

Tuvia Bielski, Commander of the Bielski Partisan Unit, Nowogródek (year unknown) (Courtesy Yad Vashem Archives, Jerusalem).

Tuvia Bielski and his wife Lilka, 1945 (Courtesy of Lilka Bielski).

Asael Bielski, Second in Command of the Bielski Partisan Unit and Commander of the Fighters, with his wife Chaja Bielski. Photo taken right after liberation in the summer of 1944 (Courtesy of Chaja Bielski and Moshe Bairach).

Sulia Wołozinski-Rubin with her husband Boris, partisans in the Bielski Unit, Nowogródek, July 1944 (Courtesy of Sulia Wołozinski-Rubin).

Zus Bielski, Head of Intelligence Operations in the Bielski and the Ordzonikidze Partisan Unit, 1942–1944 (Courtesy of Sonia Bielski).

Lazar Malbin, Chief of Staff of the Bielski Unit, 1942–1944 (Courtesy of Raja Kaplinski-Kaganowicz).

Raja Kaplinski, partisan and Secretary of the Bielski Unit, 1942–1944 (Courtesy of Raja Kaplinski-Kaganowicz).

Sewing workshop in the Budnony Partisan Unit, 1943 (Menkin Collection, Courtesy of Yad Vashem Archives, Jerusalem).

Weapon workshop in the Budnony Partisan Unit, 1943 (Menkin Collection, Courtesy of Yad Vashem Archives, Jerusalem).

Members of the Bielski Partisan Unit, Nalibocka Forest, May 1944 (Courtesy of Yad Vashem Archives, Jerusalem).

A group of Bielski partisans sent to guard a partisan airport for one month, 1943–1944 (Courtesy of Moshe Bairach and Yad Vashem Archives, Jerusalem).

Tuvia Bielski, Commander of the Bielski Partisan Unit. Photo taken by Moshe Bairach during Tuvia's last visit to Israel, 1983. Jerusalem is in the background (Courtesy of Moshe Bairach).

"Bielski also felt embarrassed by the entire business. He knew us well. He knew what kind of criminals were standing in front of him. Still, with a stern voice, he gave his verdict: 'Three days of imprisonment.' Two guards took us right away to prison."[35] In the Bielski otriad one of the ziemlankas served as a prison.

But they did not stay in prison. Instead they were immediately transferred for special duty, guarding an airport.

In addition to sending military and political men into Belorussia, the Soviets were also supplying the forest fighters with arms, ammunition, clothes, medicine, and other necessities. Usually such cargo was dropped by plane but sometimes the planes would land. If a Soviet plane landed, it would pick up wounded men who required more extensive care.

Contacts with Soviet planes called for well-coordinated arrangements. In places the partisans deemed "safe" they had built special "airports" that were, in effect, clearings. Not surprisingly, such illegal airports needed to be continuously watched and protected. Groups from different detachments would stay in the vicinity of an airport and guard it. In addition to keeping watch, those assigned to airports had to signal the approaching planes in a prearranged way. Communication with planes involved making signs with fire or light, with the particular sign either in the form of a square, a triangle, a snake, or another shape. If a pilot's mission called for a landing, these signs would inform him where and whether he could safely land.

The airport functioned as the nerve center of the partisan movement and watching the airport was a serious matter. The Soviet headquarters would request people for airport duty from different otriads.[36]

On the day the Gelers went to prison, Tuvia received an order from Platon to dispatch twenty partisans for airport duty. The Gelers were included in this group. Yudl Bielski, Tuvia's cousin, was in charge. There they met two other units from Russian otriads. Meals were prepared for all of them in one kettle. When off duty, there was much friendly talking and banter. Soon the guards became a closely knit group.

Next to the airport was a ziemlanka that housed wounded guerilla fighters who were to be flown to Moscow for medical treatment. Geler says that "One never knew if and when the plane would land. The wounded waited, some of them a long time, even a few weeks. In the ziemlanka of the wounded there were always a nurse and a doctor. . . . Sometimes the headquarters sent partisans to check on us to see if we were guarding the place well. Once the three of us who were keeping watch heard a suspicious noise. Then we noticed how an armed man was crawling in our direction on all fours. In a second the stranger was surrounded. His argument that he was a Russian partisan did not help. We disarmed him, tied him up, took him to the headquarters. There we found out that he was really a partisan. He was sent to check if the Bielski people were doing a good job."[37]

After the big manhunt, Luba and Janek Rudnicki joined the newly formed fighting detachment, Ordzonikidze. During an encounter with Germans Janek was injured in the leg. The couple came to the Bielski camp for treatment. At the same time another wounded Jewish partisan, Gedalia Tokar, was brought there as well.

Because Tokar's condition was grave he needed to go to Moscow for treatment but first had to be moved to the airport, some sixty kilometers from the Bielski base. His mouth was shattered and he was fed cream and butter through a straw. On the way someone had to attend to his needs. Luba Rudnicki, who came to the Bielski otriad with her husband, volunteered.

She says, "I did not believe that there were planes, but I decided to go. When I reached the airport there were many, many injured lying all over. It was full. Because there was little space they decided to take only those who had leg injuries. The rest would have to stay. But I felt that my patient needed help fast. I took a white sheet, tied his leg. He was small. I took him into my arms, screaming, 'Make way, the injured cannot walk, make way.' I had difficulty climbing the plane. There was some kind of a ladder. I pushed myself up, brought him there, and left. He survived the war."[38]

As in most groups, in the Bielski otriad those who came to the unit early had more privileges than newcomers. Most old timers came from Nowogródek, and most were relatives or attached to each other by friendship. Those who had come from other communities sometimes complained that "People from Nowogródek fancied themselves more important than those who had come from other places."[39]

Where one came from plus the time of arrival in the forest affected a person's social standing. And this social standing in turn was reflected in people's lifestyles. For example, those who belonged to the top echelon "could be seen riding their horses, wearing leather coats, riding breeches, with a gun stuck in their belts. Their clothes and guns served as status symbols, many were anxious to demonstrate their 'superiority' in the social hierarchy, and their membership in the upper strata.

"The members of the fighting units who were responsible for the safety and the supplies of the company were on the next level of the social ladder. Their rifles were always carried on their shoulder, inseparable. Not as well dressed as the top, they were aware of their importance and looked down on the unarmed members. These were tall, stalwart young men whose parents used to be in inferior trades, peddlers, wagoners, and artisans."[40]

The workshops and those connected to them were a very important part of the Bielski otriad. Quite early, whenever the otriad experienced a period of stability, it would set up workshops that had involved gun repair, tailoring, shoe repairs, and a smithy. These early shops came into existence without any special planning.

Malbin explains that "When we needed something built out of wood we asked a carpenter to do it. We had a bakery. In a bakery you need all kinds of special tools so we asked the carpenter to make the shovels out of wood."[41]

With the establishment of a more permanent base in the fall of 1943 in the Nalibocka forest, the Bielski otriad concentrated on its workshops. Conscious of their advantages, Tuvia insisted that, as a community, they had the potential to take care of their own needs and the needs of other partisans. He urged his people by saying, "Whatever we will need for setting up such workshops, we will find, even if we have to unearth it. The order was to bring machines and other tools that could help create all kinds of jobs."[42] Occasionally, the craftsmen themselves came to the otriad with their tools. More often than not, however, such tools had to be searched for or improvised by some of the creative mechanics.

Workshops developed slowly and continued to grow. From the beginning special rules applied to these workshops. However, these rules were not always followed. According to one basic rule, all members of the Bielski otriad were entitled to all services free of charge, and any member could have something repaired—a pair of shoes, a coat, a gun. Although there was no need for permission, people still had to wait their turn. If a member of the otriad wanted to order something new, he or she required permission from headquarters or from Tuvia.[43]

Transactions that involved other partisans had to be arranged by headquarters, often by Tuvia himself. Payments for services were usually in the form of food, medication, and, less frequently, arms. All profits belonged to the otriad, not to individual craftsmen. New rules were constantly emerging as new and specialized workshops continued to develop.

"Most workshops were situated in a very large hut. The din emanating from this hut could be heard from afar, banging of sledgehammers, sawing of wood, clatter of sewing machines, laughter, lively conversations rich in partisan slang.

"The huge hut, with its raised ceiling which looked like a large machine shop in a factory, accommodated tens of workers, who were divided up according to their trades. Large windows provided proper lighting for the various workshops located in all corners of the hut. The different workshops were separated by wooden partitions and a number of people worked in each cubicle. More workshops were spread out throughout the camp. All spheres of production were intertwined. All materials provided by nature in the forest were put to use . . . new productive areas were added, there was more variety in the type of jobs to be done, and almost everyone at camp was employed doing useful work."[44]

Headed by Shmuel Oppenheim, the gun repair shop retained its honored position. Oppenheim's son, also a capable locksmith, was his assistant. The elder Oppenheim was known throughout the region. Eventually, his fame brought Soviet newspapermen who were on a visit from Moscow to the Bielski otriad. They were amazed. One said, "This is the first time that I have seen such an otriad. When I write about it in the newspaper in Moscow no one will believe me."[45]

This workshop offered substantial earnings in food and sometimes weapons to the Bielski community. It also improved the otriad's safety by keeping its gun supply in good condition. Finally, by fixing some broken-down or defective weapons, it transformed some unhappy malbushim into proud gun owners.

The tailors did important work for partisans whose clothes were shabby or in need of patching. Partisans were often forced to wear a single shirt for many months. In addition to servicing their own members, the tailors had orders from distant companies. A special group of seamstresses was busy sewing shirts from the rough linen woven by local farmers.

Tailors would sit around a long table, and as they used scissors, needles, and thread they would sing. Sewing-machine noise would mingle with their songs and jokes. There was a lot of gossip about different commanders and their mistresses in the various isolated farms. They talked about men who at night went to the "wrong" ziemlanka and slept next to women other than their own wives. All this chatter was mixed with laughter and jokes. The entire atmosphere was friendly and warm.

Contrary to what was expected, the tailors seemed too busy to attend to the needs of the average Bielski partisan. Clients who paid with goods were served first and those with nothing to offer were kept waiting. And so, for example, after Kopold received a gun and went on a food expedition he acquired some warm clothes but they needed altering. Only when he paid a tailor with cheese and meat did the tailor agree to do it.[46]

Riva Reich was given a coat by her mother as she was running away from the ghetto. Tuvia ordered the tailors to remake it into a pair of pants; without his intervention she would have had to wait a long time.[47]

Too many people needed help and Tuvia could not possibly devote himself to them all. Hana Stolowicki and her family were malbushim, unable to offer the tailors anything. In the winter the girl caught a bad cold and coughed. There was no medication and the doctor could do nothing for her. The problem was solved as follows: "Someone took pity on me and gave me an old coat. One could make pants out of it to keep me warm. But the tailor did not want to work for nothing. We could not pay him.

"Mother decided not to eat bread all week and give it to the tailor. The whole family did not eat bread for a week. Indeed, this is how it was. I had warm pants and recovered."[48]

Like the tailors, the shoemakers were also overburdened with work. People would reach the camp barely dressed and often barefoot. Sometimes they were robbed of their good boots on the way. The distance they had to cover when coming to the forest would destroy their shoes, and in the forest itself leather would quickly rot from the mud and wetness.

Not surprisingly, "there were always long lines of people ordering shoes and in addition there were customers from other companies who came to order boots and paid for them with other essential products. Frequently there were other deciding factors. Gifts of food and other items brought by young men returning from their missions bribed the shoemakers and expedited the orders of the young ladies. In the shoemakers' workshops one could see about ten artisans bent over their benches sewing new soles, mending old, affixing soles to sandals, and making use of old tires, converting the rubber into soles.

"Tanners, some of whom came from Kołdyczewo, built a tannery in a distant part of the camp. They supplied the shoemakers with leather. Several weeks after they started operating the tannery, one could already see partisans in the camp in locally made boots. They were of a peculiar light yellow hue. The people made saddles and harnesses for the horses, as well as belts for the partisans, which put the finishing touches on the tanners' products."[49]

Occasionally, Tuvia would intervene on behalf of a needy person. When, for example, he saw Mrs. Geler arrive in the camp with wet rags on her feet he said nothing. But shortly after, she was called by the shoemaker to come and be measured for shoes. In a week she had new boots, a luxury that lifted her spirits.[50]

Tuvia also ordered a pair of boots for the nurse Riva Reich. New boots were too important not to be noticed. Among those who shared Riva's ziemlanka were Mr. and Mrs. Pupko, former owners of a brewery in Lida. It was rumored that the Pupkos had brought with them a lot of money, jewelry, and numerous valuables. They had a reputation of being misers and kept all their treasures hidden. No one, including them, benefited from their wealth.

Riva tells about Mrs. Pupko's reaction to her new boots. "Mrs. Pupko begged me to go to Tuvia and ask him for shoes for her husband. I went, in a naive way, I said: 'Commander, Pupko does not have shoes.' This was the first time that Tuvia raised his voice to me. He was always so polite with me. 'For them? They have so much money! For them you come to ask? Can't he take out some of his money and buy shoes?' I excused myself and left."[51]

In the Bielski camp very few people owned a watch. Most had never had one, while others lost theirs in different ways. For Russian partisans a watch was a precious possession and they kept the watchmaker very busy. The watch repair shop was yet another source of income.

The forest was generous with its materials. Carpenters were inventive, creating all kinds of items: wooden parts for weapons, sandals for the summer, barrels that were used for the production of leather and much, much more.

As in the past, the otriad had a smithy, located not far from other workshops. The blacksmiths were kept busy attending to the horses from the camp and surrounding otriads. They also repaired wagons, particularly wheels. They too were creative and tried their hands at nonconventional tasks.

Among the highly praised places was the sausage factory. Located in a separate building, it had an oven for smoking meat and a meat grinder, confiscated during an expedition. Although salt, pepper, and onions were hard to find, the sausage produced in the Bielski otriad was famous all over the partisan region. People would bring live cattle to exchange for sausage and smoked meats, with such transactions beneficial for all parties.[52]

A hut was also transformed into a barber shop. "Their implements were very dull and scratched unmercifully but for lack of anything better the partisans were forced to use their services. The gossip during the process sweetened the chore. Partisans who were at leisure like to come to the barber shop to exchange news of the front, to gossip, and to tell jokes. All of this was done in typical partisan spirit, in Yiddish intermixed with crude Russian expressions. The preference was for pungent male jokes, the young women were uninhibited, outdoing their men on many occasions."[53]

In principle everyone was entitled to a haircut and a shave. But since the barbers were also very busy, this rule was not always followed. Geler's experiences are revealing. "When a malbush came from whom they could expect nothing they would shave him with an unsharpened instrument. Such a customer would almost collapse. Once I felt it on my own cheeks. The dull knife took off my hair together with almost half of my cheek. From then on I stayed away from the barbers. Later on, when I stopped being a malbush and was going on food expeditions, I gave one of the barbers a piece of pork. After that my face felt the pleasant sensation of a sharpened instrument. The barber even told me elegantly, 'thank you.'"[54]

A very useful part of the otriad was the flour mill. Two old horses moved around, turning heavy grinding stones that made flour. Not far from the mill was a bakery. On winter evenings the bakery was transformed into the otriad's cultural center.

The Bielski workshops and production centers had far-reaching consequences for those who received the offered services and those who did the jobs.

Skilled workers in the various workshops enjoyed a higher social standing than unskilled workers or those whose skills were of no use to the company. Not only were the craftsmen more highly valued but they had more food than the malbushim. People who did only manual unskilled labor, such as kitchen duty, chopping wood, or taking care of the cows and horses, were part of the lowest strata in the camp. One could easily pick them out in the crowd on Main Street. Their appearance was careless, their clothes worn and patched, their faces pale and thin, their movements slow and plodding.[55]

Beyond the many visible exchanges of benefits, the Bielski production units created important connections within an unstable and hostile environment.

Despite the officially friendly posture of the Soviet partisan movement, unofficially Russian partisans continued to grumble that the Jews ate too much and did not fight. Often these critics would point to the size of the Bielski camp as proof that too much food was consumed there.[56] Perhaps the Russians envied the Bielski otriad for having family members with them while they were far away from home and relatives.[57]

Some think that Tuvia as a special individual provoked envy among partisans. He had good looks, presence, and could easily outshine most Russian commanders.[58] "Some peasants treated Bielski like a general. The Russian commanders were jealous because he was so respected by the natives, by the population at large."[59]

Several powerful Soviet partisans were opposed to Tuvia. One of them, General Dubov, saw the Jewish otriad as a disadvantage for the Soviet partisan movement. Eventually in the late fall of 1943, the commander of the Pervomajskaya brigade demanded a death sentence for Tuvia Bielski.

Hersh Smolar, a prominent partisan and a member of the Soviet headquarters, knew that "the accusation was that the Jews had been robbing the peasants. They take clothes, not only bread. Platon let me read the document.

"It was indeed true. There were some Jewish partisans who took clothes. The partisans were not allowed to take anything but food, but the Jews did take other things. This created a difficult situation. When Platon showed me the paper I said, 'Why pay attention to scraps of paper? In the forest we should not rely on such evidence, come let us go to the Bielski otriad and see what it is all about.'"[60]

Platon agreed. At the camp Tuvia invited them for a tour—he wanted them to see the workshops. They faced a hard-working, well-ordered community. When they came to the tannery they saw several older Jews pray, swaying from side to side. Platon was curious about these men. He wanted to know what they were doing. Smolar reports that Tuvia's answer to his question was: "They are studying the fourth chapter of the history of the communist movement (a chapter written by Stalin about the philosophy of Marxism). They started to laugh and that was it."[61]

When Platon saw huge pots with soaking animal skins he was amazed to hear that these would be transformed into leather. He appreciated the coordinated efforts and complimented Tuvia on these efficient enterprises.

Platon was sufficiently impressed and during the next meeting at his headquarters he said that every military body has suppliers of goods, people who instead of fighting provide all kinds of services for the fighters. This, he emphasized, was the function of the Bielski otriad. He felt that such services were indispensable for the partisans.

This changed the situation—Platon's attitude decided everything. His insistence that the Bielski otriad's contributions were essential to the partisan movement saved the life of Tuvia Bielski.[62]

Although important, Platon's protection only took care of dangers lurking from the Soviet partisan movement. Soon Jewish partisans were threatened by Polish partisans, known as White Poles. The traditional Polish-Soviet hostility was responsible for this development.

With considerable success, Stalin was undermining the legitimacy of the Polish Government in London. Churchill and Roosevelt agreed to a return to the 1939 Russian-Polish frontiers, but Stalin was not satisfied.

Eager to establish control over the rest of Poland, he was searching for Poles who would become a part of a puppet postwar government.

In Western Belorussia, as elsewhere, the Poles and the Russians were competing for control of the region. For a while, perhaps to neutralize the local Polish underground, the Soviets maintained an officially correct posture toward Polish partisans. The Poles reciprocated by cooperating with the Russian guerilla movement. During the big hunt, the Polish detachment Kościuszko, under the command of Miłaszewski, had coordinated their moves with the Russians. At that time, the Poles fought valiantly, broke the German ring, and lost many men. In September, reduced in size, Miłaszewski's group returned to the Nalibocka forest.

Soon a group of Polish officers came to join this Polish unit. They were sent by the Polish Government in Exile, with instructions to undermine and contain the Soviet power in this area. Some of these officers belonged to the Fascist NSZ (Narodowe Siły Zbrojne, National Armed Forces). Throughout the war the NSZ had waged multiple battles: against the Germans, the Russians, the Jews, and Poles who disagreed with them politically. Only in 1943 did a part of the NSZ agree to join forces with the AK (Armia Krajowa, The Home Army, the fighting unit of the Polish underground in Poland that was loyal to the Polish Government in Exile in London).[63]

The Soviet-Polish cooperation was shaky. For a while each side was not prepared to openly challenge the other. The Poles expressed their hostility toward Russian partisans in limited ways. "When they caught a Russian partisan, they would beat him up. After they took his weapon from him, they would free him."[64]

In contrast, Jews caught by these Poles were not as fortunate. Unprotected small groups of Jewish civilians in bunkers or Jews who were roaming the countryside were attacked and killed.[65]

A statement by Bor-Komorowski, the head of the AK, offers some clues for this double standard. Noting that independent forest dwellers made their appearance, he says that "They were wild bands of all sorts of refugees living by robbery and were a terrible plague to people in the neighborhood, who were visited nearly every night by bandits who gradually deprived them of their last belongings . . . in agreement with the Government Delegate I issued orders to the regional Home Army commanders to undertake the defense of the population against the violence of disturbing elements . . . my order was immediately used by Communist and Soviet propaganda as . . . an accusation against the Home Army, the Government in London, and myself for opposing Soviet partisans. . . they knew that my orders were forbidding clashes with Soviet or Communist partisans."[66]

A Pole, sympathetic to the government in London seems to be justifying and clarifying this order when he says that "Jewish bands in the forests were robbing and looting the peasants and the Home Army moved against them to protect the local population and to keep order."[67]

In line with Bor-Komorowski's directive, the White Poles were using Jews as shooting targets. A group of twelve men from the Jewish family camp Zorin was attacked on the way back from a food expedition. Only one escaped to tell the story.[68]

When I asked Tuvia about this incident, in his characteristic manner he cut through the ambiguities. "They attacked the Jews because they were anti-Semites and obviously because they were afraid to murder Russians."[69]

All along, the Soviets saw the Polish underground's loyalty to the Polish government in London as a serious interference. When the Red Army moved into Polish territory, particularly in late 1943 and 1944, it was ruthlessly destroying the non-Communist Polish resistance.[70] In the late fall of 1943, Russian headquarters in the Nalibocka forest ordered a surprise attack on the Kościuszko group. Several otriads were asked to contribute fighters. The Bielski unit sent fifty men.

At dawn the Poles were surrounded and without a single shot were taken prisoners. About fifty who were away from the base were not included among the captured.

One of the participants in this attack, Greenstein from the Ponomarenko otriad, says, "When we took these Poles prisoners, we divided the soldiers among them into small groups and sent each group into a different Russian unit. Many of them had come from the surrounding villages and towns. Soon most of them ran away."[71]

For a while these Poles were watched closely yet the ultimate fate of most of them is not known. In contrast, the Polish leaders were arrested, including Miłaszewski. Kept in a bunker, they were guarded closely until an order came to send them to Moscow.

The same plane that took these prisoners to Moscow had earlier brought Poles from the USSR for the explicit purpose of organizing pro-Soviet–Polish partisan units.[72]

Although a large Polish otriad had been disbanded, this did not eliminate Polish underground fighters from the region. They remained active in the surrounding villages. When Russian guerillas would come to such a village for food they were told that they were protected by Poles. Jewish partisans in particular tried to avoid areas dominated by White Poles. Poles preferred not to enter the forests and stayed in surrounding villages. Well-organized and strong, they continued to attack unprotected Jewish civilians.[73]

Bielski partisans who went on food expeditions would occasionally stop over at the Ordzonikidze otriad, the part of the original Bielski group now commanded by Russians. As head of intelligence, Zus would supply them with important information about road safety. Occasionally too Ordzonikidze would offer them luxury gifts like salt, vodka, eggs, fat, and more. Placed at the edge of the Nalibocka forest as a fighting unit, Ordzonikidze gave a measure of protection to its parent, the Bielski otriad.[74]

12

The Fate of Women

"I was walking alone in the forest. A man with a rifle stopped me . . . I thought it was a partisan, but I was not sure. I was frightened. He pretended to be arresting me. He was forty or so, maybe a Belorussian. He told me to follow him and took me to a tent. Then he forced himself on me. . . . He did not ask . . . he just did it. He raped me.

"Inside me something broke. I was sick. This feeling stayed and stayed, for years. I could not look at men . . . I had some kind of an inner conflict. I was excited about sex, but I was scared, turned inward. I did not like the physical part of sex. I did not put sex in an important place. I am glad I don't need it . . . I imagine that many women had similar experiences . . . shocking, this rape . . . horrible."[1]

A young teenager at the time, this girl escaped from a ghetto and lost her way. She spoke about the rape only at the very end of a lengthy interview. When she did, her voice was monotonous, halting, as if pushing back some invisible force that demanded attention.

After this hurtful experience the girl moved on, finding comfort from other fugitives to whom she became attached and with whom she joined the Bielski otriad.

I heard about the fate of some other women ghetto runaways from Hersh Smolar, a prominent communist partisan. One day Smolar received an unexpected visit from Tevel, a one-time rabbinical student, a brave partisan, and a personal adjunct to a respected head of a Soviet brigade. One glance at Tevel told his host that something terrible had happened. Smolar recalls that "Short of breath, in a strange voice, Tevel tried to tell me something. At first all I could figure out was his 'you must come.' The boy had a horse ready for me. Then I heard that we were going to Niemen (a major river). This was quite a distance from us. When I asked what it was about he answered that I would soon see for myself what 'they' do."[2]

During the long ride that followed the two barely spoke and Smolar sensed that this otherwise friendly youth wanted silence. He kept wondering what it all meant, but said nothing. They reached the river Niemen, the side identified as partisan land.

When their horses stopped, the older man saw the answer to his questions. Scattered on the ground were bodies of young women who were clearly Jewish.

"They were dead. It was obvious that these women had tried to save themselves. They had succeeded in swimming through Niemen and then they were murdered. They had come to our side. On our side there were no Germans. We stood there stunned."[3]

After what seemed like a long time, motionless, with eyes on the ground, Tevel whispered, "What should I do?" His question was addressed to the man next to him, a man who, because of his communist convictions, had spent many years in Polish jails.

Now this man was speechless. When Smolar recovered, his answer was, "Tevel, say Kaddish" (a Jewish prayer for the dead).

And Smolar did the following. "I went to our headquarters and I made a big fuss. I screamed. I told them that our people had murdered the Jewish women. I demanded justice. To this the commander, who was sent to us from Moscow, said that after all this was war and we have received a notice that the Germans were sending Jewish women to poison our good kettles. This was an excuse for the killings. Nothing changed."[4]

In the forest, physical strength, perseverance, fearlessness, and courage were all highly valued. No one associated these characteristics with being a woman. In the rough, jungle-like environment of the forest, most men were convinced that women were unfit for combat and therefore burdensome.

The Soviet government did not support such views. On the contrary, the government officially praised women's contributions to guerilla warfare. In fact, it claimed that women partisans symbolized the supreme dedication and patriotic struggle for the country.

Although widely publicized and backed politically, women's participation in the Russian partisan movement was limited. The estimated proportion of women among these forest fighters ranges from two to five percent. In contrast to approving governmental attitudes, the partisan leadership in the forest was convinced that this small number of women was all the movement could effectively absorb.[5]

Usually, women who joined Soviet partisan detachments were relegated to unimportant duties. The closest they came to combat was working as scouts and intelligence agents. But even these jobs were performed only rarely—eagerness to participate or special fitness seldom tipped the scale in their favor. Instead, most women were assigned to service jobs involving the kitchen: cooking and keeping the place clean.[6]

It has been argued by some that women's acceptance into Russian partisan detachments was due largely to their usefulness as sex partners. Indeed, officers from a brigade commander down to battalion commanders would select sexual partners from among the women enlisted in a unit. When this happened, these women became the property of the officers, which by implication gave them officer status, with the privileges that went with the position of their partners but without combat assignments.[7] Most high-ranking Russian partisans had mistresses. In recognition of this arrangement, such a woman was called a "transit wife."[8]

Most Christian women stayed in the forest because of a special relationship with a man. Only a fraction came because of a desire to oppose the Nazis and even a smaller proportion because their lives were endangered. Within the forest non-Jewish women were a definite minority.

Unlike their Christian counterparts, Jewish women came to the forest to avoid death. Before entering the woods, such Jewish fugitives knew that for a woman alone it would be much harder than for a man and that the possibility of rape and murder was real.[9] They also realized that powerful men could give them some protection, and the more powerful the man, the better the chances for staying alive. Indeed, actual acceptance into a Russian detachment usually depended on a woman's readiness to become a partisan's mistress. It was common for young girls to sleep with Russian commanders, political heads, or whoever was in power. On the other hand, if a partisan, any partisan, helped a woman, he expected to be paid with sexual favors.[10]

While most male partisans were eager to have sexual relations with women, they also accused them of promiscuity. The very women they desired as sex partners, they viewed with contempt. In male conversations, for example, whore was often substituted for the word woman.[11]

In the forest Jewish women in particular encountered dangers and refusals to become members of partisan units. Threats and rejections became more pronounced with time when more Nazi collaborators were joining Russian detachments.[12]

Jewish women were doubly disadvantaged: as Jews and as women, because of ethnicity and gender. As a result, a large but unknown proportion of Jewish women perished on the way to and inside the forests. Of those who stayed alive, some found refuge in Soviet units and some in Jewish detachments. Only a small fraction survived in small scattered family groups.

Of the Jewish women accepted into Russian otriads the majority became mistresses of partisans, usually officers. But officers with Jewish mistresses were under pressure to terminate their relationships. Sometimes such partisans would send their mistresses away to the Bielski otriad; sometimes, as in the case of Marchwinski and Shematovietz, they would go with them.[13]

There were always exceptions. Some partisans, because of their special contribution to a Soviet otriad, would be allowed to keep their women. Also here and there a Jewish woman would be allowed to join a Soviet otriad because of a connection to a powerful partisan, unrelated to sex. A brother or a father might keep his female relative if he was important enough for the otriad.[14]

A Jewish woman's acceptance into a Soviet detachment sometimes had nothing to do with her relation to a man—not all women were ready to trade sex for protection. Significantly too, only a fraction of women qualified for such transactions. Most women lacked two basic requirements: youth and good looks. Yet special skills could overcome misogynist attitudes. A physician, a nurse, even a good cook, would be accepted into a Soviet detachment without becoming anyone's mistress.[15]

In sharp contrast to most Soviet otriads, the two Jewish camps in the Belorussian forests, the Bielski and the Zorin otriads, had a policy of unconditional acceptance of all Jews. Sex, age, or any other characteristic of a Jewish fugitive made no difference. Any woman who reached the Bielski otriad was automatically accepted. The problem for Jewish women was learning about the Bielski and Zorin otriads and, once they knew about them, reaching them safely.

Most women realized that alone in the forest, their chances of reaching a haven were slim. A connection to an "appropriate" man could ease the move from the ghetto to a safe place in the forest, shielding a woman from hurtful experiences and death. Men who could do that were usually the more resourceful, lower-class youths. And because such youths could promise some safety, they had the freedom to choose their women partners. A simple, common youth had no trouble getting a socially superior girl, someone he could have only dreamt about before the war.

Everyone wanted to live. Still, for an upper-class woman to become involved with a lower-class man was not easy. Sometimes upper-class Jewish parents preferred to see their daughters dead than sexually involved with a crude and uneducated man.

These situations often resulted in painful confrontations and misunderstandings. For example, in the ghetto, one upper-class woman was courted by a crude lower-class man who had lost his wife and child during an Aktion. He was determined to save this young woman. When he heard about an impending raid, he asked her to run away with him.

The young woman had a father, mother, and sister in the ghetto and refused to leave them. The man pressured her, saying that he had consulted her father who told him they should do what was best for them. Preparations for a ghetto escape were set in motion. Shortly before the couple's departure when the daughter went to say goodbye to her father, she heard him say that he did not mind dying because this way he would not have to see her as the wife of a common man. The couple left as planned; both survived in the Bielski otriad.

They were married for over forty years. After his death the woman explained that "He became my husband only because he was going to save me, not for any other reason. Do not mention my name in this connection. I would have never married him before the war. I would have never met him. We lived in very different worlds."[16]

Often women would move out of the ghetto if they thought that on the outside they would find shelter. A man in the Bielski otriad, particularly if he was young and a fighter, would sometimes invite a woman via letters or verbal message to come to the Bielski otriad in a group led by a guide. Sometimes the partisan in question was the group leader.

Sulia Rubin joined the Bielski otriad because of an invitation from the partisan Israel Kotler. When she arrived she realized he was involved with another girl. Sulia knows that she would have never reached the woods if a male partisan had not asked her to come. Kotler's invitation had told her how to escape as well as how and where to reach the Bielski otriad. It also promised her a welcome.[17]

Other women were invited but not necessarily by a boyfriend. Sometimes the invitation came from a relative or friends. To be sure, some women like Sulia arrived at the Bielski camp in a group, but without an invitation. Some came alone. Others happened to stumble on the Bielski camp by chance—sometimes such women had not even known about the otriad's existence. Still others were picked up by Bielski scouts searching for Jewish fugitives in need of a haven.

No matter how a woman arrived, no matter who she was, when she reached the Bielski otriad she could automatically become a member. Often critical of Tuvia, the nurse Lili Krawitz tells how without any hesitation he had admitted her girlfriend with a sickly and elderly mother. Inevitably, the older woman became a burden for the otriad, yet this was by no means an isolated instance.[18] Tuvia's policy of unconditional acceptance of every Jew, man, woman, and child, no doubt accounts for the high proportion of women, a proportion that fluctuated between thirty and forty percent.[19]

Inside the camp a few basic rules applied equally to men and women. Every individual was entitled to a minimum allotment of food. Tuvia permitted no deviations from this rule. All adults, men and women, were also required to do guard duty. Only a few were excused because of physical handicaps, age, or excessive nervousness.

Tuvia wanted to see his people occupied. New working opportunities were created in the Nalibocka forest, when the people settled there for the second time, with the establishment of more permanent workshops. In addition to offering a measure of personal satisfaction, work could lead to extra food earnings and thus supplement the official diet. This, for example, was true for nursing jobs. A nurse was automatically entitled to some additional food. Many women took advantage of different job opportunities.[20]

As for overall improvement of positions, women had fewer options than men. Besides, the road that led from a malbush to an improved social position was different for men and women.

As noted earlier, a man could change his situation by acquiring a gun, which usually meant going on food expeditions and ceasing to be a malbush. And some men, after receiving a gun, distinguished themselves as scouts and fighters.[21]

No such opportunities were available to women, who were discouraged from carrying arms. They were also barred from food missions. Men felt that during an expedition the presence of women would increase the danger of these missions. The few women who possessed weapons were excluded from participation in food expeditions, except for Chaja, Asael's wife. She accompanied her husband on a few missions, and Asael's high position in the otriad made this possible.

When some women complained about these restrictions, they were told that every army needs a large supporting group and that women belong to this part.[22]

In the very rare instances when a woman with a weapon reached the Bielski otriad it would be confiscated. Guns belonged to men, not to women—this was the law of the forest. Raja, an independent woman, justifies this custom and insists that Tuvia needed these guns for men who went on food expeditions.[23]

Usually the single, unattached women were dressed in rags, with shoes that threatened to fall apart or no shoes at all. A woman without shoes had to wait for her turn with the shoemaker. But if she had nothing to bribe him with, her turn might never come.[24]

If a woman was single and young, men would court her, sometimes in unusual ways. Before Sulia Rubin got a steady boyfriend, at night, in the collective ziemlanka, men would find their way to her bunk. She had difficulties fighting them off. An older woman advised her to keep a twig next to her and hit whoever would approach her. When she did, she would notice a number of bruised faces the morning after.The men soon stopped bothering her.[25]

As a rule such unsolicited nightly advances did not result in a steady relationship. But most people felt that a woman needed someone to take care of her and this someone had to be a "proper" man.[26]

A newly acquired lover was called a "Tavo." It is a Hebrew word, a masculine address, that within the context could be translated into "come here." Because men with guns went on food expeditions, a woman who had a steady Tavo with a gun had more and better food than an unattached woman. Whatever social value a youth with a gun had, it was automatically transferred to his girlfriend. And because a woman without a man did not amount to much, most young girls looked for an appropriate Tavo.

While many women made a stable connection to one particular man, some had casual sex with different men. Others made a decision not to become attached at all.

There were no official weddings. When a man and a woman shared a tent or a bunker and acted as if they belonged to each other, others began to treat them as married. A majority of adults, over sixty percent, lived as couples.[27]

Opinions about the newly created couples vary. Some claim that forest unions were a result of sober deliberations. Some feel that love in the forest was physical and rarely anything else.[28] Others criticized these unions on the grounds that women were selling themselves to men.[29]

Women who entered into marriages that improved their lot explain their decisions in different ways. Sulia, whose father was an oral surgeon and whose mother was a dentist, automatically belonged to the elite of Nowogródek. She readily admits that when in November 1942 she reached the Bielski otriad her privileged background gained her entry into the ziemlanka of the equally privileged family, the Boldos. In the forest the Boldos continued to have a top position because of their family ties through Sonia Boldo to Zus Bielski and through Chaja to Asael Bielski. In addition, the two Boldo youths, Pinchas and Josef, were fighters.

Sulia explains that "The Boldos took me in because Arczyk Bielski brought me there and asked them to keep me. He knew who I was, he was in awe of me."[30]

While Sulia's background gave her an initial advantage, it did not nullify her camp position of a malbush. She explains that "Every ziemlanka had a 'nebich,' 'a hanger-on,' a malbush—I was it. Not fit for anything. They took me in reluctantly."[31]

In fact, as soon as her teenage protector left, some of those in the ziemlanka began to grumble. "What do we need another malbush for?" Others called her a whore. This name-calling was interrupted by the appearance of a man who, to Sulia, looked like a giant. Tuvia Bielski paid a visit to the ziemlanka. She heard him say, "'Shut your mouth, how many girls do we have like that! As many as there are, I will take them. Just let them come!' He made me feel comfortable. He made me feel that I had a place. To me Tuvia seemed like a legend, unreal. . . . To see an armed Jew who was a leader—strong. I was so impressed."[32]

While this initial meeting with Tuvia made Sulia feel more protected, it did not help her adjust to the conditions of the camp. As a malbush and a newcomer, she was assigned to low-level duties.

She had to cook, but did not know how. Nor did she know how to start a fire—to keep whatever little flame she had going she would blow and blow. As a result of this constant blowing, her face, neck, and hands were all covered with sod and ashes.

An old man in the ziemlanka took pity on her. Although himself weak, he tried to assist her. With his help came a lot of mumbling. "What will become of you? How will you manage? How?" Sulia had no answers to these questions. Instead of concentrating on her work, she was dreaming about a fur coat. She had reached the forest wearing only a sweater.[33]

Overwhelmed, Sulia became depressed. Her parents and sister were still in the ghetto. She was convinced that if she had them with her, her troubles would disappear. An appropriate boyfriend, she knew, could bring her family to the forest. This idea lifted her depression. She became actively involved in searching for a partner.

Among her new friends was an older man who had recently lost his wife and children. He became a heavy drinker and had no interest in life or women. This man agreed that Sulia's troubles could be solved by finding a Tavo. He felt that his younger brother, Boris Rubierzewicki, a brave partisan, a scout, and a constant food collector, would be a good choice.

Soon Sulia was introduced to a simple, uneducated youth who had guts and a gun. Boris had never imagined that he could have such a refined girl for a wife. Sulia, too, had never met this kind of boy. Repulsed by his common ways, she was impressed with his courage.

For Sulia, it was not a question of love or attraction. She respected his strength and expected two things from him: to improve her economic lot and to rescue her family from the ghetto. Boris was interested. As proof of his intentions, he presented her with a fur coat confiscated during one of his missions. Although the fur signaled a deal, Sulia continued to keep him at a distance.

Before taking the final step she made her future Tavo promise that as long as they would be in the forest he would not leave her. She was trying to avoid future humiliation. Boris readily agreed to her demands.

Next, through guides, Sulia notified her parents about her decision to take a lover. Her mother was opposed to this union. Sulia knew that her mother was a snob and thought that he was not good enough. Her mother's letter said, "A woman who sells her body to exist I can understand, but you sold your soul because you will never leave him."[34] Sulia did not listen to her mother's complicated message. She had reached the Bielski otriad in November 1942. By April 18, 1943, she had a steady Tavo.

Of the six months alone she says, "It was like six years of Siberian torture. But in that time I learned how to cook well. They taught me how to shoot. I knew how to clean a gun."[35]

Sulia notes that after the six months, "Right away I was dressed. Right away, I got a pair of boots. I had a fur. To have a man who did not look at anybody else and who protected you was something marvelous! Soon, we got a permit to build a ziemlanka. I told my two best friends, Raja and Lola, 'I have my Tavo, the two of you can wait till the end of the war. You will stay with me. Whatever I have I will share with you. Don't rush into any relationship.'"[36]

Almost fifty years later and married to the same man, Sulia explained that "It was enough that I did what I had to do. At least I wanted my two best friends to have the freedom to choose someone they loved."[37]

Sulia's mother never met the son-in-law she so vigorously objected to.

The family was murdered before Boris was able to arrange for their ghetto departure. Although Sulia was pushed into this union by her inability to adjust, saving her family was no doubt an important consideration.

Some women insist that in the woods male-female relationships were based on more than just an exchange or wish for services and goods. A woman, they argue, felt more secure if she had a man. Pesia Bairach explains that "Even though we women did not go out for combat and food missions, we had to deal with military actions. The Germans would attack us. A woman who had a man with a gun felt more secure."[38]

Lili Krawitz, an upper-class woman and a nurse, became attached to a lower-class man in the ghetto. She is convinced that without him she would have never left nor would she have survived. In the Bielski otriad, her husband became a fighter, which gave Lili many advantages. She thinks that women who married as she did were totally dependent on their men. Trying to understand these socially asymmetrical marriages, she points out that after all no one forced a woman to become attached to any particular man. About herself Lili says, "I was young, hungry for love like so many other young women and these men gave us love.

"For my husband it was a great thing that he got a supposedly 'superior' woman. He was grateful to me for it. He was proud of the fact that I wanted him and behaved very well toward me. I am not even sure if he loved me or not. For me his goodness was compensation for everything else.

"I don't think that a woman would have sold herself for food, more likely for security. During a raid a man would look after her. It was important. One always lived in fear about what might happen next. How does one live with this fear, all alone? A young girl needed someone.

"I do not agree that women were selling themselves, but it was not love either.

"To be sure, men rather than women would select a partner. But if a woman did not like the man no one forced her. She was free to reject a man."[39]

The Bielski otriad had many couples in which a resourceful husband protected not only his wife but also her relatives. One of these unions was that of the Russian commissar, Shematovietz, his father-in-law, an Orthodox Jew, the same age as the son-in-law, and his eighteen-year-old daughter. Somehow his pretty daughter never smiled.

As a neighbor, Tamara Rabinowicz had many opportunities to observe the Orthodox Jew at prayer. He prayed outdoors and Tamara was struck by his unusual habit. As he swayed from side to side praying he would spit. He swayed and spit, swayed and spit.[40] Intrigued, one day Tamara decided to find out what this strange combination meant, and if it was in any way related to his Russian son-in-law. She recalls: "I went out to investigate, to hear what he was saying as he spit. I passed close to him and heard, 'drop dead.' Each time he prays and spits he swears. 'He should break his legs,' 'Let the earth swallow him up,' and so on and on, adding all kinds of curses. He was saying all this in Yiddish."[41]

But some male-female relationships were shortlived affairs. In the Bielski otriad, as in most other detachments, the powerful men had easier access to women than the rest, and Tuvia and Zus took full advantage of these opportunities. In their defense some say that although the two brothers had many affairs, each tried to stay away from married women and those who had steady boyfriends.[42] Others accuse them of trying to sleep with every attractive woman.[43]

Chaja comes to her brother-in-law's defense by pointing out that Tuvia was a man any woman could easily fall in love with. She admits that he was sexually active, but insists that "the women were really after him. He was gorgeous and supposedly good in bed."[44] Others echo her statement and emphasize that "He did not force anyone. He never pushed himself on anyone. All the women in the otriad were in love with him. He was extremely attractive and he was the commander. One always had a way out. Some women even bragged that they slept with him."[45]

Some partisans are convinced that Tuvia respected women, that he gave them a chance to be self-sufficient, adding that "if any of these women became his mistress, it was because they were chasing him."[46]

Talk about Tuvia's infidelities covers the time he was married to Lilka, the most beautiful girl in the forest.[47] Not yet twenty, Lilka was madly in love with her husband. Although she suffered because of his sexual escapades, she was willing to overlook his transgressions and made an effort not to interfere.[48] Some think of her as refined, but removed from the rest of the otriad.[49] Some see her as "restrained, someone who knew how to keep secrets. She did not try to make an impression. While people might have had an inferiority complex when dealing with her, she herself did not act like the wife of the commander. As far as her husband was concerned, she tried not to see much of what was happening. Maybe it was better for her not to know too much."[50] When she talks about herself, Lilka leans toward this last assessment.[51]

Others tend to disagree and say that Lilka was trying to show that she was the wife of the commander. Some suspect that she wanted everyone to appreciate her and bow to her. Actually, she was a young, inexperienced girl, and, despite her beauty, not exceptional, just like so many other youngsters in the forest.[52]

Perhaps these diverse opinions are not as incompatible as they seem. Retiring people are often shy and give the impression of snobbishness and distance. Being Tuvia's wife in itself separated her from the rest of the people without her trying to do anything about it.

Indeed, when asked to evaluate Lilka, Tamara Rabinowicz said, "In the forest she had her clique. Her people were those that belonged to the 'Court.' They were riding horses, they would wear pants, they had boots. Each one of them had a gun. When Lilka was on a horse and I, Tamara, walked, she would look down at me. What was I to her, the wife of a man who does not have shoes? Nothing. In the forest people like us were not worth much. What counted was the gun and one's closeness to the rulers. My husband and I were removed from it. Nevertheless, they treated us very well, especially Tuvia and Asael."[53]

Despite Tuvia's infidelities, when Lilka talks about their marriage she emphasizes the positive. She insists that he taught her a lot about life, about love. Wistfully, she recalls that "When there was danger he always looked after me. He could rely on me, too. I never interfered. I did not nag him. I did not force him to take me or stay with me."[54]

She even defends Tuvia's infidelities when she says that "his private life was his business. If he went on a mission and slept with someone, I did not see, I did not know. Who cared? . . . Even in front of me women approached him."[55]

Others say that "Probably for Tuvia it wasn't a question of love. Lilka was young and beautiful. It was sex. His wife Sonia was killed. He began to drink too much after that. He wanted to forget. He stayed away for about two weeks. Before that the family had lived together, in one tent. He came back to his family, to the same tent. Lilka and her father were there."[56]

Lilka's deep commitment to her husband was atypical. Most forest unions began with a woman offering sex in exchange for goods, for protection, for security. Asael Bielski's devotion to Chaja resembled Lilka's feelings for Tuvia.

Some of those who speak about Asael and Chaja are convinced that this marriage would have never taken place before the war.[57] Although Chaja might have agreed with this assessment, she learned to see Asael in a different light. About his attitudes toward her she says that "He did not have any other women, he adored me. He felt that I was worth more than any other woman. . . . I was a virgin. This he appreciated, especially since I had quite a number of boyfriends."[58] Others also emphasize Asael's devotion and love for Chaja.[59]

Chaja came to value his warmth and generosity. Looking at her marriage from a distance of almost fifty years, she says, "I loved him very much. He was very decent. He had inborn intelligence. He was brave, daring, courageous. He was very handsome. He knew how to love. He did not touch me for a whole year. He went to non-Jewish women because he was afraid that I would become pregnant. When we were liberated, the same night I became pregnant. He wanted to make sure that I would not leave him. I would have never left him. He proved himself."[60]

Several women mentioned a time lapse between the decision to become sexually involved with a man and actual lovemaking. They seem to have a need to affirm that they did not go to bed with a man right away. Sonia Boldo, for example, insists that Zus did not leave her for another woman because when she met him she was able to resist his sexual advances. Sulia also made a point of telling me that although she agreed to become Boris' mistress, it took quite some time before she lost her virginity.[61]

I have no way of knowing whether this occurred or not and perhaps the "truth" here is not so important. Just saying that they did not easily become sexually involved with a man may have restored some of their self-esteem. These women were brought up thinking that a "good" girl had sexual relations only with a man to whom she was properly married.

Perhaps the upheavals and turmoil of war failed to erase these ideas—old traditions do resist change. And even in the forest these traditions were expressed in a variety of ways.

The kinds of jobs men and women had reflected some old values. Each group gravitated toward tasks traditionally associated with womanhood and manhood. Not only were the men in charge of the otriad, bearing arms and supplying the group with essential provisions, but even the non-leaders, those who stayed inside the camp, were busy with the so-called manly jobs. Men who were not a permanent part of a workshop would cut trees, collect building materials, and build ziemlankas. The heavy job of digging wells also belonged to men.

In contrast, women devoted themselves to traditional jobs that involved different ways of caring for people. Chaja, for example, busied herself with the welfare of others, particularly women's welfare. Washing in the winter was a problem. She taught the women that by rubbing their bodies with snow they would achieve several excellent results. Snow would clean the skin, massage it, and raise the body's temperature, all at once.[62]

Enterprising and energetic, Chaja was constantly working on new projects that benefited women. She was soothing and universally loved.[63]

The care of children was automatically left to women. The number of children fluctuated with time, ranging from ten to thirty.[64] The Germans were determined to destroy Jewish children. By eliminating the young, they expected to do away with Jews as a people. An estimated one and a half million Jewish children were murdered during World War II.[65] And this Nazi obsession made the children who survived particularly precious. On the other hand, it was harder to smuggle children out of the ghettos than adults and it was also harder to protect them on the outside. Indeed, all conditions conspired to make the rescue of Jewish children an exception.[66]

Although faced with insurmountable obstacles, Tuvia was eager to save children. He instructed his ghetto guides to bring back children, particularly orphans, and occasionally they did.

Many members of the otriad had lost their children, so they doted on those in the camp. Informal adoptions of orphans were accepted and gratifying, both to the adopting parents and the children. During the manhunt of 1943, when crossing the swamps of the Nalibocka forest, people willingly carried the orphaned children on their shoulders.[67]

Of the otriad's youngsters, only a few were younger than ten. Most were in their teens. All children were dressed in clothes inappropriate for their age and height, a combination of anything within reach. With little or no parental supervision, they adopted the crude obscenities of partisan speech, and loved to curse loudly in imitation of their elders.[68]

One of them, the Garfunk boy, four or so, would salute in front of Tuvia in the morning, and say, "Commander, allow me to report that in our ziemlanka whoring has been taking place." The adults in the otriad, including Tuvia, appreciated the boy and showered him with special favors.[69]

The children would roam around the forest in groups. They played at soldiers pretending to take revenge on the Germans. Through their games they reproduced life in the ghettos and around them. Sometimes they would become so involved playing that they would beat up those who assumed the roles of Germans. Much of their time they would spend spying on adults. They knew many secrets.

In the morning after the cows were milked, one could see children head toward the cowshed with tins in their hands. They would come for their daily milk ration. Later, they would go to the forest to collect berries. Even in cold weather, groups of youngsters would scatter in close proximity to the camp and search for the few leftover berries. Sometimes they were joined by adults who were on the lookout for mushrooms.

The adolescents were mobilized into the work force. Some tended the cows and horses, helping out in the cowshed and stables. Occasionally they would help with kitchen jobs and some youngsters found work in workshops and served their apprenticeship as shoemakers and carpenters.[70]

Toward the end of 1943, it was decided that children, particularly the younger ones, should lead a more orderly life. A woman partisan, Czesia, was in charge. In a hut, each day, she assembled about twenty youngsters. This was a special kind of school: without books, pencils, or paper. There was no formal teaching. Instead, the children listened to Czesia's stories, played group games, went on excursions, exercised, sang, and laughed.[71]

For the Purim holiday, they prepared a show. "The commander, his staff and all other guests crowded into the large hut of the workshops where a stage had been set up. It was a cold day; a strong wind carrying gusts of snow howled outside. We watched the children ascending the stage. They sang and recited in Yiddish and Russian. They were dressed in white shirts and red ties, their innocent charm lit up the faces of the dancers and gymnasts."[72]

The audience seemed mesmerized.

Suddenly, in the middle of a dance, one of the partisans jumped up, shouting, "Where are my children?! Revenge! Revenge!" He drew his gun, ran out, and shot several times into the air.

This outburst was met with total silence. People continued to sit immobilized. After what seemed like a long time, slowly, still without saying a word, they left their seats. Some quietly wiped their tears away. As if embarrassed, without looking at each other, one by one, they disappeared into the darkness of the night.[73]

The grownups loved children and would have welcomed newborn babies. But constant threats, harsh living conditions, and the severe weather all conspired against it. In fact, there was an informal rule against new births. Women had to carry the burden of unwanted pregnancies and abortions were common. These were performed under the most trying circumstances, with inadequate instruments and without medication.

Dr. Hirsh, the camp's physician and gynecologist, was kept busy performing abortions not only on women from the Bielski otriad, but also on those from the surrounding Russian detachments.[74]

The ever-smiling, thin doctor, his instrument case always in his hand, was very busy. His surgical reputation spread far and wide. In payment for his services the doctor would receive such valued provisions as fat, pork, flour, and other food articles. He demanded edibles from partisans who went on food missions and gold coins from those who could afford them. It was gossiped that his instrument bag and a special bag he had tied around his neck were each filled with gold coins. Women from the Bielski otriad who could not afford his "fees" were accommodated free of charge.[75]

Although most people agreed that no child should be born in the forest, those who refused to follow this rule were not pressured. In fact, two or three babies were born in the woods.[76]

Of the young women who were sexually active, many had one or more abortions. One woman died during the procedure. After the war some of these women might have had difficulty becoming pregnant. To this day women from the Bielski otriad find it hard to talk about these experiences. Some had several abortions and were bitter about it, blaming the men for not practicing birth control. Although still hurting, most are frank about their feelings. Basically their arguments run as follows: "After all I paid with my body, did I need pregnancies yet? It depended on him, not on me."[77]

Stories about sexual escapades were widely circulated with few facts to substantiate or deny them. Some people deplored what they saw as pervasive promiscuity, while others denied that people were sexually active. Still, within the Bielski otriad, coarse expressions about sex were part of everyday language. Men would jokingly equate having sexual intercourse with being alive. When asking a woman to become his sex partner a partisan would say, "Let me check if I am alive."[78]

Although many young women were sexually active, no one coerced them into a relationship. No woman was ever raped in the Bielski otriad. On the contrary, compared to the Soviet detachments, in the Bielski unit women had more options. They had the freedom to refuse a man—no woman was dismissed because she rejected a man. In fact, no woman was even dismissed at all.[79]

Some of the women who could have had lovers opted for celibacy. One of them explains that "There were many vulgar men. I did not go for such men. I saw how the partisans behaved sexually. I slept in the same ziemlanka . . . and I felt nauseated by them. I preferred to remain a malbush."[80]

Riva Reich, young and attractive, also refused to become involved with a man. She preferred to concentrate on her nursing job. She readily admitted that a woman with a lover was more protected, but this was not for her—she had not even considered this possibility. Besides, she felt that as a nurse she was in a more favorable situation than most other single women. About her life she said, "I did not have it difficult because I was working. I felt good because I was helping people. I knew who needed what and was able to do things for them. I was surrounded by respect, approval. Tuvia protected me and I did not care about anyone else."[81]

As a secretary of the Bielski otriad, Raja concentrated on her work rather than on men. Women who withstood the pressure and refused to become involved with a man were more highly respected by the people in the otriad than those who gave in.[82]

Frequently, these unattached women drew an unflattering picture of the men in camp as crude and sexually unappealing. At times such views were transferred to men in general and after the war some of these women were reluctant to marry. Even some who in the forest had a male protector agree with these negative assessments of males.

For example, before the war, Sulia Rubin preferred men to women as friends. The war changed her attitudes. She explains: "I did not see one man sacrifice himself and go to the grave with his children. My cousin went. She could have survived; a German wanted to save her. She was gorgeous, with blue eyes and dark curly hair. She won the beauty contest of the town Druzgieniki. Her name was Mina Bencjanowski. During a deportation a German wanted to take her to the side that was spared. She said, 'Aber meine Kinder' (But my children). 'Die Kinder kann ich nicht' (I can't take the children) was his answer. 'Da gehe ich mit den Kinder zum Tod' (Then I will go to death with my children). This was during the first big Aktion in Nowogródek. Her husband was caught at the very beginning of the occupation with fifty other prominent citizens. They were all murdered."[83]

Sulia continues. "Not only did the men not sacrifice themselves for their children, but when a man's wife was barely dead he would already look for a woman with whom to have sex. This happened to me. We were hidden during an Aktion. There was a man whose wife was taken away not long before that. He was trying to make out, first with my sister, then with me . . . I wanted a prince on a white horse to come and take me, but there was no one like this. I was attractive, young. Sure, men wanted me only for sex, not for my soul. The war cured me of all men. I tell my husband when he is jealous, 'Don't be afraid. I don't need any men. I don't want them.' Many of my girlfriends feel that way as women. We had them up to here (she places a hand above her mouth)".[84]

Yet, after the war, like Sulia, most of the women continued to stay with the men they had married in the forest. When I asked why they did not change their seemingly unsuitable partners the women had no answers. Perhaps their decisions had to do in part with the cultural climate of the times. In those days a woman who lost her virginity lost much of her self-esteem. Perhaps these women thought that "good" men would find them undesirable. Most of these young women had lost their entire families and the only people they were close to were their forest husbands. Perhaps they opted for the familiar, something that reminded them of their families and their lost world.

In the forest both Jewish men and women continued to suffer. Compared to men, women had fewer options. Their suffering was more serious if they had not come directly to the Bielski otriad. For those who reached the camp most had no one. Some lost their families while hiding in the woods; others were raped by Russian partisans on the way to the camp. For practically all these women the Bielski otriad was the last hope.

But life was harsh even in the Bielski otriad. Still, women rarely blamed this otriad for their difficulties. Instead, they repeatedly emphasized that if it were not for the Bielski camp they would have never made it.[85]

Whatever influence a woman had was usually channeled through a man to whom she was sexually attached. These male-female distinctions were a reflection of the society that had such values and the partisan movement was imbedded in it. Despite the elimination of lives and the destruction of the social fabric, some old traditions survived. Perhaps one might argue that the very devastation surrounding these people encouraged them to pursue whatever superficially might resemble normalcy. Perhaps the social, cultural, and spiritual vacuum created by the war demanded some familiar structures.

13

Keeping Order

"Put a partisan's jacket on the ground and it will begin to walk."[1] Lice, the permanent inhabitants of partisan garments, were credited with this unexpected motion. Overcrowding, no change of clothes, and limited washing—all acted as an invitation to these insidious pests.[2]

Lice are stubborn. Once they establish residence they are difficult to dislodge. Inevitably, filth and lice lead to disease. Aware of this connection, Tuvia supported those who could improve the otriad's cleanliness. Because Riva Reich qualified for such work he assigned her to double duty: nursing and sanitary inspection of the camp. Every morning she walked through the otriad checking its hygiene. When she finished this overall survey, she stepped inside the kitchen and into several ziemlankas.[3] But Riva was only one of several women who took turns watching over the camp's cleanliness.

Those who did not live up to the inspectors' standards were asked to correct the situation. If they did not change, they were given a warning. When this too produced no results, Tuvia intervened, which usually took care of the problem. People knew that if they disobeyed him there would be a public reprimand, even an arrest.[4]

While most people tried to cooperate with these efforts, Dr. Hirsh, the camp's physician, had no trust in the positive power of cleanliness. On the contrary, the historian Amarant notes that the doctor "did not avail himself of the bathing nor cleaning facilities which existed in the camp. I rarely saw him wash and his clothes were in awful condition. When we complained about the unsanitary conditions around our common food kettle, he would seriously, and in professional terms, try to convince us that dirt in food has advantages, is good for immunization. He would interject several scientific terms into his arguments, and we had to swallow our food in spite of ourselves."[5]

It was rumored that the doctor was rich, yet his personal appearance challenged this gossip. Dr. Hirsh wore dirty, shabby clothes and his boots threatened to fall apart at any minute.

The doctor seemed to have found an original, although sporadic solution for the special condition of his boots that was illustrated by the experiences of a partisan admitted into the hospital. Unable to make a diagnosis, Dr. Hirsh prescribed rest to this patient. While the man was in bed the doctor wore his new boots. When he began to feel better the doctor continued to prescribe bed rest. The boots, the patient thinks, were responsible for his prolonged hospital stay. The doctor, he suspects, used this method with other patients as well.[6]

Unlike Dr. Hirsh, his nurses devoted themselves to fighting filth and particularly its companions, lice. Nurse Krawitz encouraged people to boil their clothes which, in effect, reduced the amount of lice.[7] Nurse Reich devised another system. She told her patients to wipe themselves with their urine particularly in places favored by lice—armpits and genital areas. "Urine contains a chemical that kills lice," she insisted. The application of urine did reduce the population of lice without, however, eliminating all of them.[8]

Another approach used in cold weather, by those who had a change of clothes, was hanging garments outside their ziemlankas. When they did this, the next morning the snow underneath was covered with tiny yellowish spots—frozen lice. Some lice, however, survived even this ordeal.

The traditional approach of washing also diminished the amount of lice. But to be even partially effective, washing required soap. During the German occupation soap was a luxury. The Belorussian peasants had learned to do without it and, therefore, partisan expeditions yielded no soap.

As a substitute for soap the Bielski otriad used ashes, which were also added to pots of boiling laundry. Although a poor alternative to soap, ashes did reduce the population of lice.

In the fall of 1943, in the Nalibocka forest, Tuvia heard that one of his partisans, Sioma Pupko, knew how to make soap. This led to the establishment of a soap factory. Ashes and animal fat were the basic ingredients of this soap. The end product resembled dark brown dough. And because the quantities produced were small, they were distributed only within the otriad.[9]

The manufacture of soap was soon followed by the establishment of a bathhouse. The deserted nearby town of Nalibokī had a public bath. Like the rest of the town, the bathhouse was partially in ruins. Bielski partisans, some of whom were locksmiths and carpenters, collected and transferred stoves, pipes, and other equipment to the camp.

The otriad's finished bathhouse had two inside wells. This large structure was divided into several sections. One part was used as a fumigating room. Here, very high temperature would destroy some of the lice and no doubt other disease carriers. People were urged to bring their clothes to this place and many did.

Sometimes a garment was so well disinfected that it fell apart. The lice were killed but at the expense of their living quarters, the garment. This is how the partisan Basist lost a cherished fur jacket.[10]

Besides a regular bathing room, one room served as a Turkish bath. Heated stones made up a large part of this room; when water was poured over these stones it created steam.

The Bielski otriad was proud of this bathhouse and urged guests to visit it. One of these visitors, General Platon, was full of admiration, and was particularly impressed with the Turkish bath.[11]

The bathhouse operated from early morning until evening. Bathing was compulsory. Special times and days were allotted to residents of different ziemlankas. Men and women used it separately. One enthusiastic supporter tells of how the bathers "were luxuriating in the hot vapor, lying on the benches, pouring water over each other. But one could not enjoy it for too long, half an hour was allotted for each group."[12]

Those who went on outside missions had special access and usually, on their return, would avail themselves of their privilege and visit the bathhouse. The chief cook, Engelstern, was one of the few entitled to a daily bath. As soon as he distributed breakfast, he would head for the Turkish bath.[13]

In addition to helping people stay healthy, Tuvia and the headquarters also made provisions for death. A special piece of land was set aside for a cemetery. Of those who were buried there the majority were victims of enemy attacks.

Dr. Amarant thinks that "death took a short holiday and let the remaining refugees from ghettos bear their daily partisan tribulations and wait for the war to end."[14]

Most people avoided visits to the doctor. Some say that Dr. Hirsh "had only two diagnoses: you will live or you will die."[15] In part, people stayed away because, for no known reasons, life-threatening illnesses in the forest were practically non-existent. Despite the harsh weather conditions, very few partisans suffered from colds, coughs, or sore throats. It was as if the body learned to cope with the extreme temperature changes.

Still, the otriad did not escape from all health problems. People suffered from various kinds of skin diseases: blisters, scabies, boils, and furunculi. Inadequate hygiene and malnutrition contributed to these conditions. With a shortage of medications the doctor and nurses tried to cope with these ailments in unconventional ways. Bullets were a help. The powder inside them contains sulphur, a disinfectant that fights infections. When this powder is mixed with fat the result is a creamy substance that cures all kinds of skin diseases, including boils and furunculi.[16]

Boiled milk in the form of injections was also used for a variety of skin infections. A patient injected with milk would develop a fever and the body would then mobilize its immune system to fight the infection.[17]

In the winter of 1943–44, the Zhukov brigade was moving away from the Nalibocka forest. The commander wanted to leave behind a few wounded partisan officers for whom travel was difficult, as well as some women he considered superfluous. Tuvia agreed to accept both. As a thank you, the Soviet commander gave the Bielski otriad food supplies, horses, cows, arms, and different articles of clothing.[18]

Lili Krawitz, the nurse, was assigned the care of one of the wounded officers. When this patient did not improve, Dr. Hirsh decided that he should go to Moscow for more extensive treatment. The wounded officer had to be moved to an airport and then flown to the USSR. The patient requested that his special nurse accompany him to the airport. Krawitz refused—she was afraid to make the trip. When she did not budge from her position, Tuvia stepped in. He reiterated the order: She had to go with the patient to the airport or she would be shot for disobedience.

Krawitz stuck to her initial decision and argued that since she was the wife of a fighter she never took the extra food allotted to nurses. And because she was not receiving those special provisions, she had no obligation to follow Tuvia's order.

As the wife of a fighter she had enough food and therefore did not need the provisions. Krawitz had counted on her husband's support and influence, but at the time he was away on a food expedition. Lili ran away into the forest and waited for his return. The next day, when her husband came back he went to see Tuvia. He used the same argument: Since his wife did not get extra food for her nursing duties she should not be forced to go to the airport.

Unmoved, Tuvia repeated the order and the warning: If Krawitz refused to accompany the patient, she would be shot for disobedience. She gave in but was convinced that she had been wronged and another nurse should have taken her place. Almost fifty years later she is adamant about this incident. Still very emotional, she said, "I was crying all the way to the airport . . . I was so scared. In fact, Boris, Sulia's husband, took me and left me there. It was not that Tuvia wanted to please the Russian, it was more a question of an order which I had to follow. After a week they came to take me back. . . . Later Tuvia acted as if he had forgotten the whole incident. I never had any contact with him except for that one time. I never talked to him."[19]

Lili Krawitz never forgave Tuvia for what she saw as a personal defeat and an injustice. Krawitz's refusal to follow a commander's order was a challenge to his authority and Tuvia could not tolerate this kind of disobedience. Had he allowed this to happen, it would have undermined his ability to lead. Moreover, it is to Tuvia's credit that once the incident ended he took no punitive measures and behaved as if nothing had happened.

On joining the Bielski otriad, people were required to hand over to the nurse or doctor whatever medication they had. These drugs were then used for the sick. When Moshe Bairach came to the forest he brought some anti-infectious drugs and gave them to Dr. Hirsh.

Much later, Moshe returned from a food expedition feeling very ill. His arm was infected, swollen, and filled with pus. Pain and high fever would not let him rest. Moshe begged his wife, Pesia, to go and ask the doctor for help. Dr. Hirsh tried to dismiss her and when Pesia continued to plead for some medication, all the doctor said was, "Come back in half an hour and I will give you half a 'Kogucik' [equivalent to aspirin]."[20]

Usually gentle and soft-spoken, at that time Moshe Bairach was not in a condition to tolerate this kind of treatment. Feverish and furious, his only desire was to kill the doctor. Not fully dressed, he jumped out of bed, reached for his gun and ran out. Half-crazed, he burst into the doctor's ziemlanka. When Mrs. Hirsh saw him she screamed. In no time Bairach was surrounded by men who overpowered him and brought him back to his place.

Moshe recalls that "In half an hour all the 'high-ups' came to my ziemlanka: Tuvia, Malbin, the nurse. Tuvia said, 'Bairach, what happened to you?'"

Instead of answering, Moshe uncovered his arm—red, swollen, full of pus. It had an awful smell.

"Tuvia turned to the nurse, 'How do you allow this to happen?'"

"She said, 'I have nothing to wash it with. What can I do?'"

"'I don't care, unearth it, but find some medication, take care of it!'"

"He turned to me, 'Bairach, I give you my word that from now on they will attend to you.'"

"Then he added, 'A person in your condition should not have a gun. Give me your gun.'"

"I said, 'What are you doing to me?'"

This was a terrible blow for him.

"He repeated, 'A man in your condition should not have a gun. You see what could have happened. Give me your gun and I promise you that the day you feel well and come to me, I will return your gun.'"[21]

Moshe adds, "He spoke to me gently, like a father to a son. All was done with concern. . . . By the way, he sat on my bed despite the horrible smell that came out of my arm. All the others who had come with him disappeared as soon as they saw what it was all about."[22]

Tuvia's interest in Moshe's arm continued. He invited a doctor from a different otriad to check on his condition. This specialist thought that the arm would have to be amputated. But because of Tuvia's intervention, Dr. Hirsh and the nurses took special care and no amputation was necessary.

In the winter of 1943–44, for the second time, typhoid fever paid a visit to the Bielski otriad. People suspected that the clothes left by the Zhukov brigade were the carriers of the disease. The epidemic spread like wildfire and the estimated number of typhoid patients ranges from sixty to one hundred forty.[23] An additional ziemlanka had to be built to serve as a hospital.[24]

No medication for treating typhoid fever was available. All the doctor and nurses could do was to try to keep the patients more comfortable. And since high fever took away the patients' appetite, they were fed boiled water and melted snow.[25] Only with their fevers down, on the way to recovery, did the sick develop a ravenous appetite.

Tuvia ordered special food for all of them: meat, milk, chicken, and chicken soup. To this day some of the former malbushim remember these post-typhoid meals as veritable treats. The epidemic claimed one life, but soon allowed all others to return to their previous activities.[26]

The beginning of 1944 brought a certain measure of tranquility to the Nalibocka forest. But fewer external threats did not necessarily translate into internal peace. On the contrary, for some detachments there seemed a peculiar connection between external and internal threats. As outside dangers increased, people became more cooperative. In contrast, as soon as they felt more secure, they became more competitive, more critical, and more disobedient.

All groups have provisions for keeping disruptive behavior in line, behavior that threatens to destroy their social fabric. Among the Soviet partisans unruly conduct was punished haphazardly and severely. For example, when at a closed meeting Smolar dared to criticize what to him looked like irresponsible actions by some leaders, he was secretly sentenced to death. His superiors decided that Smolar was to be sent on a mission and shot on the way. Warned by a friend, Smolar notified General Platon about this verdict. Platon, in turn, ordered him to come and stay at headquarters.

Mazurek, Smolar's friend, was less fortunate. A brilliant lawyer, writer, and linguist, as a member of partisan headquarters Mazurek had access to a radio and listened to the BBC news. He shared this news with the people at headquarters. Since the BBC was the center for wartime information, Mazurek offered his listeners accurate news.

To a secret service agent, an NKVD man, this looked very suspicious. This agent decided to eliminate Mazurek on the grounds that he was an English spy. Both the accusation and execution happened in secret. Mazurek was sent on a mission, with a man whose job it was to shoot him. The official report stated that Mazurek was lost on a mission.[27]

In most Soviet otriads falling asleep while on guard duty resulted in a death sentence. Quite a number of partisans lost their lives this way. Others were shot for less obvious reasons. Jacov Greenstein, a partisan in the Soviet otriad Ponomarenko, tells about his commander, Shapajev, who used to shoot at random when drunk, which was quite often.

Once Greenstein lost a dear friend for a different but strange reason. It happened during an anti-German mission that called for an eighty kilometer hike. Greenstein's friend could not keep up with the group—his feet were hurting. And because this otherwise courageous partisan lagged behind, the Russian commander shot him. Greenstein is convinced that this kind of an incident would have never happened in a Jewish otriad.[28]

In the Bielski otriad human life was highly valued. But, like all other detachments, the Bielski unit had provisions for controlling deviance. In the Nalibocka forest, one ziemlanka was converted into a prison. Falling asleep while watching the camp, refusal to obey an order, or unauthorized taking of goods—all resulted in a jail sentence of a few days.

Berl Chafetz, a rabbinical student, was a shy, awkward youth who did not quite fit into the forest life. He seemed totally dependent on his older brother, a man with a gun and a frequent participant in food expeditions. One day after Berl's brother returned from a food mission he was put into prison. He was accused of taking too many provisions for personal use. Berl became frightened. He went to headquarters to find out if he too would be punished. After he explained the purpose of his visit, Tuvia laughed and said, "Don't worry, go home, and don't think about it. It does not concern you."[29] Berl stayed in the Bielski otriad while his brother chose to move to a Russian detachment. Both survived the war.

When people were accused of more serious "crimes," the lawyer Volkowyski was assigned to their cases. He was the otriad's prosecutor. His job was to investigate whether accusations and suspicions were justified. Volkowyski was a methodical man. Once he was satisfied with the collected evidence, he presented it to the headquarters of the Bielski otriad.

Among the more serious cases were Jews accused of collaboration with Nazis. In the Bielski otriad this happened to three different men. Each raised sensitive issues. Should such Jews be accepted in the first place? Does the otriad have an obligation to punish these collaborators? There was also a question of evidence and the possibility that the accusations were unjustified. Heated, emotional debates surrounded each case, and each led to a different resolution.

The first man, Lansman, came in the winter of 1942–43. His arrival was preceded by reports that he had worked in the ghetto Nowogródek as a Gestapo agent and was responsible for the deaths of many Jews. In the forest, even before his arrival, some people were eager "to get their hands on him." Just before coming to the forest, in an obvious reversal of fortune, the Germans were about to execute him. Lansman managed to run away, but only after a bullet penetrated his hand.[30]

His appearance at the Bielski base created an uproar. At a special meeting emotions ran high. Only after a long shouting contest were some people able to express their opinions. One of them, Alter Titkin, Lilka's father, said, "Every Jew who leaves the ghetto should live together with us; the trials and judgments have to wait till after the war."[31]

But Titkin was in the minority. The rest felt that Lansman was a menace; he could run away and report their whereabouts to the authorities. Then someone raised the question of how this collaborator had known about the location of the camp. Actually no one trusted him, not even Alter Titkin. The only difference between Titkin and the others was that he was against a death sentence while they were for it. The majority prevailed. Lansman became the first Jew to be officially executed in the Bielski otriad.[32]

More ambiguous was the case of Daniel Ostaszynski. As someone who was at first a high Judenrat official and later the head of the Judenrat in Nowogródek ghetto, he had many enemies. Ostaszynski was accused of supplying the German authorities with lists used for the collection and murder of Jews. But when the remnant ghetto inmates were on the verge of total annihilation, he helped organize a ghetto breakout. His defenders argued that through this action he rehabilitated himself.[33] Others disagreed.

As a ghetto runaway, Daniel entered the forest while searching for the Bielski otriad and met a few Bielski partisans. They knew about Ostaszynski's past and considered him a traitor. They began preparations for his execution by removing his boots, a partisan custom. It is much easier to take boots off a living than a dead person since a dead person's feet have a tendency to swell.[34]

By chance, Sonia and Zus Bielski came upon these preparations. Sonia and Daniel had been schoolmates and friends and she stopped her friend's execution. Later, Sonia arranged for Ostaszynski's acceptance into the otriad.[35]

Powerful protection and absence of clear incriminating evidence allowed Daniel Ostaszynski to stay in the camp. Rumors about his past continued. He must have known that some of the Nowogródek Jews wanted him dead. For safety, Daniel never left the camp, never went on food expeditions. He never became fully integrated into the life of the otriad and kept pretty much to himself.[36]

In the fall of 1943, among a group of arrivals from Lida was the barber Białobroda who came with his wife Miriam. In the Lida ghetto Białobroda had worked as an informer for Nazi collaborators, the Belorussian policemen. When Jews, employed outside the ghetto, were returning home, Białobroda would tell the Belorussian policemen who was trying to smuggle food. The policemen would then confiscate the goods and share them with Białobroda. The Jews were satisfied that no more severe punishment followed. Białobroda thought that no one knew about the part he played in these confiscations. He was wrong—people were well aware of his collaboration and hated him for it.

When Białobroda reached the Bielski otriad he was spared, possibly because he had come to the forest at a relatively stable time. Perhaps another reason he was treated mercifully was that his collaboration, while causing hardship, had not led to loss of lives. In addition, Tuvia's overall reluctance to kill, particularly to kill Jews, was of help. Tuvia appointed him chief barber of the otriad and hoped that Białobroda would rehabilitate.

In a short time, however, Białobroda's conduct became suspicious. From time to time he would ask permission to leave camp. At Tuvia's suggestion Volkowyski started an investigation. He learned that during these outside trips Białobroda used his gun for robbing natives of their valuables, gold, and jewelry. This was strictly forbidden. In itself such action called for a death sentence. Additional evidence revealed that Białobroda was also plotting with Israel Kesler who was in turn trying to undermine Tuvia's authority. There was a trial and the prosecutor, Volkowyski, demanded the death sentence. Białobroda was shot and buried. Most of the people approved of this outcome.[37]

Justice had to be done not only in cases of collaboration with the Nazis but also with the Soviets, when this was a part of internal intrigues against Tuvia. From the very beginning Tuvia was confronted by internal disobedience. As early as the winter of 1942–43 he had successfully dealt with a rebellion by temporarily disarming the opposition.[38]

One of these early rebels was Arkie Lubczanski. He continued to agitate. As a self-appointed spokesman for the communist cause, he had to be handled with care. Like so many others, he appreciated the power and privileges that came with the commander's job.[39] In short, Lubczanski wanted Tuvia's position.

He became friendly with Soviet officers, telling them about his commander's disloyalty to the communist cause, about his commander's personal collection of gold and valuables, and his lack of concern for the welfare of his people.

When these stories reached the Soviet partisan headquarters Tuvia was asked to defend himself. Tuvia succeeded in clearing his name, but Lubczanski continued to stay in the otriad. At the headquarters of the Bielski otriad Tuvia was advised to shoot the man for treason. "I don't like to kill Jews," Tuvia answered, but added, "we will throw him out."

Officially dismissed, accompanied by his girlfriend, Lubczanski left with an order never to show his face again in the Bielski otriad.

This girlfriend soon returned, begging Tuvia to readmit her. None of the Russian otriads was willing to accept her. She was readmitted on condition that she would not meet Lubczanski in the otriad—the two could get together only away from the Bielski base.

Lubczanski's girlfriend, his brother, and sister-in-law continued to stay in the Bielski camp until the arrival of the Red Army.[40]

Arkie Lubczanski was by no means the only one who tried to inject communist ideology into the Bielski camp. Opposition to such procommunist efforts was dangerous. When communist propaganda was free from personal attacks Tuvia tolerated it. Frequently he pretended not to notice.

Tuvia's patience was reflected in his treatment of Israel Kesler, a man he had earlier accepted into his otriad. As a head of a subunit, Kesler enjoyed more freedom and independence than other partisans in similar positions. For Kesler this was not enough, but Tuvia was not prepared to give more. Eventually these differences surfaced.

Israel Kesler was well adjusted to life in the forest. Despite his shady past as a professional thief and years spent in Polish jails, people appreciated his leadership qualities, his courage, and concern for the plight of the Jewish people. But Kesler's commitment to saving Jews had some limitations.

This is seen in the case of Abraham Viner, a teacher, then a school principal, and a native of Naliboki, Kesler's birthplace. While on the run from a Nazi work camp Viner had heard that Kesler had been in charge of a partisan group, in the days before the Kesler group joined the Bielski otriad. When Viner tried to join this group Kesler said, "'You cannot stay with us, you are not made of the proper material. You will not be able to kill, to fight, you are not fit to be a partisan.'"

Viner continues. "I left. I had no choice. I was rejected along with three others. We were from the same class. We were all educated. We had no arms, nothing. I dreamt about coming to the Nalibocka forest. It was not easy. After this refusal it took me about two months to reach Bielski."[41]

Despite this rejection Viner still described Kesler as very ambitious, energetic, and brave.[42] Others, who were close to Kesler, think of him as goodhearted, as someone who liked to share what he had with the needy, and as a just man.

Removed from the main base by more than a mile, Kesler's group was a well-run, orderly camp. He was totally in charge of the camp's internal activities. When it came to official dealings with the Soviet authorities, the Kesler group was represented by the leadership of the Bielski otriad, mainly Tuvia. This bothered Kesler.

Kesler's supporters minimize his aims; they insist that all he wanted was to transform his own group into an independent detachment with himself as commander.[43] Others doubt that this was all Kesler desired—they are convinced that his wish to separate was only a first step in a more ambitious journey. They argue that if allowed to happen such a breakoff would have led to Kesler's takeover of the entire otriad.[44]

No doubt Kesler had some leadership qualifications and charisma. He was well suited to life in the forest and because of his past it was particularly easy for him to confiscate goods. He had a sense for guessing which peasants had hidden jewelry and gold. Shrewd and calculating, Kesler knew how Soviet officers felt about Tuvia. Kesler stayed away from General Platon, a supporter of Tuvia, while trying to get close to General Dubov.

Dubov disapproved of Tuvia and the Bielski otriad. It bothered him that essentially it was not a fighting unit. But he was aware of Platon's protection of the Bielski camp and did not act on his opposition.

Kesler approached Dubov with gifts of jewelry, with a request to support his wish to become an independent detachment, and with a complaint about Tuvia's mismanagement of funds. Dubov agreed to act but asked for more evidence. Kesler delivered more letters in which he reiterated the accusations against Tuvia and expressed the wish to form a separate otriad.

Dubov passed on this information to General Platon who invited Tuvia, Asael, and Malbin to headquarters. The three were confronted with two issues: the establishment of a separate otriad out of Kesler's group, and accusations about the unauthorized acquisition of gold, jewelry, and money by the commander of the Bielski otriad.

For Tuvia, the request to reduce the size of his otriad, even by a small number, touched a sensitive nerve. He had given up almost two hundred people to form the Ordzonikidze otriad, but only because he realized that a refusal could lead to the destruction of his entire unit.

Now the situation was different. This was not an order from Soviet authorities but a demand made by a disobedient member of his own group. In itself this demand was a challenge to Tuvia's authority. He felt that it was something he did not have to agree to. Besides, Tuvia was convinced that if he granted Kesler's wish, others would be only too glad to follow his lead and ask for another part of the otriad and then another. This could result in a breakup and eventual elimination of the entire camp.[45]

From the very beginning Tuvia had fought for and succeeded in establishing a large detachment. He successfully argued that its size in and of itself offered safety, and history was proving him right. Scattered, small groups of Jewish runaways, even if armed, were at a disadvantage. Most were attacked and destroyed almost as soon as they were formed. Russian, Belorussian, and Polish partisans, while sometimes glad to destroy small Jewish units, were reluctant to attack a large group.[46]

Moreover, Tuvia made sure that his potential enemies were not aware of his weaknesses. They did not know how few of his people could defend themselves. He also made a point of not revealing the otriad's limited supply of weapons and ammunition and was determined always to appear strong. A reduction in size meant danger and the likelihood of a breakup. With some justification, he equated the survival of his otriad with large size.

But Tuvia did not expect Platon to be concerned about the survival of Jews. With this Russian general, he knew he would have to use different arguments and take into consideration Platon's needs. And so Tuvia argued that Kesler's independent unit would pave the way for other independent units and the eventual breakup of the entire otriad. Inevitably, the splintering of the Bielski otriad would interfere with the operations of the different workshops and lead to the disruption of production and services. In the end, the Soviet partisans who relied both on production and services would be the losers.

To Platon this made sense. He agreed that the Bielski otriad had to continue in its entirety.[47]

This took care only of a part of the problem. Denunciations about unauthorized acquisition of gold, jewelry, and money had to be dealt with next. Although easily used as weapons, charges about irregularities in money matters were hard to prove and hard to deny.

A Soviet decree further complicated matters. After the manhunt of

1943, the Russians ordered the confiscation of gold and jewelry from the local population and their delivery to partisan headquarters. As a result, in the areas controlled by the Russian partisans personal possession of such valuables became illegal.[48] Not easily enforced, this rule left much room for abuse. Partisans whose job was to collect these valuables would sometimes apply force before peasants agreed to part with their hidden treasures.

In principle each otriad was allowed to confiscate for its own use only basic food. But each needed more than just food. The Bielski otriad suffered from a chronic shortage of weapons, ammunition, medication, and more. For as long as the surrounding ghettos were in existence inmates would steal drugs from the authorities and give them to the Bielski guides.[49] Occasionally a cow would be slaughtered and a trustworthy peasant would go to town to sell the meat and buy medication. For example, Bairach's arm was saved from amputation because of Tuvia's insistence that it should be cared for regardless of cost. Money was spent for his medication.[50]

All along, members of the Bielski otriad had contributed part of the necessary money. Chaja recalls an early incident when Tuvia turned to her mother for help. Chaja's mother owned a fur-lined man's coat. Tuvia told her that in exchange for this fur coat he could receive arms from the Russian commandant, Victor Panchenko. Without a word she took off her coat and handed it to him.[51]

When in the summer of 1943 Tuvia sent guides to the Lida ghetto, a few ghetto runaways still had some money and offered it to headquarters. Some of it was used for buying guns, some for medication. Tuvia took only what people were willing to part with—no one was ever coerced into giving.[52]

There were also those who refused to part with their money. When the Garfunks ran away during the liquidation of Lida ghetto, Luba pinned dollars and gold pieces into her hair curlers. In the forest she was stripped naked and searched by Russian partisans. They found nothing. The Garfunks, their little boy, and the money reached the Bielski otriad safely. Although Luba felt grateful to Tuvia and knew that the otriad was short of money, the treasure in her hair continued to stay there untouched, until the summer of 1944, when the Red Army came to liberate them.[53]

Even though no one was ever pressured into giving money, stories about unfair financial dealings were common. Some accused Tuvia of personally amassing a fortune. Others claimed that the Bielski otriad demanded entrance fees, with the implication that the poor were refused entry. Once established, such rumors did not go away.

For example, the Sawickis reached the Bielski otriad penniless. After they were admitted, Tuvia inquired if they had any money. Cila Sawicki was offended by this question and I in turn was curious why she thought it was wrong of him to ask. After all, he had asked only after she and her husband were admitted into the otriad.

Embarrassed by my questions, Cila reluctantly backed down, saying,

"It seems that there were expenses. Certain things they had to buy for money, for example, this was true for salt. They needed the money. Tuvia certainly did not want money for himself."[54]

Another partisan was more emphatic in dispelling rumors when she said, "People told me that you had to have gold to get into the Bielski otriad. That was all baloney. I had nothing. They received us and gave a place to live."[55]

There is no evidence to support the idea that people had to pay for being admitted into the Bielski otriad. The overwhelming majority of those who reached the forest and the Bielski otriad were penniless.[56]

There was, however, an important difference between informal gossip about financial irregularities and formal accusations made by Soviet authorities. General Platon confronted the leaders of the Bielski otriad with official accusations.

Taking the lead in self-defense, Tuvia reminded Platon about the deliveries he had made of gold and jewelry to headquarters. He showed receipts to back up his statement and this evidence turned the tide. Platon dismissed the accusations about financial irregularities as unfounded. Together with the previous decision that no subgroup should be formally separated from the Bielski otriad the entire interrogation turned into a victory for the accused leaders of the Bielski otriad and a defeat for Kesler.

This confrontation also brought out into the open Kesler's denunciations and demands. Kesler was fearful of retaliation. His friends remember him as depressed, afraid for his life. Still he was unwilling to run away; instead he intensified his opposition to Tuvia as he searched for protection from the outside.[57]

Volkowyski continued to monitor Kesler's moves and learned about more letters that denounced Tuvia's conduct. Together with special gifts, these letters found their way to various Russian commanders and their assistants. In addition Kesler had someone compose a petition against Tuvia that he asked some Jewish partisans to sign. Essentially this petition repeated all the previous accusations. When Sulia heard that her brother-in-law was asked to sign this petition she reported it to headquarters.[58]

In the Bielski otriad, if a partisan wanted to leave the base, he or she had to ask permission from the commander or whoever else was in charge. One day when Tuvia was away on official business, Kesler slipped out of the camp without telling anyone. He went to see Sokolov, a Russian general. Kesler returned with a letter from Sokolov stating that Kesler was visiting him on official business and under no circumstances should he be harmed. It was in effect a letter of protection.

When Tuvia returned he was confronted with news about Kesler's circulation of a petition, about more negative letters, and about the unauthorized trip to Sokolov. Tuvia ordered Kesler's arrest. Given a choice, Kesler's wife, Rachel, came with him.

Volkowyski was stepping up his investigation. Three days later he had even more incriminating evidence. At headquarters Kesler was tried and sentenced to death. Just then Chaja and Asael returned from a food expedition. On hearing the news, Asael volunteered to do the shooting. Chaja pleaded with him not to, but he would not listen.[59] The trial and sentencing occurred without any public announcements.[60]

With about ten people, leaders and officials, Tuvia went to see Kesler in prison. In addition to Kesler and his wife, there were four more prisoners. All were ordered to step outside the jail. Each was then asked to give reasons for the arrest. The offenses ranged from failing to show up on a job to stealing a minor item. Tuvia did the asking. When Kesler's turn came, his answer was, "You know what my offense was." Then he proceeded to speak rudely and not to the point. Asael shot four bullets into him and he died on the spot.[61]

Rachel was afraid Asael would shoot her as well, but no one bothered her. She stayed in the otriad and was liberated by the Red Army.[62]

Echoes from Kesler's execution reverberated for a long time and they represented a range of reaction. Some of Tuvia's supporters argued that Kesler's denouncements would have led to Tuvia's death. They were convinced that Kesler was aware of this and wanted to become commander of the entire otriad. They were also certain that the elimination of Tuvia would have led to the breakup of the camp and the possible deaths of most members. From their perspective, Kesler's death was a welcome and inevitable development.

But not everyone felt that way. For example, the nurse Riva Reich thinks that shooting Kesler made a very bad impression on the entire otriad. She says that "To kill a Jew was not something we could have approved of. They referred to him as an enemy. But what kind of an enemy was he? Just another Jew who wanted to save some Jews. I think that Tuvia himself was not to blame in the whole affair. There were others who were spreading malicious gossip and making difficulties."[63]

Kesler's supporters were stunned by his death and saw no justification for his execution. At best they felt it was a grave error.[64] But, fearing retaliation, they were reluctant to talk about it.

Voices from a more objective direction were also heard. One of these, Kopold, argued that "Just the fact that a Jew went to a Russian, a Christian, to complain was already improper and wrong. . . . Most people in the otriad justified Kesler's death."[65]

More detached and more of an outsider, Daniel Ostaszynski views Kesler's death from a broader perspective. He emphasizes that the Bielski otriad with children, many older people, and women had a hard time maintaining itself in a hostile environment. Because of the dangerous surroundings, the otriad's existence depended on internal discipline and order. Tuvia was responsible for that discipline and order and it helped him achieve a balance. The destruction of this balance would have led to the destruction of the otriad. For the sake of the group, Kesler's relentless opposition had to be stopped. A threat to Tuvia's authority was a threat to the existence of the otriad—the two could not be separated.[66]

The Russian partisan leaders also reacted to Kesler's death in a variety of ways. In Western Belorussia, Soviet control over the partisan movement was located in three centers, around the towns of Baranowicze, Iwieniec, and Lida. Each region was headed by a Russian general: the Baranowicze region by Platon, the Lida region by Sokolov, and the Iwieniec region by Dubov.

A generous and kind man, Sokolov supported the idea of saving helpless people. He would often come to the Bielski otriad with gifts for the children and enjoyed watching them perform and sing. In fact, he arranged for one of the girl singers to travel to Moscow to study voice. Kesler's execution violated Sokolov's letter of protection and he had reason to be angry. Nevertheless, he objected to Kesler's death only informally and only to Tuvia. Very soon Sokolov dropped the matter.[67]

With General Dubov, however, it was a different story. For Dubov this unauthorized killing only strengthened his otherwise negative opinion of Tuvia and the Bielski otriad.

Abraham Viner, a clerk at Bielski headquarters, was Dubov's prewar friend. Aware of Dubov's personal dislike for Tuvia, Viner went to meet Dubov at his headquarters. Viner was confronted by a furious general ready to punish, eager to avenge. Viner pointed out that a move against Tuvia would lead to a rebellion—the Bielski partisans would stand up for their leader. He also suggested that this conflict might spill over to other otriads, creating serious problems for Soviet partisans.[68]

Apart from these arguments, General Platon's continuous support must have tipped the scale in Tuvia's favor. Although the case was debated by Soviet authorities, no official action was taken against Tuvia.[69]

Whether or not Kesler's execution could have been avoided remains a moot question. The only evidence is the history of the otriad which shows that Kesler's disappearance eliminated the threat of effective opposition. No matter how the people felt about the Kesler affair, their feelings were not expressed as an organized rebellion.

Tuvia thinks that after Kesler's death people were more careful, less prone to meddle—a positive outcome. He believes that Kesler was responsible for his own death—had he stopped plotting he would have lived. Tuvia was convinced that it had been a choice between the death of the otriad or Kesler's death.[70]

Map of the Bielski Otriad in Nalibocka Forest

During the Last Phase of Its Existence
(Fall 1943–Summer 1944)

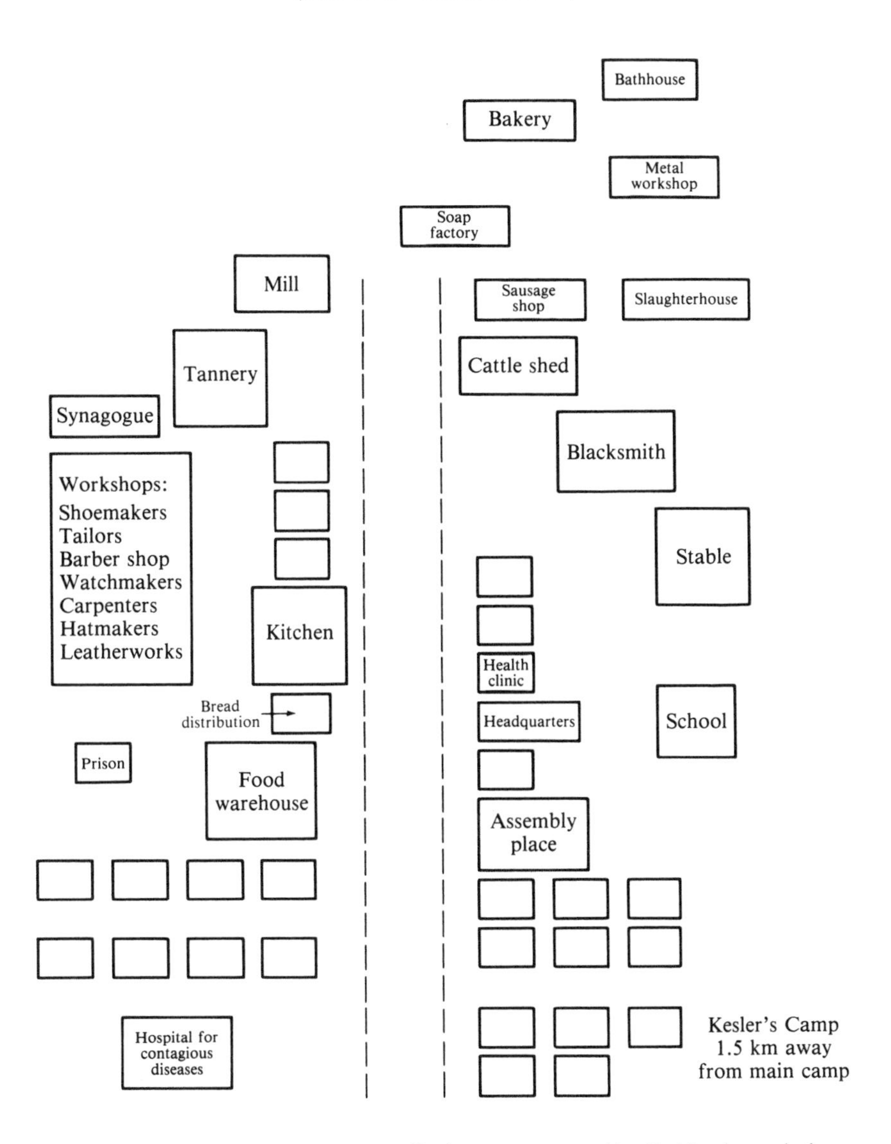

Unidentified structures are living quarters; all other structures are identified by the particular function each served.

This map is based on information provided by Chaja Bielski.

14

The End of the Otriad

By the end of 1943 close to three million Polish Jews had perished, about ninety percent of the prewar Jewish population.[1] In Western Belorussia this Jewish destruction was evident in the German liquidation of surrounding ghettos and camps. In contrast, within the Russian partisan movement, informal anti-Semitism seemed to be losing its effectiveness while official acceptance of Jews appeared to be increasing. The Bielski otriad continued to grow as more Jewish fugitives reached the camp. Some new arrivals had come after one or more stopovers at hiding places that for a variety of reasons became exposed. Some covered vast distances. For example, one family had traveled from as far away as Vilna.[2] Still others were guerillas who had left Russian detachments because as Jews they felt discriminated against.[3]

The Bielski otriad kept sending scouts into the region to find Jews who had been in need of shelter. At times this meant picking up fugitives who roamed the forests. Sometimes it involved going into a village and removing Jews from an exposed hiding place.

Within the otriad, larger size and experience led to better organization and, with time, internal improvements of the Bielski otriad had given the people a sense of a real community, a Jewish "shtetl." Fondly, some gave to this forest camp the name Bielsk, while others called it Jerusalem.[4]

Toward the middle and end of 1943 most Jews in the surrounding ghettos and camps knew about the Bielski otriad. Inmates from Nowogródek ghetto regarded it as their last hope and an expectation of a safe haven gave them the strength to organize a ghetto breakout. The number of Jews in Nowogródek had shrunk to a little over five hundred; this group was forced to live in a former courthouse.

Even at this stage, the Germans had assured inmates that their usefulness would protect them from future disturbance. But with these promises came a more vigilant watch. The ghetto was guarded day and night by twelve gendarmes. In addition, the entire place was surrounded by two walls. One wall, two meters high, was built out of wood; the other consisted of barbed wires. A distance of two meters separated them and at night a huge projector kept the area lit.[5] Despite this vigorous watch, a few groups made their way into the forest and each breakout was followed by a reinforcement of guards.

The Jewish answer to these deteriorating conditions was an organized resistance. Among its various underground leaders the two outstanding ones were Dr. Kagan and Berl Yoselewicz.[6] These leaders planned an escape into the forest and contacted Tuvia Bielski. They wanted the Jewish partisans to attack the ghetto from the outside during the breakout.

This request was made in the spring of 1943, when the Bielski otriad had no permanent base and its people were having a hard time surviving. This explains Tuvia's answer. "I am ready to accept all of you. . . . But we are not organized enough to send people to you. Whoever comes will be received well."[7]

Although not what they had hoped for, this letter pushed them into a more independent plan. The agreed-on date for the general ghetto breakout was April 15, 1943.[8]

This plot called for a group of inmates to stay close to the ghetto gate. They were to signal to another group when to move in and destroy the gate with a grenade. The rest of the population was told to be ready for a breakout on that date. They too were to be close to the gate. However, for security reasons, they were not supplied with details.

Among the active participants of this rebellion was the Jewish police commandant, Salek Jakubowicz. His brother, a physician, had injured his leg shortly before April 15. Clearly, this brother could not join the runaways, but Jakubowicz insisted that they should proceed with the breakout as planned. He argued that they had no right to sacrifice everybody because of one person.

On the day of the planned attack the man designated to destroy the gate was at his post, grenade in hand. It was evening, and he was waiting for the appropriate signal. A few moments were left. Just then a large reinforcement of guards arrived, double the usual number. Without knowing how it happened, the rebels understood—someone had exposed their plan. All the inmates dispersed.

An investigation by the underground concluded that Mrs. Jakubowicz, the wife of the doctor with the injured leg, had denounced them—she did not want her husband to be left behind and killed.[9]

The Germans became more watchful than ever. At the same time they distributed extra food, trying to give the impression that the inmates would be spared.

Angry and disappointed, the underground leaders were not about to give up. They continued to plan. Before they could act, on May 7, the Germans descended on the ghetto. Another Aktion was about to begin. One survivor recalls that "They took us to the center of town. I saw how they selected my wife. Then we heard that two hundred fifty women and forty-five men were shot that day."[10] This reduced the ghetto population to less than three hundred.[11]

With this May Aktion behind them, the underground was still busy with preparations for an escape, essentially a repetition of the exodus plan at the gate. But the Germans were determined and alert.

Each day, toward evening, the number of guards increased. The enemy's readiness made it clear that a breakout through the gate might turn into a suicidal gesture. They had to consider a different idea. Dr. Kagan and Berl Yoselewicz proposed that a tunnel with an exit to the outside should become their escape route.

Digging a tunnel called for technical know-how. But among the remaining ghetto inmates there were no engineers. In the end, all the specialized knowledge came from carpenters, locksmiths, and electricians.

Berl Yoselewicz was in charge of the actual building and had to coordinate the different parts of the job. Another person supervised the financial aspects of the project. Digging the tunnel required physical strength; extra food had to be allotted to the diggers. This called for money. In addition, the organizers had to pay for certain building supplies, unavailable in the ghetto. Ghetto inmates suspected of still having some money were asked for contributions. They were told only in general terms that a mass exodus was being planned. The response was good—people were generous.[12]

Fifty individuals were recruited for the planning and work. The actual building was to take place during the day. The silence of the night, they felt, would have made any noise suspicious. A space underneath a bed that belonged to a carpenter, a refugee from Żołudek, was selected as the tunnel's starting point. They had also decided on the place for the exit, which was separated from the nearest forest by a ten-minute walk.[13]

Small shovels were made by locksmiths. Later these were sharpened in the workshop every day.[14] One digger refers to himself as the strongest and explains that his special strength came from being small. Large men required more oxygen, something in short supply. For comfort all the diggers wore loose caftans especially made for this work.[15]

Because the tunnel was only eighty centimeters high and sixty centimeters wide, the men had to dig while crawling.

The entire project was continuously challenged by new problems. Disposal of soil was the most difficult. The courthouse had a large empty attic that the organizers selected as a disposal site. Small bags were sewn for transporting the soil. First, a row of workers, each separated by a distance of one meter, covered the area from the tunnel to the attic. Bags filled with soil moved from hand to hand until they reached their final destination. They were emptied in the attic and sent back. Then the job of filling and emptying the dug-up dirt would begin again.

When the attic could accommodate no more soil, they decided to put it underneath the wooden floors. Floors had to be pulled up and then replaced with the additional earth underneath them. When this space was used up, they came up with yet another idea.

The courtyard had a huge garbage container. Every Sunday Jews had to empty the contents of this huge box into several wagons and transport it to a special dump outside the ghetto. The craftsmen made an "invisible" hole in this box. The dug-up soil from the tunnel found its way through the opening, in with the rest of the garbage.

The ghetto workshops became places where new methods and new devices were improvised. Protected by the inmates' solidarity, these secret solutions were never discovered. One device was a wooden box that moved on wheels. Its purpose was to speed up transportation of the soil. Actually, two large wooden boxes were constructed: An empty one moved to the digging site while the other, filled with soil, moved out and its contents were emptied into special sacks. These boxes had to be pulled by ropes. Since the ghetto had no strong ropes, people cut up rags and sewed them together. Every day the diggers succeeded in lengthening the tunnel by two to three meters.

Inevitably, enlargement led to new difficulties. Increased moisture prevented oil lamps from staying lighted and for a while the diggers had to work in semi-darkness. But then two electricians came to the rescue. One day, while fixing the ghetto's projector, they decided to establish a separate electrical line for the tunnel. From then on, the tunnel had ceiling lights. Every three to four meters there was a bulb, stolen from different areas in the ghetto. Electricity improved the workers' movement and speeded up their job.

Another problem was lack of oxygen. Increased tunnel length decreased the amount of oxygen and, to cope with this, craftsmen built special pipes, five centimeters wide. On the outside, from above the tunnel, in grassy spots, they pushed these pipes into the tunnel. Openings on top of the pipes allowed air to enter the tunnel.

But then, during June and July, strong rains flooded the tunnel. At first people tried to collect the water from within. When this did not help, they dug very deep holes, actually wells. These absorbed any surplus water, preventing flooding.

Moisture also caused collapsed walls. The carpenters cut specially prepared wood that was later made into walls. Lined with stones, these walls were placed in the tunnel at intervals of a few meters, thus averting the danger of collapse.

From its starting place to its exit the tunnel had to follow a straight line and this required checking. Every eight to ten meters the workers measured the walls to see if they were straight. Sometimes they had to redo the job.[16]

The ghetto population was informed that an escape route was being prepared. They also knew that everybody would be included. However, specifics about the kind of escape and who was responsible were kept secret.[17] The entire ghetto was preoccupied with the possibility of a breakout.

Some people criticized the preparations for taking too much time. They felt that the Germans would murder them before the plan could be put into effect. Others objected to a breakout altogether. This group argued that the plan would be discovered by the Germans and they would all die because of it. Faced with heightened emotions and growing debates, the underground decided to vote on the exodus. Over seventy percent favored a ghetto breakout. Despite lingering apprehension, some think that mere knowledge that an escape was being prepared had a calming effect—it gave people an important glimmer of hope.[18]

By July the tunnel was about to reach its outside opening. The organizers felt that August would be an appropriate month for the escape; with wheat and grass still uncut, the fields were more hospitable to runaway fugitives. The tunnel was finished when news about the August 1943 big hunt reached the ghetto. Because the region was swarming with German soldiers, the move had to be postponed.

Around that time the German commissar, Traube, paid one of his rare visits to the ghetto. The German's departure was followed by a rumor that the ghetto would be liquidated and only twenty craftsmen spared.

The underground leaders were faced with a serious dilemma. Dangers on the outside and inside seemed to be mounting. They knew that any escape would result in lost lives, but they wanted to minimize them. They also knew that a German liquidation of the ghetto would mean total destruction and death. In the end they waited.

The finished tunnel extended for more than two hundred meters.[19] Without a definite date, preparations for an escape were set in motion and twenty people moved to the tunnel to rehearse the breakout. They concluded that two to three hours would be needed for the entire exodus.

The electricians temporarily disconnected the electricity in the entire camp several times. After a few hours they would reconnect it, thereby showing the authorities that the projector could be out-of-order without any grave consequences. They planned to disconnect the projector during the ghetto breakout which would give people a better chance to save themselves. The lights in the tunnel were placed away from any outside openings and were not connected to the projector. During the breakout they intended to leave the tunnel lights on.[20]

The final date for the escape was set for September 26, at 9 P.M. According to the Jewish calendar, this was to be a moonless night. As the day neared, more difficult decisions needed to be made. The question of who should leave when and in what order was debated. One woman threatened that unless she was the first to go she would reveal the entire plan.[21] Six underground leaders were busy with the organization of the actual move. In the end, people were asked to supply the names of those they wanted to leave with. On the basis of these preferences, they were told to appear at specific times at the tunnel opening. Only then were numbers distributed entitling each person to a special place in line.

At the actual time of departure, people followed orders quietly. Some participants were impressed not only with the excellent organization but also with the warm and caring attitude of the organizers. Each person followed his or her number. Some people held hands, some tied themselves to each other with belts. Most underground leaders left last.[22]

The exodus lasted over two hours. At the end something went wrong but no one knows quite what. A fire broke out at the mouth of the tunnel inside the ghetto. German guards spotted it and called for help; when no one responded, they guessed that something unusual had happened and started shooting wildly. By then all those who were supposed to pass through had left. As they exited the lighted tunnel into the gloomy rainy night, some lost their way and began to move in circles near the ghetto.

The guards intensified the gun fire and search. The estimated number of survivors from that September 26 breakout ranges from one hundred to one hundred fifty.[23]

For the runaways, the struggle for life did not end. One of them, a fourteen-year-old girl, all alone, after much hesitation, turned toward the direction of the woods and darkness. Rain was pouring, dogs were barking, and shots filled the oppressive air. From a distance she saw a house, and not far from it an outhouse. She chose the outhouse. Although uninviting, it was better than the cold, unfriendly rain and she was grateful for the dryness. For hours she sat there, listening to the shooting. When night made room for the approaching day, the girl left. Not to ruin her boots in the mud, she took them off and continued in the direction of the forest. After several painful detours, she ended up in the Bielski otriad.[24]

Kushner, one of the tunnel diggers, reached the exit, tied by a belt to his two teenage daughters. He had difficulty breathing. Kushner remembers: "I told my children: 'You are young, go save yourself, leave me here. What God wishes will happen.' 'No father, we will not leave you!' Each took me under an arm. 'I am about to faint,' I said. One of my daughters had an onion and an apple. She smeared the onion over my face and I felt better."[25]

By chance the three reached the house of a friend, a Polish settler. He directed them to the forest, to the Bielski camp. On the way they met a group of partisans. The commander of this group, a native Belorussian, both knew and liked Kushner. The commander's support led to the Kushners' acceptance into their otriad. However, after a month, this partisan detachment received an order to join another Russian otriad. At this point, with food and instructions on how to reach the Bielski group, the fugitives left. The Bielski camp became their last stop.[26]

Later, by March 1944, another special addition to the Bielski otriad came from the Kołdyczewo camp. In 1942, the estate of Koldyczewo, fifteen kilometers from the town of Baranowicze, had been transformed into a combination work and death camp. Known for its harsh conditions, Kołdyczewo had the distinction of being the only camp in Western Belorussia to gas its inmates.

This Nazi-controlled compound was first populated by Russian prisoners of war. They were soon joined by Poles and Belorussians, so-called political prisoners. Of these two groups of prisoners only a handful survived the conditions of the camp.[27]

After December 1943, with the liquidation of Baranowicze ghetto, one hundred twenty inmates were brought to Kołdyczewo. They were moved into a separate barracks that shared a wall with the camp's tannery. This tannery, in turn, was close to the barbed electrified wire surrounding the entire camp.[28]

When after a short time the one hundred twenty prisoners dwindled to ninety-five, the group decided to act. Soon, with an underground organization in place, and a man named Kushner (not related to the Kushner from Nowogródek) as its leader, they devised a breakout plan. The plot called for the creation of an opening in the wall that joined their barracks to the tannery. With hand drills they made tiny holes in the wall, thereby creating breaks. The inmates would not actually push through the wall until the day of the escape. While engaged in this job the Jews had to be careful not to arouse the suspicion of the authorities and the "political," non-Jewish prisoners who lived in an adjacent shack. In fact, while making the holes they would sing and their songs drowned out the drilling noise. Even after the job had been finished the holes surrounding the opening were invisible to the uninitiated eye.

The organizers succeeded in collecting a few guns. They also received help from a Polish policeman who gave them three grenades.[29] In preparation for the actual escape, an electrician had created a short to give the prisoners a chance to cut the electrical wires close to the tannery. To prevent noise, special covers were placed on the ground connecting the barracks with the tannery. Before they left, each person drew a number from a hat circulated among the people; according to the numbers each then moved through the opening in the wall. The exodus took place on a moonless night that was later marred by a vicious snow storm.[30]

Of the runaways, seventeen lost their way by mistakenly staying in the vicinity of the camp. Gendarmes who by then were on the lookout for fugitives shot fifteen of them. Miraculously, two of these succeeded in escaping and the rest entered the forest. Searching for the Bielski otriad, some met with rejections from Soviet partisans.[31] The majority reached the Bielski camp. Most were skilled craftsmen, some were tanners, and they ultimately improved the output and quality of leather goods in the Bielski otriad.[32]

Although preoccupied with rescue and survival, the Bielski otriad had some room left for so-called "non-essential" activities. Especially later, at a time of relative tranquillity, people seemed to take interest in leisure-type pastimes that touched on less basic aspects of life.

During the winter months of 1943–44, after working hours, small groups would go deeper into the forest in search of firewood. Old dried-out trees were special targets. Once located, such trees would be cut down, split into manageable size, transported, and placed next to the living quarters. In the evening, inside the ziemlankas, people fed the stoves with this wood and would gather around the burning fires. While some concentrated on keeping warm, others would hold onto burning wood that served as substitutes for electrical light.

Amarant recalls how the "thin flickering light would cast a shadow over the faces of men and women sitting around, over their straggly hair, the stubble of their unshaven faces, and over the pale, thin, weather-beaten faces of the women. Those huddled around the stove would enliven the evening with merry songs, or stories of adventure in the spirited fashion of the partisans. Memories of recent horrors were not welcome here. An intimate family atmosphere prevailed in the circle around the stove. Sometimes Tuvia would go from hut to hut to participate in these gatherings."[33]

Such get-togethers are remembered as being particularly heartwarming because they were neither dictated by nor approved of by the communist authorities. Indeed, because of the tolerant attitudes of the Soviet commissar, Shematovietz, and after several unsuccessful attempts, other communists also stopped their efforts to control these evenings. Here, the talking and singing was mostly in Russian and Yiddish. Like Communism, Zionism too was absent from these gatherings.[34]

In this forest community some people were drawn to the theater, music, and dance. A few acted on their interests. Sulia Rubin, for example, established a theater group that was busy with the preparation of shows. Sulia herself wrote several plays and performed as a dancer. One of the wounded Russian officers from the Zhukov brigade was a playwright who supplied the otriad with original plays.[35]

Special holidays served as outlets for these artistic endeavors and the Soviet authorities were interested in celebrating all the official communist holidays. February 23 was Red Army Day—as a gesture of loyalty the Bielski otriad was planning special festivities for that day in 1944.

The camp had many musicians but no instruments. Those who went on food expeditions were alerted to this need and, as a result, the Bielski otriad acquired a guitar, a violin, and a mandolin. These led to the creation of a trio ensemble. One participant recalls how "One could hear the sounds from the newly created trio all over the camp. All rehearsals for the celebration took place at headquarters. The chorus sang to the accompaniment of the trio of musicians. All kinds of songs were included. Many were patriotic songs that had to do with the Red Army.

"During the preparations, Tuvia would invite the musicians to his ziemlanka to play for him. He would give us drink and food. We would play, starting with popular Russian songs, and then without even planning it we moved to Jewish folk songs. We were supported by all kinds of singers. One of them was Moniek Przepiórka. He had TB. Winter and summer he wore a warm shawl around his neck. He was a wonderful singer, a troubadour."[36]

A refugee from Lódź, this singer, like so many others, had come to this part of Poland to save himself from the Nazi menace. Perhaps in part because of his fatal illness, Przepiórka was perpetually depressed. His emotions were reflected in his songs, which were filled with pain and unfulfilled longings.

In addition to the singer Przepiórka, the February 23 program included the musical trio, a children's chorus, dancing, and a brief play. The evening performance was attended by Russian partisans from different otriads, by Bielski headquarters, and by all others who were physically fit. The enthusiasm of this audience only added to the event's success.[37]

Partisan newspapers and radio supplied the forest dwellers with wartime information. By 1944, stories about Allied victories continued to spread. These stories added a hopeful air to the forthcoming May 1 holiday.

The Bielski otriad started planning festivities weeks before this important date. The actual celebration began early in the day with a military parade.

"All members of the camp gathered in Headquarters Square, red flags were raised and the surrounding huts were decorated. It was a particularly bright spring day . . . peaceful. About 1200 of us stood in the square. Three sides were flanked by the fighting units on their decorated horses, along with the reconnaissance units. The fourth side was reserved for the unarmed members who also stood in formation. Members of headquarters were in the center of the square, Tuvia Bielski, the commander, at its head. He read the Red Army order of the day, which proclaimed that fierce fighting was being crowned with success. Tarnopol had been liberated and the enemy had fled in panic. Loud applause and cheers greeted the announcement. The commander stated: 'The war will soon move into enemy country, into Germany, there the Nazi beast will be totally annihilated.'"[38]

The evening program was to continue at the neighboring Zorin otriad. Smaller, including six to seven hundred people, and much poorer than the Bielski detachment, this unit also had an open-door policy toward all Jewish fugitives. Most of its members were Russian Jews from the Minsk area. Unlike Bielski who was apolitical, the Jewish commander Zorin was a long-time loyal communist. Because he was politically active, he was supported more vigorously by Soviet headquarters. The two Jewish commanders, Bielski and Zorin, had friendly and neighborly relations.

On the evening of May 1, 1944, the Bielski performers and the rest of the otriad moved to the Zorin base. In addition to the different artists who had prepared a program of music, songs, jokes, and games, the Bielski otriad also brought food and drinks.

With the festivities well underway, Zorin himself joined the performers. Exuberant and full of life, he danced the Russian "Kozachok," a folk dance that demands great agility. Then, in high spirits, this middle-aged communist led the people from Russian to Jewish folk songs. One partisan observed, "A strange man, this Zorin, he dances like a Kozak and sings Yiddish like a Hasid."[39] When night began to change into day, an exhausted yet exhilarated crowd made its way toward the Bielski base.

Not all leisure activities were as lively. Much more subdued were what people referred to as "cultural pastimes." Ten people who saw themselves as serious intellectuals set up a cultural club. They would meet in the evening to discuss literature, philosophy, and social issues. Two of these members were former Yeshiva students. They and the others found these meetings intellectually uplifting. Yet they realized that they had elitist tastes with little or no support from the rest of the people and continued to assemble away from the public eye.[40]

Among the more accessible cultural activities were Dr. Amarant's lectures. Usually these were devoted to current political events and the meaning they had for Jews. One other important cultural undertaking was Dr. Amarant's appointment as the official historian of the Bielski otriad. Members would tell him about their wartime experiences, what happened to their towns, to the people they knew and to their families. With the assistance of his wife, Dr. Amarant tried to preserve this information by writing it down. Supportive of this project, headquarters gave the Amarants an adequate working space and paper. After the war when the Amarants were on their way to Palestine the Russians confiscated all their notes.[41]

As part of the communist partisan movement, the Bielski otriad had no official provisions for religious observances. Still, Tuvia tried to accommodate the camp's orthodox Jews. One of them, a man from Baranowicze, was a professional "shochet" (ritual slaughterer). He was allowed to kill animals in the traditional way and observant members of the otriad were allotted special food that they could cook in their ziemlankas. Those who wanted to be excused from working on the Sabbath were usually permitted to do so. Each evening some of the pious men would gather in the tannery for prayer.[42]

With the approaching Passover of 1944, a delegation of older orthodox Jews came to Bielski with a request to bake matzoh, unleavened bread traditionally eaten by the Jews during Passover. Not wishing to stir up trouble, Tuvia directed them to Shematovietz, the Russian commissar. He, in turn, gave the order to bake matzoh. And so, in the forest on Passover 1944, matzoh was available to all who wanted it.

One partisan recalls that the matzoh "was primitive, mostly burned. . . . We had a seder with children asking the traditional questions. . . . We had no 'kneidlach' (dumplings made of matzoh), no wine. Instead of talking about the departure from Egypt, we spoke about our departures from the ghettos."[43]

On June 6, 1944, the Allied forces landed in Normandy, France. On June 23 of the same year the Red Army had opened its Great Offensive on the Belorussian and Baltic fronts. As Soviet troops pressed forward, Jewish partisans played their part in disrupting the arrival of German reinforcements by destroying vital roads and bridges.[44]

News about German defeats was soon supplemented by more concrete evidence. At night, people at the Bielski base could hear dull thudding echoes of distant thunder—very welcome sounds. These war noises were followed by the appearance of groups of German soldiers on the run, pursued by Russian partisans. Fighters from the Bielski camp were eager to join forces with the Russians. Together, they searched for escaping Germans.

Rapid developments at the front created mood swings. One minute people expected to be liberated by the Red Army. The next moment they feared an attack from the retreating Germans. Then small groups of German deserters would be captured and killed, lifting the people's spirits.

Amarant reports that "the members of our armed units enjoyed the activity and looked forward to ambushes and clashes. Upon their return to the camp in the evening, they would excitedly talk about their exploits, their faces shone with enthusiasm. The taste of revenge was sweet in their mouths. The Germans were like animals, panicky and starving, dodging in and out of the thickets, their desperate resistance weakening. How wretched-looking were the remnants of an army which had purported to conquer the world!"[45]

Continual searches for runaway Germans undermined the discipline of the camp. The partisan fighters, sentries, and many others were all caught up in the fervor of going after the enemy. When sentries brought the first live Germans to the Soviet partisan headquarters, Volkowyski was asked to be the interpreter. Later, other Jewish partisans also became interrogators and interpreters. While doing the job, all the partisans were exposed to the same story: The prisoners had nothing to do with the Nazi party. They claimed not to have known what was happening. They begged for mercy.

Those in the Bielski otriad who had no opportunity to chase after the Germans felt left out. They were disappointed. Then, one day, the Bielski partisans brought back three SS men. When word about the arrival of these deserters got around, the entire otriad, more than a thousand people, descended on headquarters.[46] The crowd was agitated and the few guards had trouble keeping order.

When the situation threatened to deteriorate into complete chaos, Tuvia emerged and addressed his people: "Comrades, partisans, we have three SS murderers. At the moment we are interrogating them. Soon we will have a verdict. The bloody dogs will get their just punishment. We have lived to avenge ourselves. Be quiet. Do not disturb us in our work." Tuvia went quickly back inside.[47]

One of the few partisans admitted inside headquarters recalls that "I will never forget it, how on their knees the SS men were begging for their lives. They pleaded for mercy because they had children and wives at home. They swore that they were not to blame for anything that had happened to the Jews."[48]

On the outside the crowd was impatient. People seemed to be grinding their teeth. Tense yet quiet, they stood, waiting. Then two partisans came out leading a tall blond German, his hands tied in back.

The mass of people darted toward the prisoner, shoving each other, stepping over one another. The first to reach the SS man was Pupko, the oldest man in the otriad, described by some as seventy-five years old, by others as eighty. With knife in hand, Pupko screamed, "God, my grandfather was not a murderer, my father was not a murderer, but I will be a murderer."

Something in the old man's voice stopped the crowd's advance. It halted with eyes glued to Pupko's knife. As if in a trance, this main actor began the job of cutting up the SS man. A murmur of approval escaped from the mass of people. This wordless conversation urged Pupko to continue, to finish the job. In a few intense, highly charged moments, the SS man was unrecognizable and dead. The two other prisoners were shot.[49]

More scattered groups of German soldiers continued to roam the forest. Propelled by fear, suspicious, often hungry, they tried to shoot their way through the woods. As Russian partisans intensified their search-and-destroy missions, Bielski fighters continued to join them enthusiastically. Relentlessly pursuing their enemy, the Jewish fighters became less inclined to think about the safety of their own people.

Then something happened on July 9, 1944. Many Bielski fighters were away on an anti-German mission. On that day, at dawn, members of the otriad were awakened by shots. This was followed by a strange commotion coming from the direction of headquarters. A witness recalls that "We clearly heard German voices, loud orders, screams and curses, all the while shots whistling over our heads. We burst out of our huts and plunged into the forest under a barrage of bullets. We descended into the swamp and hid among the bushes. We were petrified . . . [the Germans] had thrown hand grenades into huts, and shot anyone who came their way."[50]

Although tired-looking, unshaven, and hungry, these Germans still had their excellent weapons and discipline. When Tuvia heard about the raid of the retreating army, he quickly assembled a few men and launched a counter-attack. Scattered groups of remaining Bielski partisans followed suit. Their counter-attacks were haphazard; they aimed at giving the people, the non-fighting majority, a chance to run deeper into the forest and hide.

Soon the shooting attracted Russian partisans from neighboring detachments. They rushed to the Bielski otriad. In the end, none of the Germans survived.[51] When it was over eleven Jewish partisans were dead and several others wounded.[52] After everyone returned to the camp, a few partisans dug a common grave at the top of a hill, not far from headquarters. In a brief and solemn ceremony, Tuvia eulogized the dead. A volley of shots bid them farewell.[53] Just then the first, dust-covered group of Red Army soldiers reached the Bielski camp. People rushed to greet them with hugs, kisses, and handshakes.

Inevitably these expressions of welcome were muted by recent events. The Russian soldiers seemed to marvel at the very presence of Jewish partisans. Nudging each other, they could be heard saying: "Look, look, Jewish fighters!" All were friendly. "One of them, a major, also a Jew, looked in wonderment around him. Then quietly he said: 'I passed from Stalingrad to Belorussia, I freed many towns and villages. I did not meet any Jews at all. I am happy to see real Jews, alive.' Fondly, he embraced the children. 'I left my wife and children in Kiev, now they are all in Babi-Yar.' It was hard to watch how from the eyes of this officer tears kept rolling down his cheeks."[54]

Glad to have met, the two groups mingled. They exchanged impressions, cigarettes, and vodka. And then the Red Army visitors were on their way.

This was July 1944—the war was still on. The same day an official order came: The next day the Bielski otriad was to evacuate their base and go to Nowogródek.

With some ziemlankas in ruins, the camp was partly in disarray. People were frightened, reluctant to enter the ziemlankas from which, only a short while ago, they had been forced out. But since the coming move would be stressful, they had to rest. People continued roaming aimlessly through the camp, not daring to go inside. They obeyed only when Tuvia ordered them to retire for the night.[55]

Even in their own quarters many had a hard time relaxing. Sleep refused to come as tension continued to mount. Some tried to cope by turning to alcohol. Tuvia was one; that night he drank heavily. Some suspect that he continued to drink in the morning.[56]

During this last night, in the Bielski camp, the majority had to cope with conflicting emotions. Tuvia's feelings had to have been even more burdensome, no doubt more oppressive. One partisan muses that "on the one hand Tuvia was happy that we were going to be free, on the other, he was losing his authority. For him, it had to be a crisis . . . after all, till then, Tuvia had had dictatorial power!"[57]

But no matter what Tuvia's inner struggles were, he also had a difficult job ahead. Until the very end he felt responsible for his people.[58] He was determined to bring them safely to Nowogródek where he knew they each would be supplied with a protective document. The communists saw an enemy in anyone who did not have proper identification, and the trip to Nowogródek would take care of this.

Because the Russians feared that the Germans and some locals might use the forest as a base for anti-Communist sabotage, they ordered Tuvia to destroy whatever could be of value to these potential opponents. And so, early in the morning, the Bielski partisans broke windows, wrecked bunks and doors, filled in wells with soil, and buried tools and utensils. For hours, the camp resounded with destruction and explosion. With this work behind them, the people assembled in front of and around headquarters.[59]

They were preparing for a long and perilous trip. The order was to travel light, with only small hand luggage. No one was to take goods that belonged to the otriad.[60] One partisan explains that "We wanted to arrive in town as fighters. We wanted to make the proper impression. We did not want to enter the town as robbers with property."[61]

Indeed, they planned to return cattle to their rightful owners or to the authorities.

When assembled for departure the otriad extended for almost two kilometers.[62] "First came the scouts on horseback, followed by marching fighters, then came the carts pulled by horses, followed by all the walking survivors. Only the sick and very weak were allowed to ride in the carts. A herd of cows came next to a unit of armed fighters on foot. Other horse riders came last."[63]

More than 1200 people took part in this exodus.[64] The move began in a subdued and orderly fashion. On both sides of this mass of people and goods were fighters. Their job was to protect this move from enemy attacks.

Polonecki, a member of the Bielski otriad, was one of the assembled. A blacksmith by profession, Polonecki had joined the Bielski otriad a few months before the forest exodus. He had arrived with a wife and baby boy.

Those who mention Polonecki describe him as ungrateful, manipulative, and dangerous.[65] Two different stories are told about his arrival in the otriad. One says that a peasant who had been sheltering Polonecki's family changed his mind, denounced them, or threw them out. When the news reached Bielski headquarters, Tuvia sent a group of fighters to rescue the Poloneckis.[66] Another version tells how Polonecki roamed the countryside, unprotected, with wife and child. By chance, Bielski scouts encountered the family and brought them to the base.[67]

Some say that Polonecki was friendly with many locals, with whom he had been able to leave leather or other goods. One week before the liberation Polonecki had gone to his Christian friends and collected his belongings. Hours before the company was to leave the forest he put all his possessions into a wagon. Leaving with the rest of the otriad, he pulled the loaded wagon while his wife walked next to him with the baby in her arms.[68] Another account says that Polonecki's wagon was filled with leather owned by the otriad. Polonecki took it thinking that the high value of leather would give his family an economic start.[69]

Whether he had leather or other goods, whether these goods belonged to him or not, would not have made a difference in what happened. Just taking a wagon filled to capacity was a violation of Tuvia's order. In fact, his action was the only flagrant violation.

On horseback, riding among the moving mass of people, Tuvia spotted Polonecki's wagon. The commander's request to leave the wagon met with a refusal. Angry, Tuvia repeated the order. "Leave the wagon, take the child in your arms, and enter the row together with all the people!"[70]

Instead of obeying, the man said, "You finished your career, you are not the commander any more."

"If you don't leave the wagon, I will shoot you!" Tuvia warned.[71]

After a defiant "I am not leaving it!" a shot was heard and Polonecki's body hit the ground. The forest had claimed its last victim.

Most people say they did not witness the incident and most doubt that Tuvia actually killed the man.[72] Two partisans admit to seeing the shooting. Although Tamara Rabinowicz says she saw Tuvia shoot the man, she does not know precisely what precipitated the deed. After the death of Polonecki she was overcome by a feeling of deep sadness. She valued Tuvia Bielski, approved of all he had done for the Jews, and was grateful that he had saved her and so many others. But somehow this last act partly overshadowed Tuvia's great achievements. Still, she adds that it is difficult to judge, especially since she did not know what really happened.[73]

Pinchas Boldo, the second witness to the shooting, elaborates. "I remember that the wagon was filled with stuff, I do not remember what was on it, how he got it. I do not know who he was. Tuvia told him to leave the things, I remember it clearly. Tuvia was on the horse, he took out his gun, and he shot the man. I think that he reacted too fast. He could have arranged it so that the man would have left the things and stayed alive. It all happened too fast. Tuvia was under pressure. Perhaps he was under tension because he was losing his leadership position.

"I did not hear what the man told him. I was a few feet away but I did not hear what was said. It was clear that the man did not want to listen but if two or three of us could have interfered we would have taken care of it.

"I don't know what they did with the body. I was there and I don't know. Had they brought the dead man to the town I would have known it. For me it was terrible, many people were shaken when they heard about it."[74]

No one defends Polonecki. Yet all were sad that the incident took place. Unable, perhaps unwilling to accept Tuvia's responsibility for the man's death, some partisans have woven farfetched stories.[75]

However, many use the incident as an opportunity to defend and express admiration for Tuvia's overall wartime conduct. One partisan insists that Tuvia was an angel not to have lost his patience more often with his undisciplined people. Indeed, they marvel at his restraint and admire him for it.[76]

Specifically, Baruch Lewin, a Jewish partisan heavily decorated by the Soviets, on hearing about Polonecki's death, said to Moshe Bairach, "If I were your commander I would have shot half of you."[77]

Someone else, putting the incident in context, justifies the act. These were perilous times. They had to travel light, there had to be discipline—

their very survival depended on listening to Tuvia who assumed the responsibility of bringing them safely to their destination. Polonecki's rebellion had endangered the entire group.[78]

Although understood and justified by some, Polonecki's death nevertheless lowered the already low spirits of the people. For days, a constant reminder—the crying of the dead man's widow—followed the mass of subdued marchers. Upset about the present, anxious about the future, overcome by sadness, they moved on.

Here and there they came upon decomposing corpses of German soldiers and animals. Only after several days did they leave behind the Nalibocka forest. They passed a village where the peasants came out to greet them. Their surprised eyes and equally surprised voices said over and over again, "So many Jews! So many Jews!?"

One of the partisans recalls that "At first we marched in military order with the armed units leading the way and closing the column. Then the marchers dispersed and walked in groups at their own pace in the oppressive heat. We made slow progress. There was no joy in our hearts. As we neared the areas of Jewish settlements, we realized the extent of the disaster that had befallen us. It looked as if our very lives had been consumed by flames, we were walking into a wasteland. The houses of Nowogródek appeared on the horizon. We stayed overnight on a neighboring estate, in the barns, granaries and other structures. . . . Travel weary, we lay on the straw."[79]

The next day many of them visited Nowogródek. There they were faced with Jewish houses now occupied by new tenants, none of them Jewish. The Jewish community had disappeared. The Bielski partisans made up only a fraction of the thousands of partisans swarming the town and its surroundings. All these guerillas were treated to propaganda, speeches, and parades. They were supposed to rejoice—it was practically an order.

Later, at an estate close to Nowogródek, the Bielski partisans had their last official parade. Tuvia addressed the crowd briefly. To each he gave an appropriate document. With the ceremony and the distribution of documents, the Bielski otriad officially ceased to exist. Parades and festivities continued into the middle of July. During most of them the achievements of the Bielski brothers were praised.[80]

But underneath this officially friendly posture loomed danger. Shortly after Tuvia's arrival in Nowogródek he was denounced to the Secret Service (NKVD) by several of his former partisans who in the past had not dared to act on their dissatisfaction. A few were former exiles from the Bielski otriad, but it is not clear what their accusations were. Disloyalty to the communist cause, financial irregularities, or practically anything else could have had disastrous consequences. Before the authorities took action, Tuvia, Zus, and their wives escaped to Rumania. From there, with the help of a Jewish agency, they reached Palestine.[81]

Ironically, for Tuvia as for many other partisans, these former allies and liberators, the Soviets, became a real threat. Stalin saw all those who were independent and free in spirit as foes, as competitors for power. Tuvia fit into this potentially "dangerous" category, as did many guerilla fighters.

Indeed, the Soviet authorities had adopted a tough policy toward all partisans. One of them explains that "All kinds of elements came out of the forests. The Russians did not want to bother selecting or differentiating between them. Whoever was of a military age they sent to the front. Those who were better educated and needed for administrative jobs they retained. My brother and I were in this last group. I became the secretary of the prison. The director of the prison was an illiterate Russian. He had lived fifteen years in Siberia. Because he worked for the NKVD they sent him to Nowogródek to be the director of the prison. He did not know how to sign his name. I had to sign it for him.

"Those who ended up in the army were sent to the front line. Most of them fell on the front."[82]

Indeed this happened to Asael. Asael and Chaja and part of her family went to Lida. Immediately Asael was taken into the Red Army where he fell in battle, in Marienburg, Germany. Chaja and Asael were expecting their first child and Chaja, together with her family, settled in Israel. She gave birth to a daughter, Asaela.[83]

Palestine, soon Israel, became the home of the two brothers, Tuvia, Zus, and their wives. Lilka's and Tuvia's daughter, Ruthie, was born here. Both brothers served in the Israeli army.

In the forest Tuvia was a powerful leader widely admired, worshipped by some. He rose to his position of power during extraordinary times. Like most charismatic leaders, unbound by tradition, he improvised. Tuvia's charisma, intelligence, and special talents led to his success. Each time he performed a new miracle that gained a measure of security for his followers his power and authority grew. But as soon as his historical moment passed, the authority of this charismatic leader faltered.[84]

When he came to regular society he did not fit in. Tuvia had no formal education. He was apolitical and did not push, yet he had a family to support. When some leaders in Palestine inquired how they could help him, Tuvia asked for a taxi. Driving and owning a cab was a profitable business.

His partisans, however, were shocked to see their leader as a cab driver.[85] In different ways they say that "the minute he became a cab driver he already stepped down to a level that led nowhere. . . . Had he insisted that he was a hero, that he had to be recognized as such, he would have been given an important position."[86] Agreeing with much of what has been said, another voice adds that Tuvia was "a tribal leader, not a regular leader. . . . The war was made for his kind of leadership. But nothing else. I believe that till the last day of his life he missed this time, longed for it. Tuvia could not have been a leader of a political party. His brain was not built for political things. He came to Israel with an extraordinary record. After all, a man in the times of the Nazis had fought and saved over 1200

Jews. There were no people like him. No one came with his record, and he did not use it because he did not know how."[87]

One of Tuvia's admirers, Hersh Smolar, a prominent journalist, regrets that things did not work out for Tuvia. Like others, Smolar seems to be blaming Tuvia when he says that he "did not recognize the Israeli reality, he did not know the Israeli bureaucracy. Israelis did not pay much attention to Tuvia. It had to be a terrible disappointment for him."[88]

During one of their meetings Tuvia explained to Smolar that he did not know how to beg and that he could not behave differently. One day Smolar delivered a lecture on the meaning of Tuvia Bielski with Tuvia present. Smolar said that Tuvia "was a glow in the Jewish history, an exceptional phenomenon." He emphasized that what Tuvia had accomplished was unprecedented and extraordinary. After this lecture Tuvia's comment to Smolar was, "I have not even thought about it this way."[89]

Zus and his family moved to the United States. Disappointed and ill, Tuvia and his family followed. In America the Bielskis had an older brother, a rich factory owner. This was the brother who had emigrated to America in the 1920s and Tuvia counted on his help. Some think he expected the American Jews to make a fuss over him, to support him until the end of his life. The Bielski partisans see the move to America as another grave mistake.[90]

Disappointment followed Tuvia to this new country as well. The rich brother was not as forthcoming as Tuvia wished. Here, too, Tuvia had to support a growing family, a daughter and two sons. He owned one truck. Then two trucks. At eighty-one, partly forgotten and disillusioned, he died.

15

From Self-Preservation to Rescue

People who are exposed to extreme dangers may be paralyzed into inaction. Whether this occurs is in part contingent on the extent to which they define a situation as hopeless. As a rule, fighting requires hope, yet hope tends to fade with grave dangers. Those who have been sentenced to death tend to give up hope. Yet hope dies reluctantly. For some individuals condemned to death even a slim chance of survival turns the wish to live into an all-consuming passion. During the Nazi occupation, among the Jews in Western Belorussia, a clinging to hope and life was expressed in a variety of ways.

In the summer of 1941, when the Germans occupied Western Belorussia they were already experienced in the murder of Jews and seemed less concerned with keeping their crimes secret. And since in this part of Poland most mass shootings happened close to home, Jews who had eluded the killings had a hard time denying the grim reality.

As special targets, faced with overpowering destructive forces, many Jews obeyed German orders. Their compliance was based on a host of arguments. Some reasoned that opposition to the Germans, although personally gratifying, would only hasten the destruction of all Jews. Others claimed that by conforming to the German demands they would gain time. They had hoped that in the meantime the war would end and thus interfere with the Nazi plan to murder them. Others, especially older people, rejected the idea that the Germans intended to annihilate all Jews. They pointed out that it was to the Germans' advantage to keep some of them alive, as part of the labor force. Still others argued that in view of the Nazi superior power all Jewish opposition would be a suicidal gesture, resulting in immediate death rather than the possibility of survival.

Some Jews had hoped to avoid death through compliance. In the end practically all of them perished as did most Jews who opposed the enemy. Those who refused to submit to German terror were the rebels; they were more independent and often endowed with leadership qualities. United in their refusal to become victims, they were preoccupied with opposition to the enemy through self-preservation. As these rebels continued to elude the Germans, they began to feel more self-assured. While they came to feel personally less threatened, the leaders among them were ready to consider other issues.

Some of these rebels switched their attention from self-preservation to revenge. Although they knew that death was a real possibility, they opted for revenge through armed resistance. The more successful they were, the more absorbed they became in this nearly impossible struggle.

Indeed, many Jewish partisans preoccupied with revenge had perished. Among them were such heroic leaders as Atlas, Kaplinski, and Dworecki.[1] If given the opportunity, these fallen heroes and others like them would have argued that their deaths had been different from the deaths of defenseless ghetto inmates. Indeed, they were. And yet, by fighting the Germans, these Jewish partisans speeded up their own deaths and the deaths of their followers.

Tuvia Bielski's opposition to the Germans was different. Whatever feelings of revenge he had, these took a back seat to his determination to save lives. Refusing to become a victim, rejecting the role of avenger, Tuvia Bielski concentrated on gathering Jewish fugitives and protecting their lives.

Tuvia's opposition to the Germans had the support of his entire family. The Germans, however, succeeded in murdering many of them. Defiance did not work for some of the Bielski family any better than it worked for the majority of Jews.

Clearly, survival in a hostile and devastated setting was unlikely. Moreover, German terror led to the destruction of Jewish traditions, making Jewish prewar leaders ineffective. But a strong and able leader could help improve the chances for survival. Tuvia Bielski was such a leader. His skills developed gradually, keeping pace with the community he headed.

A one-time peasant, a resident of a small provincial town, Tuvia Bielski was an unlikely leader, an unlikely hero. He was both different from and more independent than most of his fellow Jews. Social upheavals propelled him into a position of prominence. Extraordinary times require extraordinary leaders, unbound by old traditions.

And so, independent, set apart from traditional leaders, Tuvia's reactions to the German onslaught were unusual. With his two brothers, Asael and Zus, Tuvia became a part of the small minority of Jews who, from the very start, vowed never to be ghetto inmates. Their refusal to become victims was linked to a determination to stay alive. The three brothers joined forces, enlarged their group to include more relatives and friends, and elected Tuvia their commander. From the outset Tuvia was concerned with saving lives and argued that the survival of their group depended directly on its enlargement and hence on bringing more Jews into their camp.

Time and survival of the expanding group led to feelings of confidence. With some initial success came greater freedom to consider the welfare of others—at first, family and friends. The group grew. The group's growth and survival made Tuvia even more sensitive to the needs of other Jews, those who were not necessarily relatives or friends.

Tuvia's concerns about the Jewish plight were further strengthened when the Germans stepped up their persecutions of Jews in 1942. That summer witnessed the liquidation of many ghettos and the resultant murder of most Jews.

The longer the Bielski otriad continued to oppose the Germans, the more involved it became with the rescue of Jews. While saving many lives, the Bielski partisans also participated in military moves against the Germans. But rather than inflicting damage on the enemy, preservation of life remained their major, all-absorbing mission.

As an independent charismatic leader, untouched by political ideology, Tuvia made no distinction among different kinds of Jews: the old and the young, the weak and the strong. Quite naturally, the rescue efforts of the Bielski otriad came to include all Jews, no matter who they were.

As an effective leader, Tuvia was able to persuade others to listen to him and to obey. With time, he converted more and more people to his open-door policy of taking in every Jew. The growing support for Tuvia's policies and the success of these policies were intricately connected.

But Tuvia's achievements were sprinkled with criticism. Some of his partisans had argued that, after all, without the cooperation of his brothers, of Lazar Malbin, of the young fighters who followed his orders, he could not have made it. True. Still, Tuvia's leadership and vision brought these people together in the first place. Moreover, it was his ability to control, to organize, and to protect that made the people follow his orders.

A general needs an army and an army needs a general. Tuvia was a superb general and he led his people to victory. Most prominent among Tuvia's victories was the rescue of more than 1200 Jews, all condemned to death. But Tuvia's achievements went beyond such rescues. The Bielski partisans stood for justice, attacking and killing local collaborators. By punishing informers they helped reduce the Jewish death toll and at the same time intimidated other potential collaborators. This resulted in safer roads which, in turn, encouraged more ghetto escapes.

Eventually the local population realized that the Bielski otriad would punish all those who mistreated Jews. Such knowledge prevented some peasants from dismissing or denouncing their Jewish charges. It also prevented some from refusing help.

Moreover, whenever Jewish partisans in Soviet detachments felt threatened by Russian partisans they could count on finding protection in the Bielski otriad. Indeed, on the eve of the manhunt of 1943, twenty-two Jewish fighters from the Orlanski detachment found refuge in the Bielski camp. They came because they had been exposed to anti-Semitic threats.

In a different way the Bielski otriad saved the lives of the Kesler group. They had come to the Bielski detachment reluctantly, but if they had refused to join they would have been killed by the Victor Panchenko group.[2] These examples represent only a fraction of the Jews who were protected from Russian partisan threats.

The Bielski otriad initiated other forms of rescue. Special guides were sent into ghettos to help people escape, and scouts searched the roads for Jewish fugitives in need of protection. While most of those collected by the guides and scouts ended up in the Bielski otriad, some did not. A few joined Soviet detachments, while others found refuge among the local population. Still the activities of these guides and scouts added to the number of Jews saved by the Bielski otriad.

The degree of rescue was also affected by the losses of life. It is not easy to determine exact casualties of an expanding community like the Bielski otriad. Nor is it easy to compare the attrition rates of different partisan groups. Precise estimates are hard to find, yet those available for the Bielski otriad list specific names of places and names of the dead. Most other estimates seem to lack such supporting facts.

Under conditions in which attrition occurred it was extremely difficult for people fighting for survival to make precise records. In an environment so hostile, devastated, and devastating, people with limited hope for survival and with minimal resources are not likely to make systematic tabulations.

Attrition had many components. The major part of attrition was probably death from enemy attacks. But death also resulted from disease, suicide, unintended injuries, punishment mandated by the group's authorities, and from incidents among the armed group members. Desertions, expulsions, and accidental separation also occurred.

Thus, in the absence of systematic records, one cannot know either exact numbers or causes of the attrition that led to a reduction in the number of partisans who survived the war. Of course, merely to compare the survivors with the initial recruits is inadequate, since there usually was a continuing and poorly recorded influx of new partisans.

It might also be inaccurate to make simple direct comparisons of the losses of the Bielski otriad and those of other partisan groups. However, by reviewing the most reliable estimates and taking into account the many sources of uncertainty and the nature of the estimates, we can learn much about the relative attrition of these groups.

Independently, two former Bielski partisans estimated that fifty people had died in their otriad.[3] One of the partisans, Chaja Bielski, followed up her statement with a thorough search. The end results are very similar to the initial estimate.

Listing specific names, places, and dates, the number of people who died in the Bielski otriad came to fifty-five. Included in this group are four people whose death was ordered by the otriad's authorities, one suicide, and one person who drowned. When these special individuals are excluded from the fifty-five, this leaves forty-nine people who died because of enemy attacks.

Based on the conservative figure of the total number of Bielski group members of 1200, the forty-nine deaths represent an attrition rate of less than five percent. When offering these estimates, Chaja cautions that her figures are approximate.[4]

In contrast to figures from the Bielski otriad, one frequently quoted source says that the size of the entire Russian partisan movement rose "from 30,000 in January 1942 to 175,000 by June 1944. The personnel turnover resulting from casualties, sickness and desertions over a three-year period brought the total of men who at one time or another participated in the partisan movement to about 400,000 or 500,000. . . . These figures represent the number of partisans present in regular, permanently, organized combat units."[5]

One can make a rough estimate of the attrition rate for the Russian partisan movement by assuming that the participation was uniformly spread out over the three-year period (thirty-six months) for an average number of participants of about 14,000 per month. Thus the estimated total number of incoming participants over the eighteen-month period from January 1942 to June 1944 was 250,000. This adds to the starting value of 30,000 in January 1942 and gives an intake of 280,000, of which only 175,000 were left in June 1944. This gives a conservative estimated attrition rate of forty percent.[6]

Other estimates for partisan deaths in different Soviet otriads vary. For example, one partisan from the Ponomarenko detachment tells that they lost one-third; out of one hundred fifty partisans, fifty died. This is an attrition rate of thirty-three percent. For most detachments such information is not available.[7] An article in the *Encyclopedia Judaica* states that one-third of the Jewish partisans who served in Russian otriads perished.[8] Another study of Jewish partisans in Lithuania estimates that fifty-three percent of them died.[9]

More vulnerable than guerilla units were small, unprotected family groups. Indeed, of the single individuals who reached the Bielski otriad, many were sole survivors of small family camps. Also, in some forests, family groups were abandoned by fighting partisans before a raid. Unprotected, such camps were easily destroyed.[10] But even "properly" protected family camps suffered heavy losses. For example, in the Parczow forest in Poland, of the 4000 Jewish ghetto runaways who found their way into this forest only two hundred survived the war. This shows an attrition rate of ninety-five percent.[11] In contrast, the Bielski otriad lost only five percent of its people.

Indeed, no matter what group the Bielski detachment is compared to, it seems to have had by far the smallest losses. Much of the credit for this successful protection of lives belongs to the group's commander, Tuvia Bielski.

Perhaps the Bielski otriad and its charismatic leader can serve as a model showing the link between self-preservation and the selfless protection of others.

In times of upheaval among the ruins of established society, those who are independent and removed from the mainstream of tradition are likely to see hope where there is none.[12] Threatened by overpowering forces of destruction, those with hope will concentrate more vigorously than others on overcoming death. When this struggle yields a semblance of success, self-preservation makes room for concern about the welfare of others. While at first these others are an extension of self, such as close family members, with time and with further success, they come to include friends. Feeding on their own achievement and gaining more support, the protectors and the protected soon come to include anyone threatened by destruction and death. Each becomes transformed from a prospective victim to a rescuer, blurring the distinctions between the rescuer and the rescued. What had begun as an unrealistic glimmer of hope turned into a chance to survive, a cooperative effort. Armed with a trust in a better future, those who are independent and free from social constraints can more easily transfer their hopes, their possibilities, and their successes to the less hopeful and less resourceful. This transfer creates its own reality, a reality that opposes the life-threatening environment and death.

And so, in times of crisis, when old, established leaders fail, the uninitiated, the independent who are free from traditional constraints have the opportunity to develop their leadership skills and their strategies of survival. At first equipped only with hope and a feeling of self-worth, they soon translate these hopes into actual gains. Success may carry them into the position of leadership and power. Hope, independence, ability to organize, and the resultant success lead to more opportunities and greater achievements.

When the upheavals show signs of receding to previous societal realities, the skills of the independent charismatic leader become obsolete. In part by changing conditions, in part by their own inability to fit into these conditions, such leaders are pushed aside. Although the leaders retreat, the fruits of their achievement remain.

As commander of a forest community dedicated to the preservation of life, Tuvia Bielski gave to the Jewish people many precious gifts: hopes, dreams, and their very lives.

Notes

Chapter 1

Entries containing information from people interviewed for this book are identified as "Personal Interview." I have conducted these interviews in Israel and the United States. Places identified as Tel Aviv and Haifa refer also to surrounding communities of the particular town. Each interview lasted from two to six hours. I met some individuals several times, a few as many as ten times. When more than one date appears, this means the person was interviewed several times during the stated period. All interviews were taped and were conducted in Yiddish, Hebrew, Polish, or English. Sometimes people would switch languages. The majority spoke Hebrew and Yiddish. When transcribing, I have translated all the information into English. Both the original taped interviews and the transcripts are available. I hope that in the future other researchers will find them useful.

1. In identifying various geographic locales, for the purpose of consistency I have used the Polish names. I am relying on *Miasta Polskie W Tyśięcoleciu* (*Polish Towns During a Millennium*) (Kraków: Zakład Narodowy Imienia Ossolińskich, 1965) and *Skorowidz Mjejscowości Rzeczypospolitej Polski* (*Index of Polish Locales*) (ed.) Tadeusz Bytrzycki (Warszawa: Wydawnictwo Książnicy Naukowej, 1930s. Exact year not stated).
2. Moshe Bairach, Personal Interview, Tel Aviv, Israel, 1988–1989. Moshe Bairach wrote a book of memoirs entitled *Vzot Ltuda: Bgetaot Vbiarot Belorussia* (*And This Is To Witness*) (Israel: Ghetto Fighter's House Ltd., 1981).
3. Raja Kaplinski, Personal Interview, Tel Aviv, Israel, 1987–1989.
4. Chaim Basist, Personal Interview, Tel Aviv, Israel, 1988.
5. Eljezer Engelstern, Yad Vashem Testimony, No. 3249/233.
6. Chaja Bielski, Personal Interview, Haifa, Israel, 1987–1991.
7. Tuvia Bielski, Personal Interview, Brooklyn, New York, 1987.
8. Michael Bielski, Personal Interview, Brooklyn, New York, 1992. This is Tuvia Bielski's son. He shared with me material from tapes he had made of his father.
9. Jacob Lestchińsky, "The Industrial and Social Structure of the Jewish Population of Interbellum Poland," *YIVO Annual of Jewish Social Science* 2 (1956–1957), p. 246; Antony Polonsky, *Politics in Independent Poland, 1921–1939* (Oxford: Clarendon Press, 1972), pp. 42–44.
10. "Belorussia," in *Encyclopedia Judaica,* vol. 4 (Jerusalem: Keter Publishing House Ltd., 1971), p. 445.

11. Ibid., p. 446; Norman Davies, *Heart of Europe: A Short History of Poland* (Oxford: Clarendon Press, 1984), p. 120.

12. Nechama Tec, *In the Lion's Den: The Life of Oswald Rufeisen* (New York: Oxford University Press, 1990), pp. 68–78.

13. Yehuda Bauer, *The Holocaust in Historical Perspective* (Seattle: University of Washington Press, 1978), p. 61; Nicolas P. Vakar, *Belorussia, The Making of a Nation* (Cambridge: Harvard University Press, 1956), p. 187.

14. Zus Bielski, Personal Interview, Brooklyn, New York, 1989.

15. Tuvia Bielski, Personal Interview; Zus Bielski, Personal Interview, has a slightly different version. He says that they attacked the man with rocks.

16. Michael Bielski, Personal Interview.

17. Chaja Bielski, Personal Interview; Raja Kaplinski, Personal Interview; Lilka Bielski speaks about Tuvia's time in the army with pride, emphasizing that he stood up to all anti-Semites, Personal Interview, Brooklyn, New York, 1989.

18. Tuvia Bielski, Personal Interview.

19. Raja Kaplinski, Personal Interview; Luba Rudnicki, Personal Interview, Tel Aviv, Israel, 1988, was Rifka's cousin. All her life she remained a good friend to both Rifka and Tuvia.

20. Herzl Nachumowski, Personal Interview, Tel Aviv, Israel, 1987.

21. Tuvia Bielski, Personal Interview.

22. Shmuel Amarant, *Nvo Shel Adam* (*Expressions of a Man*) (Jerusalem: Published privately with the support of the Ministry of Culture and Education of Israel, 1973). A chapter in this book deals with the history of the Bielski camp. Its title is "The Tuvia Bielski Partisan Company" and was translated from the Hebrew into English by R. Goodman. I am relying on the typed version of this translated chapter. From now on, when using material from this source I will cite only the name of the author and the title of the chapter; Sonia Bielski, Personal Interview, Brooklyn, New York, 1989; Cwi Isler, Yad Vashem Testimony, No. 1706/113; Raja Kaplinski, Personal Interview.

23. Pinchas Boldo, Personal Interview, Haifa, Israel, 1990.

24. On the political climate toward the Soviet Union and the Communist party in prewar Poland, see Norman Davies, *God's Playground: A History of Poland*, Volume II (New York: Columbia University Press, 1984), pp. 393–434; Jan Karski, *The Great Powers and Poland, 1919–1945* (New York: University Press of America, Inc., 1985), pp. 69–103. In prewar Poland the membership of the Communist party did not exceed 20,000. By 1938 the party ceased to exist as an official entity. See Yisrael Gutman and Shmuel Krakowski, *Unequal Victims: Poles and Jews During World War II* (New York: Holocaust Library, 1986), p. 355; Paul Lendvai, *Anti-Semitism Without Jews* (New York: Doubleday, 1971), p. 203.

25. Chaja Bielski, Personal Interview.

26. Estelle Bielski succeeded in running away to Russia in 1941. She survived the war there and came to the United States after the war. Personal Interview, Miami, Florida, 1991.

Chapter 2

1. For centuries Poland had to defend itself against its surrounding neighbors: Austria, Prussia, and Russia. By 1795 each of these three countries occupied a part

of Poland. As a result, from 1795 till 1918, Poland ceased to exist as a political entity.

Only with the end of World War I did Poland regain its sovereignty and its lands. Its neighbors, however, were upset by this state of affairs. Russia and Germany, in particular, were eager to regain what they came to view as their own territories.

Because of the Nazi-Soviet pact, signed on August 23, 1939, the USSR soon occupied more than half of Poland. For a discussion of the conditions surrounding some of the Nazi-Soviet agreements see Jan Karski, *The Great Powers and Poland, 1919–1945: From Versailles to Yalta* (New York: University Press of America, Inc., 1985), pp. 339–363; Norman Davies, *Heart of Europe: A Short History of Poland* (Oxford: Clarendon Press, 1984), p. 129.

2. These accusations are noted and discussed in many sources. For a few examples, see Yehuda Bauer, *The Holocaust in Historical Perspective* (Seattle: University of Washington Press, 1978), p. 86; Jan T. Gross, *Polish Society Under German Occupation* (Princeton: Princeton University Press, 1979). In footnote 3 on p. 186 Gross cites several Polish documents that accuse the Jews of disloyalty to the Poles and loyalty to the Russian Communists; Yisrael Gutman, "Historiography on Polish-Jewish Relations," pp. 184–188, in Maciej Jachimczyk, et al., (eds.), *The Jews in Poland* (Oxford: Basil Blackwell, 1986); David Engel, *In the Shadow of Auschwitz: The Polish Government in Exile and The Jews* (Chapel Hill: The University of North Carolina Press, 1987), p. 168; Kazimierz Iranek-Osmecki, *He Who Saves One Life* (New York: Crown Publishers, Inc., 1971), p. 185.

3. Yisrael Gutman and Shmuel Krakowski, *Unequal Victims: Poles and Jews During World War II* (New York: Holocaust Library, 1986), pp. 36–39.

4. Chaja Bielski, Personal Interview, Haifa, Israel, 1987–1991.

5. Alexander Dallin, *German Rule in Russia, 1941–1945: A Study of Occupation Policies* (New York: Octagon Books, 1980), p. 486; Thomas Fitzsimmons, Peter Malof, John C. Fiske, *USSR* (New Haven: HRAF Press, 1960), p. 450; Raymond Pearson, *National Minorities in Eastern Europe, 1848–1945* (New York: St. Martin's Press, 1983), pp. 180–182.

6. Yitzhak Arad, *Ghetto in Flames: The Struggle and Destruction of the Jews in Vilna in the Holocaust* (New York: Holocaust Library, 1982), pp. 20–21.

7. "Belorussia" in *Encyclopedia Judaica,* Vol. 4 (Jerusalem: Keter Publishing House, Ltd., 1971), p. 447.

8. Indeed, from Communist-occupied Poland, including Belorussia, people "were deported in four vast railway convoys . . . in February, April and June 1941. . . . The vast majority were convicted for no known offense but simply because the Polish nation was seen as the inveterate enemy of its Russian master." See Norman Davies, *God's Playground: A History of Poland* (New York: Columbia University Press, 1984), Vol. 2, pp. 448–451; Kazimierz Plater Zyber, "In Defense of Poles in the USSR," *Polish Ex-Combatant Association,* 1982, pp. 3–19.

9. "Belorussia," in *Encyclopedia Judaica,* p. 447; Engel, *In the Shadow of Auschwitz,* pp. 125, 156; Gutman and Krakowski estimate that of the deported 50% were Poles, 30% were Ukrainians and Belorussians, and 20% were Jewish. Depending on the source, there are some variations in these estimates. See: *Unequal Victims,* pp. 36–37.

10. Arad, *Ghetto in Flames,* pp. 25–27; Eljezer Engelstern, Yad Vashem Testimony, No. 3249/233; "Nowogrudok," in *Encyclopedia Judaica,* Vol. 12, p. 1238;

"Lida," in *Encyclopedia Judaica,* Vol. 11, p. 212. Deportations to Siberia often led to broken families. This happened to Raja Kaplinski who was left behind when the Russians deported her parents. Personal Interview, Tel Aviv, Israel, 1987–1989.

11. Chaja Bielski, Personal Interview.

12. Because most historical publications use Vilna, the Russian name for the city, I will use it as well. Other versions are (1) Wilno—Polish; (2) Vilinius—Lithuanian; and (3) Wilna-German. See Davies, *God's Playground,* p. 505.

13. Sonia Bielski, Personal Interview, Brooklyn, New York, 1989.

14. Lilka Bielski, Personal Interview, Brooklyn, New York, 1989.

15. Raja Kaplinski, Personal Interview.

16. Chaja Bielski, Personal Interview.

17. Esia Lewin-Shor, Personal Interview, Bronx, New York, 1989.

18. Motl Berger, Personal Interview, Brooklyn, New York, 1989.

19. Lilka Bielski, Personal Interview.

20. Luba Rudnicki, Personal Interview, Tel Aviv, Israel, 1988.

21. Tuvia refers to Sonia as his wife. Lilka does not think that Tuvia and Sonia were officially married. Lilka Bielski, Personal Interview; Tuvia Bielski, Personal Interview, Brooklyn, New York, 1987.

22. See Gutman and Krakowski, *Unequal Victims,* pp. 35–36.

23. For discussions about Jewish discrimination in prewar Poland, see Gutman and Krakowski, *Unequal Victims,* p. 37; Celia S. Heller, *On the Edge of Destruction* (New York: Columbia University Press, 1977), pp. 85–136; Jacob Lestchiński, "Economic Aspects of Jewish Community Organization in Independent Poland," *Jewish Social Studies 9, no. 1–4* (1947), pp. 319–338; Emmanuel Ringelblum, *Polish-Jewish Relations During the Second World War* (Jerusalem: Yad Vashem, 1974), pp. 10–22; Edward D. Wynot, Jr., "A Necessary Cruelty: The Emergence of Official Anti-Semitism in Poland, 1936–1939," *The American Historical Review 76, no. 4* (October 1971), pp. 1035–1058.

24. In 1921, with postwar Poland newly sovereign, after more than a hundred years of Polish subjugation by others, the commander of Poland's army, Józef Piłsudski, ordered the takeover of Vilna, then part of Lithuania. Later, over the objections of Lithuania, Poland conducted a plebiscite in which the inhabitants of that city had to state their national preference. Sixty-four percent of the voters wanted to become a part of Poland. Interpreting the results as a green light, Poland annexed the city—an act unpopular with everyone but the Poles. Nevertheless, at the 1923 Conference of Ambassadors, Vilna was formally recognized as a Polish city. For a discussion of the Lithuanian-Polish conflict, see Karski, *The Great Powers and Poland,* pp. 71–74.

25. This last act of generosity masked the Soviets' determination to annex Lithuania and other Baltic states. Both these future plans were described in the secret sections of the Nazi-Soviet pact. See Davies, *God's Playground,* p. 443.

26. Yehuda Bauer, *A History of the Holocaust* (New York: Franklin Watts, Inc., 1982), pp. 282–283.

27. Of the Jews who came to Lithuania some succeeded in obtaining British visas for Palestine. Others moved to Curaçao and Costa Rica. Still others found an unexpected ally in Sempo Sugihara, the Japanese consul in the city of Kovno. Singlehandedly, and over the objections of his superiors, Sugihara issued 6000 transit visas. Holders of these precious documents were entitled to a three-week stay in Japan. In reality most of them remained in the Far East, at least for the duration of the war. Eventually the Japanese authorities caught up with Sugihara

and removed him from the diplomatic service. See Arad, *Ghetto in Flames,* pp. 18–20; Marvin Tokayer and Mary Swartz, *The Fugu Plan* (New York: Paddington Press Ltd., 1979), p. 79; Leni Yahil, *The Holocaust: The Fate of European Jewry* (New York: Oxford University Press, 1990), pp. 616–617.

28. On August 3, 1940, Lithuania officially became a part of the Soviet Union. See Arad, *Ghetto in Flames,* pp. 20–24; Nechama Tec, *In the Lion's den: The Life of Oswald Rufeisen* (New York: Oxford University Press, 1990), pp. 23–24.

29. Moshe Bairach, Personal Interview, Tel Aviv, Israel, 1988–1989.

30. Zus Bielski, Personal Interview, Brooklyn, New York, 1989.

31. Arad, *Ghetto in Flames,* p. 29.

32. Tuvia Bielski in *Yehudei Yaar* (*Forest Jews*). Narratives of Jewish Partisans of White Russia, Tuvia and Zus Bielski, Lilka and Sonia Bielski, and Abraham Viner, as told to Ben Dor (Tel Aviv: Am Oved, 1946). Translated from Hebrew into English by R. Goodman and available in typed form, less than 200 pages. I am relying on the English translation of this book. When quoting from this typed version I will identify the specific narrator. For convenience I will exclude the long subtitle of the book and the page numbers. Needless to say, the page numbers of the published version do not correspond to the page numbers in the typed version.

33. Tuvia Bielski, in *Forest Jews.*

34. Riva Reich, Personal Interview, Tel Aviv, Israel, 1989.

35. Eljezer Engelstern, Yad Vashem Testimony, no. 3249/233.

36. Zorach Arluk, Personal Interview, Tel Aviv, Israel, 1988; "Russia," in *Encyclopedia Judaica,* Vol. 14, pp. 433–506.

37. Hersh Smolar, Personal Interview, Tel Aviv, Israel, 1989–1990.

38. Nicholas P. Vakar, *Belorussia: The Making of a Nation* (Cambridge: Harvard University Press, 1956), pp. 170–174.

39. Chaja Bielski, Personal Interview. Also in neighboring Lithuania native Lithuanians were murdering Jews even before the Germans arrived. See Raul Hilberg, *The Destruction of European Jews* (New York: New Viewpoints, 1973), p. 205.

40. Esia Lewin-Shor, Personal Interview.

Chapter 3

1. Chaja Bielski, Personal Interview, Haifa, Israel, 1987–1991. Martin Gilbert reports that, arrested with her friend, Chaja was forced to clean a hall. See *The Holocaust: A History of the Jews of Europe During the Second World War* (New York: Holt, Rinehart & Winston, 1985), p. 169. Gilbert received this information from Idel Kagan, in 1984, in England. When asked about it, Chaja denies that she was ever seized or was made to clean a hall.

2. In 1944, after the Germans were evicted from this region, a Soviet commission came to investigate Nazi crimes. They dug up graves to check how the victims died. At one of those sites Riva identified her husband by the pajamas he wore. Riva Reich, Personal Interview, Tel Aviv, Israel, 1989.

3. Tuvia Bielski in *Yehudei Yaar* (*Forest Jews*) (Tel Aviv: Am Oved, 1946).

4. For examples of discussions about the start of the Final Solution and the part played by the Einsatzgruppen, see: Yehuda Bauer, *The Holocaust in Historical*

Perspective (Seattle: University of Washington Press, 1978), pp. 14–15; Lucy S. Davidowicz, *The War Against the Jews, 1933–1945* (New York: Holt, Rinehart & Winston, 1975), pp. 119–128; Raul Hilberg, *The Destruction of European Jews* (New York: New Viewpoints, 1973), pp. 177–219; Nora Levin, *The Holocaust and the Destruction of European Jewery, 1933–1945* (New York: Schocken Books, 1973), pp. 234–267; Leni Yahil, *The Holocaust: The Fate of European Jews* (New York: Oxford University Press, 1990), pp. 253–287.

5. Michael R. Marrus, *The Holocaust in History* (Hanover: The University Press of New England, 1987), pp. 64–65.

6. Zus Bielski in *Forest Jews.*

7. The *Encyclopedia Judaica* gives the date as the 10th of July and the *Encyclopedia of the Holocaust* as the 11th of July. See "Novogroduk," in *Encyclopedia Judaica,* Vol. 12 (Jerusalem: Keter Publishing House, Ltd., 1971), pp. 1237–1238; and "Novogrudok," in *Encyclopedia of the Holocaust,* Vol. 3 (London: Macmillan Publishing Co., 1990), pp. 1072–1073.

8. News about the fate of the Jews in Horodyszcze reached Chaja Bielski and her family in Duża Izwa. Chaja Bielski, Personal Interview; Yehuda Szimszonowicz, Yad Vashem Testimony, No. 3312/296. When the area was retaken by the Red Army in 1944, a Soviet commission examined the mass grave of the Horodyszcze Jews. Some of the victims could be recognized by their clothing. See Major W. Goldsey, "Dos Tor Nit Zain Geshenkt!" ("This Should Not Be Forgiven"), *Einikait,* December 14, 1944, p. 2.

9. For a description and discussion of such mass murders, see Vassily Grossman, "The Story of an Old Man, Shmuel Dovid Kugel," pp. 182–223, in Ilya Ehrenburg and Vasily Grossman (eds.), *The Black Book* (documents the Nazis' destruction of 1.5 million Jews) (suppressed by Stalin) (New York: Holocaust Library, 1980); Nechama Tec, *In the Lion's Den: The Life of Oswald Rufeisen* (New York: Oxford University Press, 1990), pp. 71–75, 91–95, 121–124.

10. Lilka Bielski, Personal Interview, Brooklyn, New York, 1989.

11. Zus Bielski in *Forest Jews.*

12. Ibid.

13. Yitskhok Rudashevsky, *The Diary of the Vilna Ghetto, June 1941–1943* (Israel: Ghetto Fighters' House, 1973), p. 32. This youth was later murdered by the Germans. The book is based on his diary left behind in an attic.

14. Tamara Rabinowicz, Personal Interview, Haifa, Israel, 1990.

15. "Lida," in *Encyclopedia of the Holocaust,* Vol. 3, pp. 868–870.

16. Eljezer Engelstern, Yad Vashem Testimony, No. 3249/233; "Anton Schmidt," in *Encyclopedia of the Holocaust,* Vol. 4, p. 1222. Schmidt helped with a plan to escape from Vilna to Sweden across the Baltic Sea. The plan had to be given up because the Germans became aware of this route. See Yitzhak Arad, *Ghetto in Flames: The Struggle and Destruction of the Jews in Vilna in the Holocaust* (New York: Holocaust Library, 1982), pp. 187, 197–198.

17. Tuvia Bielski, in *Forest Jews.*

18. Aharon Bielski, called Arczyk, lives in the United States. Here he changed his name to Bell. During the war he was a young teenager and survived with his brothers in the forest. I tried to interview him without success. Twice he made an appointment with me in New York City. Once he failed to appear; the second time he refused to answer my questions. When I asked why he had made the appointment if he had no intention of giving me information, he offered no explanation. Occasionally in the forest he acted as a guide. Those I spoke to agree that his

participation and impact on the life of the Bielski otriad was minimal, almost nonexistent. I will refer to him only very rarely.

19. Zus Bielski, Personal Interview, Brooklyn, New York, 1989.

20. Cwi Isler, Yad Vashem Testimony, No. 1706/113; Luba Rudnicki, Personal Interview, Tel Aviv, Israel, 1988; Both Isler and Rudnicki were inmates of the Nowogródek ghetto. Some of the information they give is repeated in "Novogrudok," in *Encyclopedia of the Holocaust,* Vol. 3, pp. 1072–1073.

21. "Lida," in *Encyclopedia Judaica,* Vol. 11, pp. 212–213; "Lida," in *Encyclopedia of the Holocaust,* Vol. 3, pp. 868–870. There is some discrepancy about the number of Jews murdered and the date. *The Holocaust Encyclopedia* gives July 8 as the date and the number killed as eighty.

22. Tuvia Bielski, in *Forest Jews.* Lilka Bielski, Personal Interview, talks about Tuvia's insistence that her family should not go to a ghetto and about his offer of help.

23. Asael told Chaja the details of his encounter with the pharmacist; Chaja Bielski, Personal Interview.

24. The Holocaust literature is filled with examples of parents who urged their children to save themselves and leave them behind. For such illustrations, see Tec, *In the Lion's Den,* pp. 15, 142.

25. Chaja Bielski, Personal Interview.

26. Tuvia Bielski, *Forest Jews.*

27. Ibid.; Chaja Bielski, Personal Interview.

28. Ibid.

29. Lilka Bielski, Personal Interview.

30. Ibid. According to one source during this Aktion, 200 people ran away from the scene of slaughter. See "Lida," in *Encyclopedia Judaica,* Vol. 11, pp. 212–213. Another source states that 5670 Jews were murdered that day; see "Lida," in *Encyclopedia of the Holocaust,* Vol. 11, pp. 878–880.

Chapter 4

1. Chaja Bielski, Personal Interview, Haifa, Israel, 1987–1991.

2. Tuvia Bielski in *Yehudei Yaar* (*Forest Jews*) (Tel Aviv: Am Oved, 1946).

3. Lilka Bielski, Personal Interview, Brooklyn, New York, 1989.

4. Eljezer Engelstern, Yad Vashem Testimony, No. 3249/233.

5. Chaja Bielski, Personal Interview.

6. Eljezer Engelstern, Yad Vashem Testimony.

7. Dr. Isler describes one of Kozłowski's visits to ghetto Nowogródek and how the Belorussian tried to persuade the doctors to escape to the forest, to join the Bielski otriad. They were all amazed by his faultless Yiddish and his dedication to Jewish rescue. On leaving the group, Kozłowski gave them his address. Later on, when looking for shelter, this helped Dr. Isler find his way to the Bielski unit. See Cwi Isler, Yad Vashem Testimony, No. 1706/113.

8. Motl Berger, Personal Interview, Brooklyn, New York, 1989.

9. The name of the Zhukov otriad was not officially recognized by the Soviet Partisan Movement. With the 1943 reorganization of the Russian partisans they were given the name Kalinin. See Tuvia Bielski in *Forest Jews.*

10. Pesach Friedberg, Yad Vashem Testimony No. 3332/320, suggests that he switched positions because the group grew and food supply distribution became

a central issue. Other partisans suspect that Lazar Malbin's arrival, a more capable man who became the Chief of Staff, was responsible for the change in Friedberg's position. Among the partisans who expressed these views are Pinchas Boldo, Personal Interview, Haifa, Israel, 1990; Raja Kaplinski, Personal Interview, Tel Aviv, Israel, 1987–1989; Riva Reich, Personal Interview, Tel Aviv, Israel, 1989.

11. Raja Kaplinski, Personal Interview, as the secretary of the otriad worked closely with Malbin at the headquarters. She speaks highly about his qualifications as Chief of Staff.

12. Ibid.; Daniel Ostaszynski, Personal Interview, Tel Aviv, Israel, 1989. Ostaszynski knew Malbin well and admired Malbin's organizational skills and his ability to stay in Tuvia's shadow.

13. Lazar Malbin, Personal Interview, Tel Aviv, Israel, 1987–1988. I have interviewed Malbin twice. At that time he lived in an old-age home on the outskirts of Tel Aviv. Although a stutterer, he did not stutter during our meetings. I have examined Malbin's Yad Vashem testimonies, but will rely on my own interviews because they cover the same areas as the testimonies and include the additional information I was particularly interested in eliciting.

14. Chaja Bielski, Personal Interview. Chaja is devoted to the preservation of Asael's memory. An honest and decent person, she emphasizes Asael's positive qualities without compromising the truth.

15. Pinchas Boldo, Personal Interview.

16. Lazar Malbin, Personal Interview, insisted that only Tuvia qualified as commander.

17. Raja Kaplinski, Personal Interview.

18. Pinchas Boldo, Personal Interview.

19. Motl Berger, Personal Interview.

20. Sonia Bielski, Personal Interview, Brooklyn, New York, 1989.

21. Chaja Bielski, Personal Interview.

22. Shmuel Amarant, "The Tuvia Bielski Partisan Company," in *Nvo Shel Adam* (*Expressions of a Man*) (Jerusalem: Misrad Hahinuch v Tarbut, 1973), translated by R. Goodman.

23. Pinchas Boldo, Personal Interview. I was particularly impressed with Dr. Boldo's honesty and courage to express his ideas openly.

24. In the forest Jashke Mazowi did not care about Tuvia's emphasis on saving lives—he considered fighting more important. Jashke Mazowi, Personal Interview, Tel Aviv, Israel, 1989. Other partisans also recall their ambivalence or disapproval of Tuvia's position. For example: Hanan Lefkovitz, Personal Interview, Tel Aviv, Israel, 1988; Jacov Greenstein, Personal Interview, Tel Aviv, Israel, 1984–1990.

25. When asked if they themselves ever objected to taking in more people, without exception each denies this was ever the case. Chaja Bielski, for example, objects strongly to the rumor that her father was opposed to accepting more fugitives. Zus Bielski also insists that he was always in favor of bringing in more people. It is likely that people changed their minds about it. Once they saw that they could manage even with a large group they might have been won over to the idea of saving lives. Chaja Bielski, Personal Interview; Zus Bielski, Personal Interview, Brooklyn, New York, 1989.

26. Eljezer Engelstern, Yad Vashem Testimony.

27. Ibid.

28. Philip Friedman, "Jewish Resistance to Nazis," pp. 387–408 in Ada June

Friedman (ed.), *Roads to Extinction* (Philadelphia: The Jewish Publication Society of America, 1980). Friedman notes that some Jewish leaders succeeded in escaping to the West in 1939–1940. Among them was Samuel Zygelbojm. In London he became the Jewish representative in the Polish Government in Exile. He met with Jan Karski, the Polish underground courier who delivered messages from Jewish underground leaders. See Jan Karski, *Story of a Secret State* (Boston: Houghton Mifflin Company, 1944); Zygelbojm was unable to stir up support for the Jewish plight. Feeling helpless, hurt by the world's indifference, he committed suicide in protest. See Samuel Zygelbojm, "The Conscience of the World," pp. 329–331 in Jacob Gladstein, Israel Knox, Samuel Margoshes (eds.), *Anthology of Holocaust Literature* (New York: Atheneum, 1982). This is a public letter Zygelbojm wrote before he killed himself. Henry L. Feingold, *The Politics of Rescue* (New York: Holocaust Library, 1970), p. 207. Martin Gilbert, *The Holocaust: A History of the Jews of Europe During the Second World War* (New York: Holt, Rinehart & Winston, 1985), pp. 557–565.

29. Shalom Cholawski, "Jewish Partisans—Objective and Subjective Objectives," in *Jewish Resistance During the Holocaust,* Proceedings of the Conference on Manifestations of Jewish Resistance, 1968 (Jerusalem: Yad Vashem, 1972), p. 326; Yehuda Bauer, "The Judenräte—Some Conclusions," in *Patterns of Jewish Leadership in Nazi Europe, 1933–1945,* Proceedings of the Third Yad Vashem International Historical Conference, 1977 (Jerusalem: Yad Vashem, 1979), pp. 393–405; Raul Hilberg, "Conscious or Unconscious Tool," pp. 31–44 in *Patterns of Jewish Leadership in Nazi Europe, 1933–1945.*

30. "The Judenräte," pp. 235–287 in Lucy S. Dawidowicz (ed.), *A Holocaust Reader* (New York: Behrman House, 1976); Philip Friedman, "Social Conflict in the Ghetto," pp. 131–152 in Ada June Friedman (ed.), *Roads to Extinction;* Isaiah Trunk, *Judenrat: The Jewish Councils in Eastern Europe Under Nazi Occupation* (New York: Stein & Day, 1977).

31. For examples of the variety of ghetto underground organizations and the obstacles in communication, see Yitzhak Arad, *Ghetto in Flames: The Struggle and Destruction of the Jews in Vilno in the Holocaust* (New York: Holocaust Library, 1982), pp. 221–270, 373–400; Chajka Grossman, *The Underground Army* (New York: Holocaust Library, 1987). Chajka was an underground courier and the book is filled with stories about her travels and efforts to contact different ghetto undergrounds; Israel Gutman, "Youth Movements in the Underground and the Ghetto Revolts," pp. 260–284 in *Patterns of Jewish Leadership in Nazi Europe, 1933–1945;* Chaim Lazar, *Muranowska 7: The Warsaw Ghetto Rising* (Tel Aviv: Massada P.E.C. Press, Ltd., 1966).

32. See Max Weber, *The Theory of Social and Economic Organization* (Glencoe, Ill.: The Free Press, 1947) who says that "charismatic authority repudiates the past and is in this sense a specifically revolutionary force," p. 362. In his fascinating book Aron Hirt Manheimer describes another unconventional Jewish leader who during World War II in Rumania devoted himself to saving Jews; see *Jagendorfs Foundry* (New York: Harper Collins, 1991).

33. Pinchas Boldo, Personal Interview.

34. Daniel Ostaszynski, Personal Interview.

35. Abraham Viner, Personal Interview, Haifa, Israel, 1990.

36. Hersh Smolar, Personal Interview, Tel Aviv, Israel, 1989–1990.

37. Tuvia Bielski, Personal Interview, Brooklyn, New York, 1987.

38. Pesia Bairach, Personal Interview, Tel Aviv, Israel, 1990.

39. Pinchas Boldo, Personal Interview.
40. Zorach Arluk, Personal Interview, Tel Aviv, Israel, 1988.

Chapter 5

1. With collective responsibility, punishment was applied not only to the transgressors, but also to their families or even an entire community. For some illustrations of the implications of the principle of collective responsibility see Arad, *Ghetto in Flames,* pp. 387–400; Nechama Tec, *When Light Pierced the Darkness: Christian Rescue of Jews in Nazi-Occupied Poland* (New York: Oxford University Press, 1986), pp. 52–69.

2. Sonia Boldo quotes her mother in *Yehudei Yaar* (*Forest Jews*) (Tel Aviv: Am Oved, 1946).

3. Chaja Bielski, Personal Interview, Haifa, Israel, 1987–1991.

4. Independently, in Israel and the United States, Chaja and Sonia repeated the same story, Sonia Bielski, Personal Interview, Brooklyn, New York, 1989.

5. Sonia Bielski, Personal Interview.

6. Ibid.

7. Survival in the forbidden Christian world depended in part on how a Jew looked and behaved. Those who could not pass for Christians because of their "Jewish looks" along with the old and the disabled would often not even try to save themselves. For a discussion of these issues, see Nechama Tec, *When Light Pierced the Darkness: Christian Rescue of Jews in Nazi-Occupied Poland,* pp. 27–39.

8. Pesia Bairach, Personal Interview, Tel Aviv, Israel, 1990.

9. Ibid.

10. Husband and wife, independently, told me the same story. Moshe Bairach, Personal Interview, Tel Aviv, Israel, 1988–1989; Pesia Bairach, Personal Interview.

11. Sulia Wołożhinski-Rubin, *Against the Tide: The Story of an Unknown Partisan* (Jerusalem: Posner & Sons Ltd.), p. 91; Sulia Wołożhinski-Rubin, Personal Interview, Saddle River, New Jersey, 1988. (She will be referred to as Sulia Rubin throughout.)

12. Ibid.

13. Berl Chafetz, Personal Interview, Boston, Massachusetts, 1989.

14. Motl Berger, Personal Interview, Brooklyn, New York, 1989.

15. "Novogrudok," pp. 1237–1238, in *Encyclopedia Judaica* (Jerusalem: Keter Publishing House Ltd., 1971), Vol. 12; "Novogrudok," pp. 1072–1073, in *Encyclopedia of the Holocaust* (London: Macmillan Publishing Co., 1990). Both sources report 4000 victims for the December 1941 Aktion. There is only a slight difference on dates, with the *Encyclopedia Judaica* giving December 7 and the *Encyclopedia of the Holocaust* December 8.

16. The Nazis were in a rush to apply their policies of Jewish annihilation to all of Poland including Western Belorussia. The push for these policies was given by the January 1942 Wannsee Conference. See Lucy S. Dawidowicz (ed.), *A Holocaust Reader* (New York: Behrman House, Inc., 1976), pp. 294–316; Filip Friedman, "Zagłada Żydów Polskich W Latch, 1939–1945" (Destruction of Polish Jewry, 1939–1945), *Biuletyn Głownej Komisji Badan Zbrodni Niemieckiej W Polsce, No. 6* (1946), pp. 196–208; Martin Gilbert, *The Holocaust: A History of the*

Jews of Europe During the Second World War (New York: Holt, Rinehart & Winston, 1985), pp. 280–293; Raul Hilberg, *The Destruction of the European Jews: Revised and Definitive Edition* (New York: Holmes and Meier, 1985), pp. 482–523. This stepped-up annihilation of Jews went hand in hand with barriers to rescue. For a discussion of these issues, see Nechama Tec, *When Light Pierced the Darkness,* pp. 27–39.

17. Motl Berger, Personal Interview.

18. Almost until the very end the Mir ghetto was guarded by Jewish policemen, which facilitated the organization of a general ghetto breakout. See Nechama Tec, *In the Lion's Den: The Life of Oswald Rufeisen* (New York: Oxford University Press, 1990), pp. 119–148.

19. Luba Rudnicki, Personal Interview, Tel Aviv, Israel, 1988. Ironically, the Pole Jarmałowicz had rescued a group of Jews, among them the brother of Perale Ostaszynski-Hirschprung; Personal Interview, Tel Aviv, Israel, 1988.

20. On the other hand, those who had blue eyes and blond hair had a definite advantage when they lived illegally in the forbidden Christian world. See Tec, *When Light Pierced the Darkness,* pp. 27–39; Nechama Tec, *Dry Tears: The Story of a Lost Childhood* (New York: Oxford University Press, 1984).

21. Luba Rudnicki, Personal Interview.

22. Ibid.

23. Ibid.

24. See Michael Temchin, *The Witch Doctor: Memoirs of a Partisan* (New York: Holocaust Library, 1983).

25. Luba Rudnicki, Personal Interview.

26. Ibid.

27. Ibid. Reference to this group was made by Chaja Bielski, Personal Interview. Essentially, Luba's and Chaja's stories coincide.

28. Lazar Malbin, Personal Interview, Tel Aviv, Israel, 1987–1988.

Chapter 6

1. With very slight variations, this encounter was independently described by several people. Here I am relying mainly on the following sources: Chaja Bielski, Personal Interview, Haifa, Israel, 1987–1991; Tuvia Bielski in *Yehudei Yaar* (*Forest Jews*) (Tel Aviv: Am Oved, 1946); Zus Bielski, Personal Interview, Brooklyn, New York, 1989; Lazar Malbin, Personal Interview, Tel Aviv, Israel, 1987–1988.

2. Obedience to rules does not fit well into the lives of guerilla fighters. In contrast to regular soldiers, partisans are freer and more independent. And since they only very reluctantly bow to authority they are much harder to control than regular army men.

Because partisans are usually well acquainted with the surroundings, they also have much greater freedom of movement than regular soldiers. In part greater mobility and familiarity with the environment tend to counteract the disadvantages that stem from small size and inadequate military equipment. By definition, guerilla fighters are not as big nor as well equipped as a conventional army that they oppose. See John A. Armstrong and Kurt DeWitt, "Organization and Control of the Partisan Movement," pp. 73–139 in John A. Armstrong (ed.), *Soviet Partisans in World War II* (Madison: The University of Wisconsin Press, 1964), p. 73; Henri Michel, "Jewish Resistance and the European Resistance Movement," pp.

365–375 in *Jewish Resistance During the Holocaust,* Proceedings of the Conference on Manifestations of Jewish Resistance (Jerusalem: Yad Vashem, 1972); Henri Michel, *The Shadow War: European Resistance, 1939–1945* (New York: Harper & Row, 1972), p. 13; Jack Porter (ed.), *Jewish Partisans, A Documentary of Jewish Resistance in the Soviet Union During World War II* (New York: University Press of America, 1982), p. 9; Tennenbaum, *Underground: The Story of a People* (New York: Philosophical Society, 1952), p. 385.

3. Violence also becomes a part of the guerilla fighter's life. Use of violence is often backed up by moral rationalizations. See J. K. Zawodny, "Guerrilla and Sabotage: Organization, Operations, Motivations, Escalation," *The Annals of the American Academy of Political and Social Science,* 341 (May 1962), pp. 8–18.

4. While some of these Soviet soldiers might have been willful deserters others were left behind because of special circumstances. The Russians were retreating fast, in a chaotic way, and pockets of the army might have inadvertently stayed on. In a sense, as a part of the Red Army, the Bielski brothers were abandoned by the military. Arad describes the Red Army's limited means of transportation. See Yitzhak Arad, *Ghetto in Flames: The Struggle and Destruction of the Jews in Vilna in the Holocaust* (New York: Holocaust Library, 1982), p. 30. Also in *The Minsk Ghetto Soviet Jewish Partisans Against the Nazis* (New York: Holocaust Library, 1989), pp. 4–8, Hersh Smolar talks about the difficulties in eluding the advancing German army.

5. It has been estimated that by the summer of 1942 the entire Soviet Partisan Movement consisted of 150,000 guerillas. See Earl Ziemke "Composition and Morale of the Partisan Movement," pp. 141–196 in John A. Armstrong (ed.), *Soviet Partisans in World War II.* For all Belorussia a Soviet source offers the following estimated figures for the partisans under Russian control: by late 1941, 5000; by 1942, 73,000; by 1943, 243,000; by 1944, 374,000. In Belorussia the Soviet Partisan movement included Belorussians, Russians, Jews, Poles, Slovaks, and other groups. See "Partisans," in *Encyclopedia of the Holocaust,* Vol. 3 (London: Macmillan Publishing Co., 1990), p. 1113.

6. It has been estimated that in the first six months of the war the Germans took over 3,000,000 Soviet soldiers as prisoners of war. See Ziemke, "Composition and Morale of the Partisan Movement," p. 143.

Basically, Nazi policies toward Russian POWs consisted in economic exploitation and murder. Of the two, economic exploitation was an intermediary step that led to death. Economic exploitation in turn was closely related to political and economic conditions. See Reuben Ainsztein, *Jewish Resistance in Occupied Eastern Europe* (New York: Barnes & Noble, 1974), p. 243; Martin Gilbert, *The Second World War: A Complete History* (New York: Henry Holt & Co., 1989), p. 373.

7. Nicholas P. Vakar, *Belorussia, The Making of a Nation* (Cambridge: Harvard University Press, 1956), pp. 174–175.

8. Ainsztein, *Jewish Resistance,* p. 279; Vakar, *Belorussia,* p. 192.

9. J. K. Zawodny, "Soviet Partisans," *Soviet Studies,* 17, No. 3 (January 1966), pp. 368–377.

10. Ainsztein, *Jewish Resistance,* pp. 307–338; Yehuda Bauer, *A History of the Holocaust* (New York: Franklin Watts, 1982), p. 271; Bryna Bar-Oni, *The Vapor* (Chicago: Visual Impact, Inc., 1976); Shalom Cholawski, *Soldiers from the Ghetto* (New York: The Herzel Press, 1982), p. 147; Shmuel Krakowski, *The War of the Doomed: Jewish Armed Resistance in Poland, 1942–1944* (New York: Holmes & Meier Publishers, Inc., 1984), p. 28; Dov Levin, *Fighting Back: Lithuanian*

Jewry's Armed Resistance to the Nazis, 1941–1945 (New York: Holmes & Meier, Publishers, Inc. 1985), pp. 206–207.

11. For statements about limited Belorussian nationalism see Thomas Fitzsimmons, Peter Malof, and John C. Fiske, *USSR* (New Haven: HRAF Press, 1960), pp. 17, 50; Nicholas P. Vakar, *Belorussia,* p. 185.

12. Alexander Dallin, *German Rule in Russia, 1941–1945: A Study of Occupation Policies* (New York: Octagon Books, 1980), p. 199, cites several German SD reports written at different times that illustrate the Belorussians' change of attitudes toward the German occupation. (1) July 1941: "We are enthusiastically received on all sides. (2) August 1942: "The basic attitude is one of deep resignation." (3) October 1943: "In truth the bulk of the population is hostile." Some historians feel that in large measure German cruelty was responsible for the expansion of the partisan movement. See Michel, *The Shadow War,* p. 185; Vakar, *Belorussia,* p. 191. Moreover, in a secret report the Nazi, Major Daven, writes, "The natives don't think that there is a Jewish problem at all. This is due to the communist influence that perceives no racial differences." YIVO, the Berlin Collection, Box 30, Occ E3a-14.

13. Seventy-seven percent of Polish Jews lived in urban centers. See Jacob Lestchińsky, "Economic Aspects of Jewish Community Organization in Independent Poland," *Jewish Social Studies 9,* No. 1–4 (1947), pp. 319–338; Jacob Lestchińsky, "The Industrial and Social Structure of the Jewish Population of Interbellum Poland," *YIVO Annual Social Science 2* (1956–1957), pp. 243–269; Antony Polonsky, *Politics in Independent Poland, 1921–1939* (Oxford: Claredon Press, 1972), p. 42.

14. Yehuda Bauer, *The Jewish Emergence from Powerlessness* (Toronto: The University of Toronto Press, 1979), p. 30; Yisrael Gutman and Shmuel Krakowski, *Unequal Victims, Poles and Jews During World War II* (New York: Holocaust Library, 1986), pp. 103–139; Shmuel Krakowski, *The War of the Doomed,* pp. 27–42.

15. Ideological and political concerns came later, when the partisan ranks were augmented by arrivals from the Soviet Union and by Poles connected to the Polish underground. Pinchas Boldo, Personal Interview, Haifa, Israel, 1990; Hersh Smolar, Personal Interview, Tel Aviv, Israel, 1989–1990; Vakar, *Belorussia,* p. 191.

16. For examples of how ghetto inmates dreamt about joining the partisans see Nechama Tec, *In the Lion's Den: The Life of Oswald Rufeisen* (New York: Oxford University Press, 1990), pp. 119–148.

17. Jashke Mazowi, Personal Interview, Tel Aviv, Israel, 1989. After Mazowi became a partisan he went several times into the Lida ghetto, acting as a guide for ghetto inmates. See also *Sefer Hapartisanim Hajehudim* (*The Jewish Partisan Book*) (Merchavia: Sifriath Poalim, Hashomer Hatzair, 1958), Vol. 1, p. 421.

18. Pesia Bairach, Personal Interview, Tel Aviv, Israel, 1990.

19. Several people have independently referred to these meetings. They were: Zorach Arluk, Personal Interview, Tel Aviv, Israel, 1988; Chaim Basist, Personal Interview, Tel Aviv, Israel, 1988; Moshe Bairach, Personal Interview, Tel Aviv, Israel, 1988–1989.

20. Such fears were justified. The Germans often relied on the principle of collective responsibility, punishing those who were close to or connected to the "guilty." See Arad, *Ghetto in Flames,* pp. 373–395; Tec, *When Light Pierced the Darkness,* pp. 27–39.

21. Zorach Arluk, Personal Interview.

22. Sometimes part of a Russian otriad—a hospital or headquarters—had a stable base inside the forest while the rest moved around the countryside. Chaja Bielski, Personal Interview, Haifa, Israel, 1987–1991; Raja Kaplinski, Personal Interview, Tel Aviv, Israel, 1987–1989.
23. Tec, *In the Lion's Den,* p. 192.
24. Ibid., pp. 201–202.
25. Ibid., pp. 193–194.
26. Ibid., p. 194.
27. Tuvia Bielski, in *Forest Jews.*
28. Ibid.
29. Ibid.
30. Ibid.
31. Ibid.; *The Jewish Partisan Book,* p. 431.
32. Tuvia Bielski, in *Forest Jews; The Jewish Partisan Book,* p. 438; Chaja Bielski, Personal Interview; Lazar Malbin spoke with a great deal of contempt about Israel Kesler, Personal Interview.
33. Tuvia Bielski, in *Forest Jews.*
34. Ibid.; Chaja Bielski, Personal Interview.
35. Tuvia Bielski, in *Forest Jews;* Pesach Friedberg Yad Vashem Testimony, No. 3332/320.
36. Pinchas Boldo, Personal Interview; Tuvia Bielski in *Forest Jews.* Although women spoke about these acts of revenge, none of them had actually participated. Chaja Bielski, Personal Interview; Lilka Bielski, Personal Interview, Brooklyn, New York, 1989; Raja Kaplinski, Personal Interview.
37. Tuvia Bielski, in *Forest Jews;* Tuvia Bielski, Personal Interview, Brooklyn, New York, 1987.
38. Lazar Malbin, Personal Interview; Chaja Bielski, Personal Interview.

Chapter 7

1. Pinchas Boldo, Personal Interview, Haifa, Israel, 1990.
2. The literature is filled with descriptions of abusive behavior toward the Jews by Russian and other partisans. For a few examples see Reuben Ainsztein, *Jewish Resistance in Occupied Eastern Europe* (New York: Barnes & Noble, 1974), pp. 307–308; Nachum Alpert, *The Destruction of Słonim Jewry* (New York: Holocaust Library, 1989), pp. 290–298; Shalom Cholawsky, "Jewish Partisans—Objective and Subjective Difficulties," *Jewish Resistance During the Holocaust,* Proceedings of the Conference on Manifestations of Jewish Resistance (Jerusalem: Yad Vashem, 1971), pp. 323–334; Shmuel Krakowski, *The War of the Doomed: Jewish Armed Resistance in Poland, 1942–1944* (New York: Holmes & Meier Publishers, Inc., 1984), pp. 28–58; Dov Levin, "Baltic Jewry's Armed Resistance to the Nazis," in Isaac Kowalski (ed.), *Anthology of Armed Resistance to the Nazis, 1939–1945* (New York: Jewish Combatants Publishing House, 1986), Vol. 3, pp. 42–48; Dov Levin, *Fighting Back: Lithuanian Jewry's Armed Resistance to the Nazis, 1941–1945* (New York: Holmes & Meier, Publishers, Inc. 1985), pp. 206–227; J. Tennenbaum, *Underground: The Story of a People* (New York: Philosophical Society, 1952), p. 292.
3. Jashke Mazowi, Personal Interview, Tel Aviv, Israel, 1989.
4. For activities of these leaders see *Sefer Hapartisanim Hajehudim* (*The Jewish*

Partisan Book) (Merchavia: Sifriath Paolim Hashomer Hatzair, 1958), Vol. 1, pp. 375–382, 337–343; Samuel Bornstein, "The Platoon of Dr. Atlas," pp. 217–240, in Meyer Barkai (ed.), *The Fighting Ghettoes* (New York: J. B. Lippincott Company, 1962); Lester Eckman and Chaim Lazar, *The Jewish Resistance* (New York: Shengold Publishers, Inc., 1977), pp. 51–58; Leonard Tushnet, "The Little Doctor—A Resistance Hero," pp. 253–259 in Yuri Suhl (ed.), *They Fought Back, The Story of the Jewish Resistance in Nazi Europe* (New York: Schocken Books, 1975); *Pinkas Zetel* (*A Memorial to the Jewish Community of Zdzienciół*, Baruch Kaplinski (ed.) (Tel Aviv: Zetel Association in Israel, 1957) contains several articles about Lipiczańska and other forests and talks about the two leaders Dworecki and Kaplinski.

5. Family camps varied in size, ranging from a handful of individuals to the largest single group, the Bielski otriad, that grew to more than 1200 individuals. For a discussion of the instability and precarious position of these different camps, see Yitzhak Arad, "Jewish Family Camps in the Forests: An Original Means of Rescue," pp. 333–353 in *Rescue Attempts During the Holocaust,* Proceedings of the Second Yad Vashem International Historical Conference (Jerusalem: Yad Vashem, 1977); Eckman and Lazar, *The Jewish Resistance,* pp. 83–99; Tennenbaum, *Underground,* p. 404; Lea Garber Kowenska, a member of a small family group in the Lipiczańska forest, touchingly describes how great suffering and mutual caring intermingled in the lives of this group. See Lea Garber Kowenska "Dos Fos Hot Sich Fargidenk of Aibik" (What Is Remembered Forever) *Journal Fu Sovietisher Heimland* (*Journal of the Soviet Homeland*), No. 4, 1971, pp. 92–102; Lea Garber Kowenska also describes her life in the forest, in Yiddish, in an unpublished memoir. In her group of fifteen there were seven children. Some were orphans whom she and others picked up on the way to and inside the forest.

6. *Jewish Partisan Book,* pp. 346–394.

7. Hersh Smolar, Personal Interview, Tel Aviv, Israel, 1989–1990.

8. Raja Kaplinski, Personal Interview, Tel Aviv, Israel, 1987–1989.

9. Pinchas Boldo, Personal Interview.

10. Raja Kaplinski, Personal Interview.

11. Jacov Greenstein, Personal Interview, Tel Aviv, Israel, 1984–1990.

12. Even after the Russian-German war turned in favor of the USSR, it took quite a while before Soviet partisans became an effective force. Some are convinced that the partisan fighting was much less extensive than officially claimed. Pinchas Boldo, Personal Interview, Oswald Rufeisen, a member of the Ponomarenko otriad, is convinced that the partisan battles and heroism have been highly exaggerated. See Nechama Tec, *In the Lion's Den: The Life of Oswald Rufeisen* (New York: Oxford University Press, 1990), pp. 201–202.

13. Henri Michel, *The Shadow War: European Resistance, 1939–1945* (New York: Harper & Row Publishers, 1972), pp. 278–279; Jack N. Porter (ed.), *Jewish Partisans, A Documentary of Jewish Resistance in the Soviet Union During World War II* (New York: University Press of America, 1982), p. 9; J. K. Zawodny, "Guerrilla and Sabotage: Organization, Operations, Motivations, Escalations," *The Annals of the Academy of Political Science,* Vol. 341 (May 1962), pp. 8–18.

14. Tuvia Bielski, in *Yehudei Yaar* (*Forest Jews*) (Tel Aviv: Am Oved, 1946).

15. Chaja Bielski, Personal Interview, Haifa, Israel, 1987–1991; Raja Kaplinski, Personal Interview. At one point, at the start of 1943, German planes dropped leaflets announcing monetary rewards for information leading to Tuvia

Bielski's capture. This leaflet, aimed at reaching the Christian population, claimed that whoever would give the Nazis information leading to the capture of Tuvia Bielski would receive a reward of 50,000 marks. This figure was later doubled. Because of these leaflets many Jews learned about the existence of the Bielski otriad. This information, in turn, gave them courage to escape and search for the camp. See *Sefer Lida* (*Lida Book*) (Tel Aviv: Published by the Association of Lida Jews, 1970), p. xv. The official currency exchange rate at the time was 2.5 marks per $1. For this and other exchange rates see Leni Yahil, *The Holocaust: The Fate of European Jewry* (New York: Oxford University Press, 1990), pp. 661–662.

16. This anti-German move is described by several Bielski partisans. Chaja Bielski, Personal Interview; Tuvia Bielski, *Forest Jews;* Pinchas Boldo, Personal Interview; Zus Bielski, Personal Interview, Brooklyn, New York, 1989; Raja Kaplinski, Personal Interview. Boldo and the Bielski brothers participated in the move.

17. This is a quote from a personal interview with Pinchas Boldo. He used this case as an example of a failed anti-German move. While Tuvia admits that the incident was not a great success, he is less critical. Tuvia says that this "combined attack on the Germans by the Jewish partisans and Panchenko's group brought us no loot, but we killed seven to eight Germans. Indeed, our aim was the annihilation of the entire guard and to take their machine guns and mortar. We did not succeed in realizing the entire plan. It was fortunate that we got away without sustaining casualties." Tuvia Bielski in *Forest Jews.*

18. Tuvia Bielski in *Forest Jews.* He also talked about these things during our meeting in his house, Personal Interview, Brooklyn, New York, 1987.

19. Ibid.

20. Eljezer Engelstern, Yad Vashem Testimony, No. 3249/233.

21. Ibid.

22. Ibid.

23. Chaja Bielski, Personal Interview.

24. The historian, Dr. Amarant, feels that, if allowed, this step would have become a dangerous precedent that could eventually lead to a breakup and destruction of the otriad. See Shmuel Amarant, "The Bielski Partisan Company," in *Nvo Shel Adam* (*Expressions of a Man*) (Jerusalem: Misrad Hahinuch v Tarbut, 1973), translated by R. Goodman.

25. Tuvia Bielski in *Forest Jews.*

26. Ibid.

27. Ibid.

28. Ibid.

29. Ibid.

30. Ibid.

31. Ibid.; depending on who tells the story, the size of this group ranges from twenty to forty.

32. Chaja Bielski, Personal Interview.

33. Motl Berger, Personal Interview, Brooklyn, New York, 1989.

34. Esia Lewin-Shor, Personal Interview, Bronx, New York, 1989. Among these fugitives was Esia's pregnant neighbor, someone she was very fond of.

35. Sonia Bielski, Personal Interview, Brooklyn, New York, 1989. Except for the Jewish partisans who accompanied the group, all the rest were unarmed ghetto runaways, women, children, and older people.

36. "Ivije," in *Encyclopedia Judaica,* Vol. 9 (Jerusalem: Keter Publishing House, Ltd., 1971), pp. 1155–1156.
37. Baruch Kopold, Personal Interview, Haifa, Israel, 1990.
38. Ibid.
39. Chaja Bielski, Personal Interview; *The Jewish Partisan Book,* p. 424, also refers to the help offered by the Bielski otriad to the Jews of Iwje. According to this source, because of this help, 150 Jews left the ghetto and came to the Bielski otriad.
40. Baruch Kopold, Personal Interview.
41. Tuvia Bielski in *Forest Jews.*
42. When I interviewed Tuvia two weeks before his death he spoke about this incident with much regret. He felt he should not have allowed them to go. Personal Interview.
43. Chaja Bielski, Personal Interview.
44. Ibid.; Sonia Bielski essentially tells the same story, Personal Interview.
45. Tuvia Bielski in *Forest Jews.*
46. A Belorussian policeman gave Chaja Bielski, Personal Interview, these details.
47. Ibid. For just such an eventuality Asael and Chaja had decided to meet at the home of Chaja's Belorussian friend, Piotrus, which is what happened.
48. Baruch Kopold, Personal Interview.
49. Chaja Bielski, Personal Interview.
50. Raja Kaplinski, Personal Interview.
51. For the cold weather Sonia's parents, the Boldos, stayed with Belorussian peasants. Sonia joined them there, Personal Interview.
52. Tuvia Bielski, Personal Interview.
53. Lilka Bielski, Personal Interview, Brooklyn, New York, 1989.
54. Chaja Bielski, Personal Interview.
55. Lilka Bielski, Personal Interview.
56. Ibid.

Chapter 8

1. Jashke Mazowi, Personal Interview, Tel Aviv, Israel, 1989.
2. The German defeat at Stalingrad was a turning point in the Russian-German War. See Martin Gilbert, *The Second World War: A Complete History* (New York: Henry Holt & Co., 1989), pp. 398–410; Yisrael Gutman, *Fighting Among the Ruins, The Story of Jewish Heroism During World War II* (Washington, D.C.: B'nai B'rith Books, 1988), p. 219; Henri Michel, *The Shadow War: European Resistance, 1939–1945* (New York: Harper & Row Publishers, 1972), p. 185.
3. Tuvia Bielski in *Yehudei Yaar* (*Forest Jews*) (Tel Aviv: Am Oved, 1946).
4. Ibid.
5. Tuvia Bielski, Personal Interview, Brooklyn, New York, 1987; Lazar Malbin, Personal Interview, Tel Aviv, Israel, 1987–1988; *Sefer Hapartisanim Hajehudim* (*The Jewish Partisan Book*) (Merchavia: Sifriath Poalim, Hashomer Hatzair, 1958), Vol. 1, p. 446; Hersh Smolar, Personal Interview, Tel Aviv, Israel, 1989–1990, sees the arrival of Platon as an important step in the reorganization of the partisan movement.
6. Tuvia Bielski, Personal Interview.
7. Norman Davies, *God's Playground: A History of Poland* (New York: Co-

lumbia University Press, 1984), Vol. II, pp. 466–472; Jan Karski, *The Great Powers and Poland, 1919–1945: From Versailles to Yalta* (New York: University Press of America, 1985), pp. 403–411.

8. John A. Armstrong and Kurt De Witt, pp. 73–139 in J. A. Armstrong (ed.), *Soviet Partisans in World War II* (Madison: The University of Wisconsin Press, 1964); Tadeusz Bor-Komorowski, *The Secret Army* (London: Victor Gollancz, Ltd., 1951), pp. 119–120; Michel, *The Shadow War,* p. 219.

9. Nechama Tec, *In the Lion's Den: The Life of Oswald Rufeisen* (New York: Oxford University Press, 1990), p. 181.

10. Michel, *The Shadow War,* p. 185; Nicholas P. Vakar, *Belorussia: The Making of a Nation* (Cambridge: Harvard University Press, 1956), pp. 193–194. Both Michel and Vakar maintain that German cruelty contributed to the growth of the Soviet partisan movement by forcing people into the forests.

11. Chaja Bielski, Personal Interview, Haifa, Israel, 1987–1991; Hersh Smolar, Personal Interview.

12. The Mir ghetto was liquidated on August 13, 1942. See Tec, *In the Lion's Den,* pp. 134–148. This liquidation happened two days after a massive breakout during which 300 Jews succeeded in escaping. The liquidation of Nieśwież occurred on July 22, 1942; of Zdzienciół, on August 6, 1942; see Dov Levin, "The Fighting Leadership of the Judenräte in Small Communities," pp. 133–149 in *Patterns of Jewish Leadership in Nazi Europe, Proceedings of the Third Yad Vashem International Historical Conference* (Jerusalem: Yad Vashem, 1979); The liquidation of Żołudek ghetto took place on May 9, 1942, Pesia Bairach, Personal Interview, Tel Aviv, Israel, 1990; Iwje ghetto was destroyed on January 20, 1943, see "Ivije," pp. 1155–1156 in *Encyclopedia Judaica* (Jerusalem: Keter Publishing House, Ltd., 1971), Vol. 9.

13. On the eve of the German invasion Lida had a Jewish population of 9000. By May 8, 1942, this number was reduced by about 6000 people. See "Lida," pp. 868–870 in *Encyclopedia of the Holocaust* (London: Macmillan Publishing Co., 1990), Vol. 3; another source says that after the May 8, 1942, Aktion only 1250 inmates were left in the Lida ghetto. See "Lida," pp. 212–213 in *Encyclopedia Judaica,* Vol. 11; on the eve of World War II the town of Nowogródek had a Jewish population of 6000. After the second Aktion of August 7, 1942, only a few hundred were left. "Novogrudok," in *Encyclopedia Judaica,* Vol. 12, pp. 1237–1238; the *Encyclopedia of the Holocaust* lists the Jewish population at the start of World War II as 7000, Vol. 4, pp. 1072–1073.

14. Because natives could easily identify who was a Jew, sometimes the Jews feared local denouncers more than the Germans. For a discussion of these issues, see Nechama Tec, *Dry Tears: The Story of a Lost Childhood* (New York: Oxford University Press, 1984); Nechama Tec, *When Light Pierced the Darkness: Christian Rescue of Jews in Nazi-Occupied Poland* (New York: Oxford University Press, 1986), pp. 40–51.

15. Cila Dworecki, Personal Interview, Tel Aviv, Israel, 1988; Luba Dworecki Personal Interview, Tel Aviv, Israel, 1988.

16. Tamara Rabinowicz, Personal Interview, Haifa, Israel, 1990. This guide was an unsavory character, Itchinke, who at one point robbed two Jewish fugitives of their possessions. When the two complained, the Bielski partisan, Ben-Zion Golkovitz, a good, brave fighter, pursued Itchinke and his party and brought them to the Bielski otriad. He made them return the goods and placed them under arrest. Itchinke and his wife ran away that night and no one heard from him again. His

friends apologized and were let free. They claimed that Itchinke had initiated the robbery. Tuvia Bielski in *Forest Jews*.

17. Many women with and without children were picked up by Bielski scouts. One example is Cila Kapelowicz, a Mir ghetto inmate who ran away during the August 1942 ghetto breakout. She was first a member of a small family group. Cila was absent during an attack by Russian partisans and this way became the sole survivor of her group. Eventually she ended up in the Bielski otriad where she survived the war. She now lives in South Africa but during her 1987 visit to Israel I interviewed her in Jerusalem. Some other examples of women who were roaming the forest alone and ended up in the Bielski otriad were Ester Krynicki Godorejski Berkowitz, "Sichrojnes Fun Der Deitscher Okupacje" ("Memoirs from the German Occupation"), pp. 587–602 in *Mir* (ed.) N. Blumenthal (Jerusalem: Memorial Books, Encyclopedia of the Diaspora, 1962); Cirl Rojak, Yad Vashem Testimony, No. 3597/240; Hana Stolowicki, Yad Vashem Testimony, No. 3439/191.

18. Hana Berkowitz, Yad Vashem Testimony, No. 2086/195.

19. Abraham Viner, Personal Interview, Haifa, Israel, 1990.

20. Chaja Bielski, Personal Interview.

21. Bela is a fictitious name, because the real person did not want to be identified in relation to this affair.

22. Ibid.

23. Ibid.

24. Arad, *The Partisan,* p. 138; Raja Kaplinski, Personal Interview, Tel Aviv, Israel, 1987–1989.

25. Tuvia Bielski in *Forest Jews*.

26. Ibid.

27. Cwi Isler, Yad Vashem Testimony, No. 1706/113.

28. Moshe Bairach, Personal Interview, Tel Aviv, Israel, 1988–1989.

29. This incident is described by different people essentially in the same way. The slight variations refer to the kind of animal. Some speak about chickens, some about cows, still others about pigs. A few of these sources are Chaja Bielski, Personal Interview; Tuvia Bielski in *Forest Jews;* Raja Kaplinski, Personal Interview.

30. Isler attended to Oppenheim's wound and diagnosed it as a superficial injury. Cwi Isler, Yad Vashem Testimony. Both Chaja Bielski and Raja Kaplinski describe this incident the same way.

31. Moshe Bairach, Personal Interview.

32. Tuvia Bielski, in *Forest Jews*.

33. Bielski partisans spoke independently about the letters sent to ghettos urging people to come to the forest. Two such examples are Chaja Bielski, Personal Interview; Lazar Malbin, Personal Interview.

34. Lilka Bielski, Personal Interview, Brooklyn, New York, 1989; Tuvia Bielski, Personal Interview.

35. Chaja Bielski, Personal Interview.

36. Tuvia Bielski in *Forest Jews*.

37. Cwi Isler, Yad Vashem Testimony; Zus Bielski, Personal Interview, Brooklyn, New York, 1989.

38. The attack on Belorus has attracted much attention. Most of those I spoke to have mentioned it and parts of the story have appeared in print. See Tuvia Bielski in *Forest Jews;* also, Tuvia Bielski, Personal Interview. There are only very slight discrepancies in these accounts. For example, while all sources agree that a

son went to the police to denounce the sleeping partisans, one Bielski partisan says it was a daughter. See *Korelitz: Kium un Hurbn Fun a Yidishe Kehile* (*Korelicz: The Life and Destruction of a Jewish Community*) (ed.) M. Walzer-Fass (Tel Aviv: Korelitz Societies in Israel and the USA, 1973), pp. 217–225.

39. Luba Rudnicki, Personal Interview, Tel Aviv, Israel, 1988.

40. Pinchas Boldo, Personal Interview, Haifa, Israel, 1990.

41. Eljezer Engelstern, Yad Vashem testimony, No. 3249/233.

42. Tuvia Bielski, in *Forest Jews.*

43. Ibid.; Chaja Bielski, Personal Interview. I also spoke to some of the people who were a part of this group in Tel Aviv and they reconfirm the story. There is a slight variation in the size of this group with Tuvia's twenty-five and twenty-two individuals in another source. Those who left the Orlanski otriad were a part of 120 partisans who joined this Russian detachment after their leader, Hirsh Kaplinski, fell in battle. See *The Jewish Partisan Book,* pp. 384, 438.

44. Tuvia Bielski in *Forest Jews.*

45. Eljezer Engelstern, Yad Vashem Testimony. Hersh Smolar, Personal Interview, says that the Russian partisans had resented the Bielski otriad because its people consumed too much food.

46. Cwi Isler, Yad Vashem Testimony.

47. Ibid. Chaja Bielski, Personal Interview, tells how her mother gave Tuvia her fur coat because he needed it for a bribe.

48. Ibid.

49. The size of this forest is sometimes stated in miles, sometimes in kilometers; *The Columbia Lippincott Gazetteer of the World* (New York: Columbia University Press, 1961), p. 1279, gives 160 square miles as the size of the Nalibocka forest. In sharp contrast in "Partisans," *Encyclopedia of the Holocaust,* Vol. 3, p. 1114, it is stated that the Nalibocka forest "covers 1158 square miles, 3000 square kilometers." These last figures seem questionable.

50. The estimated size of the otriad for that time was 600 to 700 individuals, Chaja Bielski, Personal Interview; Tuvia Bielski, in *Forest Jews;* Zus Bielski, Personal Interview; Lazar Malbin, Personal Interview.

51. Chaja Bielski, Personal Interview; Tuvia Bielski in *Forest Jews.*

52. Moshe Bairach, Personal Interview.

53. Ibid.; Chaja Bielski, Personal Interview; Tuvia Bielski in *Forest Jews.*

54. Raja Kaplinski, Personal Interview.

55. Moshe Bairach, Personal Interview.

Chapter 9

1. Chaja Bielski, Personal Interview, Haifa, Israel, 1987–1991, thinks that there were about 15,000 partisans in the Nalibocka woods; Jacov Greenstein, Personal Interview, Tel Aviv, Israel, 1984–1990, doubles the figure to 30,000; in contrast, Hersh Smolar's, Personal Interview, Tel Aviv, Israel, 1989–1990 estimate is 20,000. This last estimate is the same as in "Partisans," *Encyclopedia of the Holocaust* (London: Macmillan Publishing Co., 1990), Vol. 4, p. 1114.

2. Determined to increase the number of partisans and to deprive the Germans of manpower, the Soviets devised all kinds of unorthodox ways to achieve this goal. See Henri Michel, *The Shadow War: European Resistance, 1939–1945* (New York: Harper & Row Publishers, 1972), pp. 184–185; Nicholas P. Vakar, *Be-*

lorussia: The Making of a Nation (Cambridge: Harvard University Press, 1956), pp. 193–194. Also for a discussion of the effects of this open-door policy on the forest Jews, see Yitzhak Arad, *The Partisan: From Valley of Death to Mount Zion* (New York: Holocaust Library, 1979), p. 138; Yehuda Bauer, *The Jewish Emergence from Powerlessness* (Toronto: The University of Toronto Press, 1979), p. 32; Shalom Cholawski, *Soldiers from the Ghetto* (New York: The Herzel Press, 1980), p. 147; Dov Levin, *Fighting Back: Lithuanian Jewry's Armed Resistance to the Nazis: 1941–1945* (New York: Holmes & Meier, Publishers, Inc., 1985), pp. 183–185; Raja Kaplinski, Personal Interview, Tel Aviv, Israel, 1987–1989; Hersh Smolar, Personal Interview, Tel Aviv, Israel, 1989–1990.

3. In 1943 special teams of well-trained officers were sent from Moscow into the Belorussian forests. These new arrivals introduced strict codes of conduct, prohibiting drinking, card playing, and stealing from civilians. In part these new measures improved the attitudes toward Jews. They also resulted in more safety and protection for the Jews. However, these positive changes failed to eliminate subtle forms of anti-Semitism. Nor did they succeed in establishing strict military discipline. See Reuben Ainsztein, *Jewish Resistance in Nazi-Occupied Eastern Europe* (New York: Barnes & Noble, 1974), pp. 333–335; Nachum Alpert, *The Destruction of Słonim Jewry: The Story of the Jews of Słonim During the Holocaust* (New York: The Holocaust Library, 1989), pp. 317–323; J. Tennenbaum, *Underground: The Story of a People* (New York: Philosophical Society, 1952), p. 393. Still, the presence of these official measures gave Tuvia a chance to threaten Soviet partisans that they would later pay a price for their discrimination against the Jews. See *Sefer Hapartisanim Hajehudim* (*The Jewish Partisan Book*) (Merchavia: Sifriat Poalim, Hashomer Hatzair, 1948), Vol. 1, p. 448.

4. Jashke Mazowi, Personal Interview, Tel Aviv, Israel, 1989, is convinced that, on balance, and particularly in the latter part of 1943, the Soviet partisan movement played an important role in saving Jewish lives.

5. Freedom from traditional constraint is a characteristic associated with charismatic leaders. See Max Weber, *The Theory of Social and Economic Organization* (Glencoe, Ill.: The Free Press, 1947), p. 360.

6. As for medical services there was much cooperation among the different otriads. One rule, for example, specified that an injured partisan was entitled to medical treatment in the nearest detachment. For example, when Hanan Lefkowitz was wounded in battle he was brought to an unfamiliar otriad for medical treatment because it was close. Hanan Lefkowitz, Personal Interview, Tel Aviv, Israel, 1988; Riva Reich, Personal Interview, Tel Aviv, Israel, 1989; Lili Krawitz, Personal Interview, Tel Aviv, Israel, 1989.

7. Chaja Bielski, Personal Interview.

8. Eljezer Engelstern, Yad Vashem Testimony, No. 3249/233, became the chief cook of the otriad.

9. Tuvia Bielski, in *Yehudei Yaar* (*Forest Jews*) (Tel Aviv: Am Oved, 1946).

10. For examples see Nachum Alpert, *The Destruction of Słonim Jewry,* pp. 288–291; Arad, *The Partisan,* p. 115; Cholawski, *Soldiers from the Ghetto,* p. 147; Levin, *Fighting Back,* pp. 183–185; J. Tennenbaum, *Underground: The Story of a People,* p. 392.

11. Jacov Greenstein was married. Several Russian otriads refused him entry because of his wife. Eventually he came to the Ponomarenko unit, a Soviet detachment with more liberal policies toward Jews and women. Personal Interview, Zorach Arluk, Personal Interview, Tel Aviv, Israel, 1988, though recognized as

a brave fighter in the Iskra detachment, could only keep his wife. He had to place his wife's parents in the Bielski otriad where they survived the war.

12. Ibid.; Moshe Bairach, Personal Interview, Tel Aviv, Israel, 1988–1989; *The Jewish Partisan Book,* Vol. 1, p. 454; Hersh Smolar, Personal Interview.

13. Ester Marchwinski, Yad Vashem Testimony, No. 03/3567; Josef Marchwinski, Yad Vashem Testimony, No. 03/3568.

14. Ibid.

15. Among those who objected to Marchwinski's presence were Chaja Bielski, Personal Interview; Tuvia Bielski in *Forest Jews;* Lazar Malbin, Personal Interview, Tel Aviv, Israel, 1987–1988; Raja Kaplinski, Personal Interview, Sulia Rubin, Personal Interview, Saddle River, New Jersey, 1988.

16. See Chapter 7.

17. Raja Kaplinski, Personal Interview.

18. Chaja Bielski, Personal Interview; Raja Kaplinski, Personal Interview; Lazar Malbin, Personal Interview, unlike the others, spoke about Kesler with great contempt. He was particularly objecting to his past.

19. Kesler's wife lives in Israel. She refused to be interviewed.

20. The Soviet regime imposed different chains of command on the partisan movement, the party, the secret service, and the army. The party played a powerful but not well-defined role. The secret service played a significant covert part. The authority of the Red Army was precise and direct. See John A. Armstrong and Kurt DeWitt, "Organization and Control of the Partisan Movement," pp. 132–134 in J. A. Armstrong (ed.), *Soviet Partisans in World War II* (Madison: The University of Wisconsin Press, 1964); Earl Ziemke thinks that party men remained the elite of the Russian partisan movement. See E. Ziemke, "Composition and Morale of the Partisan Movement," in Armstrong, *Soviet Partisans In World War II,* p. 144. The party influence within the Soviet partisan movement was also noted by Tadeusz Bor-Komorowski, *The Secret Army* (London: Victor Gollancz Ltd., 1951), pp. 119–120.

21. Tuvia Bielski in *Forest Jews.*

22. Ibid.

23. Ibid.

24. Ibid.

25. Many of those around Tuvia aspired to his position. This opinion was expressed by many, including Raja Kaplinski, Personal Interview, and Lazar Malbin, Personal Interview.

26. Josef Marchwinski, Yad Vashem Testimony. To compete with the groups supported by the Polish Government in Exile, London, the Soviets created the Union of Polish Patriots. See Jan Karski, *The Great Powers and Poland: 1919–1945, from Versailles to Yalta* (New York: University Press of America, Inc., 1985), pp. 432–433; Bor-Komorowski, *The Secret Army,* p. 123; Michel, *The Shadow War,* pp. 299–300.

27. Stalin worked hard at undermining the legitimacy and authority of the Polish Government in Exile. By 1943, as a part of Stalin's anti-Polish campaign, he cut off official ties to the Polish Government in London. This happened because of the supposedly "false" accusation about the Katyn affair, the murder of thousands of Polish officers by the Russians. For a discussion of the Katyń affair, see Norman Davies, *Heart of Europe: A Short History of Europe* (Oxford: Clarendon Press, 1984), pp. 67–68; Karski, *The Great Powers and Poland, 1919–1945,* pp. 424–

429, 432–435; J. K. Zawodny, *Death in the Forest, The Story of the Katyń Forest Massacre* (New York: Hippocrene Books, 1988).

28. Most likely the Ukrainians in the partisan movement in this area were former Nazi collaborators; Hersh Smolar, Personal Interview; Josef Marchwinski, Yad Vashem Testimony, thinks very highly of Miłaszewski; Jacov Greenstein, Personal Interview, feels that it was to the advantage of the Russian partisan movement to tolerate the Polish group of partisans. In fact, until the fall of 1943, the Poles and the Russians cooperated closely in their opposition to the Germans.

29. Tuvia Bielski, Personal Interview, Brooklyn, New York, 1987.

30. Ibid.

31. Referred to as the "Unternehmen 'Hermann'" (Hermann Undertaking), the document, in the form of a directive, states that the average partisan group consists of 200 individuals, but some are as big as 1000. Yad Vashem Document JM/10394, ff 648.

32. Memories diverge, depending on who does the telling, and such variations reflect real but different experiences. In the end, although coming from a variety of directions, the pieces to the puzzle tend to fall into place. Hersh Smolar, Personal Interview, thinks that at least two divisions of German soldiers, from the front, were committed to the attack. A division consisted of 10,000 soldiers. Nazi documents list different brigades, special units, comando groups, and others, without, however, offering specific figures. Yad Vashem Document JM/10394, ff 650.

33. Tuvia Bielski in *Forest Jews.*

34. Ibid.

35. Jacov Greenstein, Personal Interview.

36. Nazi documents state that the German forces intended to encircle and destroy all those inside the forest. Yad Vashem Document JM/10394, ff 694–695.

37. *The Jewish Partisan Book,* p. 446.

38. Chaim Basist, Personal Interview, Tel Aviv, Israel, 1988; Moshe Bairach, Personal Interview, also talks about a bridge.

39. Lazar Malbin, Personal Interview. Chaja Bielski, Personal Interview, also denies that such a bridge was built.

40. This is suggested by Chaja Bielski, Personal Interview; Tuvia Bielski in *Forest Jews;* Eljezer Engelstern, Yad Vashem Testimony.

41. Tuvia Bielski in *Forest Jews.*

42. Ibid.

43. Moshe Bairach, Personal Interview.

44. Tuvia Bielski in *Forest Jews.*

45. Ibid.

46. Moshe Bairach, Personal Interview; Chaja Bielski, Personal Interview; Eljezer Engelstern, Yad Vashem Testimony.

47. Chaim Basist, Personal Interview. Moshe Bairach, Personal Interview, echoes Basist's comment when he says, "Tuvia ordered us not to go on the bridge, but in a different direction. There is talk that the bridge is too dangerous."

48. Moshe Bairach, Personal Interview.

49. Tuvia Bielski in *Forest Jews.*

50. Ibid.

51. Ibid.

52. German conversation has been reported by many of the Bielski partisans. Among them were Pesia Bairach, Personal Interview, Tel Aviv, Israel, 1990; Raja Kaplinski, Personal Interview; Lazar Malbin, Personal Interview; and others.

53. Chaja Bielski, Personal Interview.

54. Ibid.

55. Moshe Bairach, Personal Interview.

56. Ibid.; Chaja Bielski, Personal Interview; Tuvia Bielski in *Forest Jews;* Lazar Malbin, Personal Interview.

57. At the time Cila and Luba Dworecki were teenagers. Luba Dworecki, Personal Interview, Tel Aviv, Israel, 1988.

58. Herzl Nachumowski, Personal Interview, Tel Aviv, Israel, 1987.

59. Tuvia Bielski in *Forest Jews.*

60. Ibid.

61. Moshe Bairach, Personal Interview.

62. Tuvia Bielski in *Forest Jews.*

63. When it was over, about 20,000 natives were deported to Germany for slave labor. See *Jewish Partisan Book,* p. 455. The attack on the Nalibocka forest lasted less than two weeks. In one of their documents the Germans order their forces to return to their previous positions by August 10, 1943 (Yad Vashem Document JM/10394). In addition to destroyed property, many people were killed. Some otriads incurred heavy losses, among them the Polish unit Kościuszko and the Jewish otriad headed by Zorin. The Bielski otriad lost only one man who drowned while trying to cross the river Niemen. According to Hersh Smolar, the Germans wanted to achieve yet another objective with this big manhunt. They parachuted spies into the Nalibocka forest during the raid. There was a special school in Berlin that trained such spies, its aim was to destroy the Russian partisans from within. After the war Smolar came across one such spy in prison, Rokosovky. In the Nalibocka forest Smolar knew him as a good Soviet partisan officer. Hersh Smolar, Personal Interview.

64. Zus Bielski, Personal Interview, Brooklyn, New York, 1989.

65. Tuvia Bielski in *Forest Jews.*

66. Cwi Isler, Yad Vashem Testimony, No. 1706/113.

67. This incident was reported by Abraham Viner in *Forest Jews.*

68. Tuvia Bielski in *Forest Jews.*

69. Moshe Bairach, Personal Interview; Chaja Bielski, Personal Interview; Shmuel Geler, Yad Vashem Testimony, No. 1556/112.

70. Tuvia Bielski, in *Forest Jews.*

71. Pesia Bairach, Personal Interview.

72. The death of one person by drowning is reconfirmed by many, e.g., *Forest Jews,* Chaja Bielski, Shmuel Geler, Raja Kaplinski, and others. This is quite extraordinary considering that most detachments incurred heavy losses. The name of the drowned person varies, sometimes it is Gwenofelski at other times Gornofelski.

73. Luba Dworecki, Personal Interview.

74. Chaja Bielski, Personal Interview.

75. Eljezer Engelstern, Yad Vashem Testimony.

76. The incident with Kaplan reappears in different sources: Sonia Bielski, Personal Interview, Brooklyn, New York, 1989; Zus Bielski in *Forest Jews; The Jewish Partisan Book,* p. 448. Moshe Bairach, Personal Interview, said that "Even in the 1960s, when I met Zus, half drunk, he told me that this shooting did not let him rest."

Chapter 10

1. Tuvia Bielski in *Yehudei Yaar* (*Forest Jews*) (Tel Aviv: Am Oved, 1946).
2. Ibid.
3. Cwi Isler, Yad Vashem Testimony, No. 1706/113.
4. Zorach Arluk, Personal Interview, Tel Aviv, Israel, 1988.
5. A German document specifies the destruction of the communities and the removal from the area of all natives. This document is dated August 1, 1943, signed by Von Gotterberg (SS Gruppenführer und Generalleutenant der Polizei), Yad Vashem Document JM/10394. After the big hunt, the Nalibocka forest contained a variety of family camps, even a Gypsy camp. Tuvia Bielski in *Forest Jews;* Moshe Bairach, Personal Interview, Tel Aviv, Israel, 1988–1989; Hersh Smolar, Personal Interview, Tel Aviv, Israel, 1989–1990; Abraham Viner, Personal Interview, Haifa, Israel, 1990. According to Soviet sources, of the 374,000 partisans in Belorussia in 1944 91,000 were in family camps. See *Encyclopedia of the Holocaust* (London: Macmillan Publishing Co., 1990), Vol. 4, p. 1113.
6. Tamara Rabinowicz, Personal Interview, Haifa, Israel, 1990; Chaja Bielski, Personal Interview, Haifa, Israel, 1987–1991; Eljezer Engelstern, Yad Vashem Testimony, No. 3249/233.
7. Eljezer Engelstern, Yad Vashem Testimony; Raja Kaplinski, Personal Interview, Tel Aviv, Israel, 1987–1989.
8. Shmuel Geler, Yad Vashem Testimony, No.1556/112.
9. Chaja Bielski, Personal Interview; Eljezer Engelstern, Yad Vashem Testimony; Shmuel Geler, Yad Vashem Testimony; Lazar Malbin, Personal Interview, Tel Aviv, Israel, 1987–1988; Abraham Viner, Personal Interview.
10. Perale S. Szlossberg, "Zichrojnes Un Iberlebungen," in *Arajateinu Nalibok: Hajeja Ve Churbana* ("Memories and Experiences" in *Our Town Naliboki: Its Life and Destruction*) (Tel Aviv: Irgun Yozei Nalibok, 1967), p. 124.
11. A slight discrepancy about the distance between the two camps appears with Chaja Bielski, Personal Interview, saying that it was a kilometer and a half while Shmuel Geler, Yad Vashem Testimony, says it was three kilometers.
12. Shmuel Amarant, "The Tuvia Bielski Partisan Company," a chapter in *Nvo Shel Adam* (Expressions of a Man) (Jerusalem: Misrad Hahinuch v Tarbut, 1973) translated by R. Goodman; Chaja Bielski, Personal Interview; Raja Kaplinski, Personal Interview.
13. Amarant, "The Tuvia Bielski Partisan Company."
14. Luba Garfunk, Personal Interview, Tel Aviv, Israel, 1989.
15. Amarant, "The Tuvia Bielski Partisan Company"; Chaja Bielski, Personal Interview; Sulia Rubin, Personal Interview, Saddle River, New Jersey, 1988.
16. Amarant, "The Tuvia Bielski Partisan Company"; Tamara Rabinowicz, Personal Interview, Haifa, Israel, 1990; Rosalia Gierszonowski-Wodakow, Personal Interview, New York, 1988.
17. Jashke Mazowi, Personal Interview, Tel Aviv, Israel, 1989.
18. Ibid.
19. Ibid.
20. The book, *Forest Jews,* is filled with examples of Jews who were partisans in Soviet detachments and who found shelter in the Bielski otriad. Also, *Sefer Ha-*

partisanim Hajehudim (*The Jewish Partisan Book*) (Merchavia: Sifriat Poalim, Hashomer Hatzair, 1948) is punctuated with many such examples.

21. Zorach Arluk, Personal Interview.

22. Ibid.

23. Chaja Bielski, Personal Interview, says that Asael stayed in the Kirov brigade for a few weeks. He was there without Chaja. Because of safety he wanted her to stay in the Bielski otriad. Tuvia Bielski in *Forest Jews;* Lazar Malbin, Personal Interview.

24. Amarant, "The Tuvia Bielski Partisan Book." Dr. Amarant speaks with a great deal of affection about Asael Bielski as do many others. For examples see "Our Town Naliboki: Its Life and Destruction," *Be Geto Nowogrodek* (*Ghetto Nowogródek*) (Tel Aviv: Irgun Iozeh Nowogródek, 1988), pp. 333–348.

25. Esia Lewin-Shor, Personal Interview, Bronx, New York, 1989.

26. Chaja Bielski, Personal Interview.

27. Tamara Rabinowicz, Personal Interview.

28. Raja Kaplinski, Personal Interview.

29. Max Weber, *The Theory of Social and Economic Organization* (Glencoe, Ill.: The Free Press, 1947), p. 362.

30. Chaim Basist, Personal Interview, Tel Aviv, Israel, 1988.

31. Chaja Bielski, Personal Interview; Raja Kaplinski, Personal Interview; Hersh Smolar, Personal Interview; Sulia Rubin, Personal Interview. All those mentioned concur in their high esteem of Tuvia Bielski. Smolar, in particular, is amazed at Tuvia's ability to clarify and simplify matters.

32. For a discussion of the different sources of power in the Russian partisan movement see John A. Armstrong and Kurt DeWitt, "Organization and Control of the Partisan Movement," in John A. Armstrong (ed.), *Soviet Partisans in World War II* (Madison: The University of Wisconsin Press, 1964), pp. 73–139.

33. All agree that Shematovietz was a fine person, free of prejudices. They also agree that for the Bielski otriad he was an important asset. A few examples of people who feel this way are Amarant, "The Tuvia Bielski Partisan Company"; Tamara Rabinowicz, Personal Interview; Chaja Bielski, Personal Interview; Lazar Malbin, Personal Interview; Hersh Smolar, Personal Interview; Abraham Viner, Personal Interview. Grisha, the Russian partisan and commissar of the Bielski otriad, was transferred to Ordzonikidze. Later he fell in battle.

34. Max Weber, *The Theory of Social and Economic Organization,* pp. 361–362.

35. Raja Kaplinski, Personal Interview, as secretary of the otriad, was aware of Malbin's bureaucratic skills and emphasizes his love for new bureaucratic procedures. Two more partisans praise him in particular for his organizational skills: Daniel Ostaszynski, Personal Interview, Tel Aviv, Israel, 1989, and Riva Reich, Personal Interview, Tel Aviv, Israel, 1989.

36. Pesach Friedberg, Yad Vashem Testimony, No. 3332/320.

37. Especially during the winter months, hungry wolves would come close to the camp. Amarant, "The Tuvia Bielski Partisan Company"; Raja Kaplinski, Personal Interview.

38. The Bielski partisans were aware of the many discipline problems Tuvia had to face and cope with. This example was offered by Raja Kaplinski, Personal Interview.

39. Herzl Nachumowski, Personal Interview, Tel Aviv, Israel, 1987.

40. These are of course rough estimates. They were offered by Tamara Ra-

binowicz, Personal Interview; Chaja Bielski, Personal Interview; Pinchas Boldo, Personal Interview, Haifa, Israel, 1990; Raja Kaplinski, Personal Interview.

41. In most communities the first victims were prominent Jewish men. For example, in Lida, on July 5, 1941, the Germans collected 200 men, all members of the Jewish elite, and murdered them. "Lida" in *Encyclopedia Judaica,* Vol. 11, pp. 212–213; in Nowogródek, too, the first victims were leaders of the Jewish community, "Novogrudok" in *Encyclopedia Judaica,* Vol. 12, pp. 1237–1238. Philip Friedman notes that many Jewish leaders were deported to Russia, others escaped as the Germans were occupying the region. See Philip Friedman, "Jewish Resistance to Nazism," pp. 387–408, in Ada J. Friedman (ed.), *Roads to Extinction: Essays on the Holocaust* (Philadelphia: The Jewish Publication Society, 1980). Kahanowicz writes that the initial massive killings of Jewish leaders created a leadership gap. See Moshe Kahanowicz, "Why No Separate Jewish Partisan Movement Was Established During World War II," pp. 25–40, in Isaak Kowalski (ed.), *Anthology of Armed Resistance to the Nazis, 1939–1945* (3 Volumes) (New York: Jewish Combatants Publishing House, 1986).

42. Over seventy percent of the Jewish population lived in urban centers. For urban, rural, and occupational breakdowns of the Jewish population in prewar Poland, see Jacob Lestchińsky, "The Jews in the City of the Republic of Poland," *YIVO Annual of Jewish Social Science 1* (1946), pp. 156–177; Jacob Lestchińsky, "The Industrial and Social Structure of the Jewish Population of Interbellum Poland," *YIVO Annual of Jewish Social Science* (1956–1957), vol. 2, p. 246; L. Lifschutz, "Selected Documents Pertaining to Jewish Life in Poland, 1919–1938," in *Studies on Polish Jewry* (ed.), Joshua A. Fishman (New York: YIVO Institute for Jewish Research, 1974), p. 280; Antony Polonsky, *Politics in Independent Poland, 1921–1939* (Oxford: Clarendon Press, 1972), p. 40.

43. These distinctions were noted by most Bielski partisans. A few examples are Shmuel Geler, Yad Vashem Testimony; Raja Kaplinski, Personal Interview; Riva Reich, Personal Interview; Cwi Isler, Yad Vashem Testimony, was very sensitive to these conditions and tried to participate in activities associated with the working class.

44. Chapter 12 concentrates on the situation of women.

45. Cwi Isler, Yad Vashem Testimony.

46. Tamara Rabinowicz, Personal Interview.

47. Luba Garfunk, Personal Interview; Cila Sawicki, Personal Interview, Tel Aviv, Israel, 1989.

48. Sulia Rubin, Personal Interview.

49. Shmuel Geler, Yad Vashem Testimony.

50. Chaja Bielski, Personal Interview; Pinchas Boldo, Personal Interview; Abraham Viner, Personal Interview; Lili Krawitz, Personal Interview, Tel Aviv, Israel, 1989.

51. Lilka Bielski, Personal Interview, Brooklyn, New York, 1989.

52. Sulia Rubin, Personal Interview.

53. This ranking order was suggested by Tamara Rabinowicz, Personal Interview; Chaja Bielski, Personal Interview, Pinchas Boldo, Personal Interview; Baruch Kopold, Personal Interview, Haifa, Israel, 1990, and anyone else I spoke to. Disagreements appeared when it came to the proportion of people who fit into a particular segment. For example, while the *Jewish Partisan Book,* p. 441, states that the malbushim made up 67% of the Bielski partisans, Chaja Bielski, Personal Interview, estimates that 50% of the otriad belonged to this category. Other esti-

mates go as high as 80% and as low as 20%. While the percentage of a special category of people varied with time, by and large the malbushim, the lowest ranking members, made up the majority of the otriad.

Chapter 11

1. Zorach Arluk, Personal Interview, Tel Aviv, Israel, 1988.
2. Ibid.
3. Hanan Lefkowitz, Personal Interview, Tel Aviv, Israel, 1988.
4. Ibid.
5. Chaja Bielski, Personal Interview, Haifa, Israel, 1987–1991, was opposed to this arrangement.
6. Hersh Smolar, Personal Interview, Tel Aviv, Israel, 1989–1990; Raja Kaplinski, Personal Interview, Tel Aviv, Israel, 1987–1989, admits that as a secretary of headquarters she had not tasted the food from the general kitchen.
7. Riva Kagonowicz-Bernstein, Personal Interview, New York, 1988. Before coming to the Bielski otriad she stayed a few weeks in the Zorin camp where she emphasized that everybody ate the same food, including the commander. This, however, has been denied by Hersh Smolar, Personal Interview, who was very close to Zorin and his otriad. Smolar says that in Zorin's camp those who went on food expeditions had more and better food. Lili Krawitz, Personal Interview, Tel Aviv, Israel, 1989, a Kesler supporter, claims that in his group too everybody ate the same food.
8. Pesia Bairach, Personal Interview, Tel Aviv, Israel, 1990.
9. Riva Reich worked as a nurse and has a great deal of admiration for Tuvia Bielski. Personal Interview, Tel Aviv, Israel, 1989.
10. Sulia Rubin, Personal Interview, Saddle River, New Jersey, 1988.
11. Raja Kaplinski, Personal Interview.
12. Shmuel Geler, Yad Vashem Testimony, No. 1556/112. Pesia Bairach in "Zmanim Modernim" (Modern Times) pp. 6–7, *Idiot Achronot,* December 21, 1988, refers to the kosher meat. She is probably talking about the very last phase shortly before liberation. Even then, it was not universally the case.
13. Tamara Rabinowicz, Personal Interview, Haifa, Israel, 1990.
14. Shmuel Geler, Yad Vashem Testimony.
15. Among those who say that they were hungry are Riva Kagonowicz-Bernstein, Personal Interview; Hana Stołowicki, Yad Vashem Testimony, No. 3439/191; Cila Sawicki, Personal Interview, Tel Aviv, Israel, 1989; Hanan Lefkowitz, Personal Interview, also told me that when he visited the otriad people complained to him that they did not have enough food. Still it is true that in the Bielski otriad no one ever died of hunger.
16. Cila Sawicki, Personal Interview.
17. Shmuel Geler, Yad Vashem Testimony.
18. Luba Garfunk, Personal Interview, Tel Aviv, Israel, 1989.
19. Riva Reich, Personal Interview.
20. Tamara Rabinowicz, Personal Interview.
21. Shmuel Amarant, "The Tuvia Bielski Partisan Company," a chapter in *Nvo Shel Adam* (*Expressions of a Man*) (Jerusalem: 1973), translated by R. Goodman.
22. Ibid.

23. Cila Sawicki, Personal Interview.
24. Hana Stołowicki, Yad Vashem Testimony.
25. Luba Garfunk, Personal Interview. Baran was a legendary figure, fearless, brave, not very intelligent. Pesia Bairach, Personal Interview, talks about him in the same way. See also *Sefer Hapartisanim Hajehudim* (*The Jewish Partisan Book*) (Merchavia: Sifriat Poalim, Hashomer Hatzair, 1958), Vol. 1, p. 402.
26. Riva Kaganowicz-Bernstein, Personal Interview; Cila Sawicki, Personal Interview; Rosalia Gierszonowski-Wodakow, Personal Interview, New York, 1988.
27. As a rule, women were not included in food expeditions. One exception was Chaja Bielski because Asael would occasionally take her with him.
28. Shmuel Geler, Yad Vashem Testimony.
29. Cila Sawicki, Personal Interview.
30. Riva Kagonowicz-Bernstein, Personal Interview.
31. Ibid.
32. Pinchas Boldo, Personal Interview, Haifa, Israel, 1990; Shmuel Geler, Yad Vashem Testimony.
33. Cila Sawicki, Personal Interview.
34. Shmuel Geler, Yad Vashem Testimony.
35. Ibid. It was unusual to steal milk because quite independently others have mentioned this incident. Raja Kaplinski, Personal Interview, was one of them. Without divulging the person's name, she referred to it as an unusual case of theft.
36. Lazar Malbin, Personal Interview, Tel Aviv, Israel, 1987–1988, emphasizes the importance of the airports; for a discussion of some related issues, see Nechama Tec, *In the Lion's Den: The Life of Oswald Rufeisen* (New York: Oxford University Press, 1990), p. 197; *The Jewish Partisan Book,* p. 450; Gerald L. Weinberg, "Airpower in Partisan Warfare," pp. 361–385, in John A. Armstrong (ed.), *Soviet Partisans in World War II* (Madison: The University of Wisconsin Press, 1964).
37. Samuel Geler, Yad Vashem Testimony.
38. Luba Rudnicki, Personal Interview, Tel Aviv, Israel, 1989. Indeed, the wounded partisan whom Luba Rudnicki took to the airport searched for her after the war. When he found her he thanked her for saving his life.
39. People from towns other than Nowogródek were especially irritated by these privileges. Among them were Cila Dworecki, Personal Interview, Tel Aviv, Israel, 1988; Luba Dworecki, Personal Interview, Tel Aviv, Israel, 1988; Luba Garfunk, Personal Interview.
40. Amarant, "The Tuvia Bielski Partisan Company."
41. Lazar Malbin, Personal Interview, tends to emphasize the gradual unpremeditated development of the workshops.
42. Eljezer Engelstern, Yad Vashem Testimony, No. 3249/233. In contrast to Malbin, Engelstern stresses Tuvia Bielski's determination and conscious effort to develop the workshops.
43. Chaja Bielski, Personal Interview; Raja Kaplinski, Personal Interview; Lazar Malbin, Personal Interview.
44. Amarant, "The Tuvia Bielski Partisan Company."
45. Chaja Bielski, Personal Interview; some people say that the parts Oppenheim made were more precise than the originals. Eljezer Engelstern, Yad Vashem Testimony.
46. Most agree that the tailors were a hard-working group who demanded

special payment for their services. Amarant, "The Tuvia Bielski Partisan Company"; Eljezer Engelstern, Yad Vashem Testimony; Shmuel Geler, Yad Vashem Testimony; Baruch Kopold, Personal Interview, Haifa, Israel, 1990.

47. Riva Reich, Personal Interview.

48. Hana Stołowicki, Yad Vashem Testimony.

49. Amarant, "The Tuvia Bielski Partisan Company."

50. Shmuel Geler, Yad Vashem Testimony.

51. Riva Reich, Personal Interview.

52. Amarant, "The Tuvia Bielski Partisan Company"; Eljezer Engelstern, Yad Vashem Testimony.

53. Amarant, "The Tuvia Bielski Partisan Company."

54. Shmuel Geler, Yad Vashem Testimony.

55. Amarant, "The Tuvia Bielski Partisan Company."

56. Hersh Smolar, Personal Interview.

57. Pinchas Boldo, Personal Interview.

58. Baruch Kopold, Personal Interview; Raja Kaplinski, Personal Interview.

59. Chaim Basist, Personal Interview, Tel Aviv, Israel, 1988.

60. Hersh Smolar, Personal Interview.

61. This incident was reported by different people in a variety of ways. Among those who spoke about the encounter were Chaja Bielski, Personal Interview; Lilka Bielski, Personal Interview, Brooklyn, New York, 1989; Lazar Malbin, Personal Interview; Hersh Smolar, Personal Interview. Malbin says that during General Platon's visit Tuvia was absent and he played the host. About the praying Jews Malbin says, "They were standing there and praying. The general had seen this for the first time. He asked: 'What is this?' So I answered: 'These are rabbis. They are too old to go for operations.'

"'So what do they do here?' he wanted to know.

"'They sit and pray,' I explained. 'They pray to God so that the Russian army would win the war.'"

"'What do they have on the arm and on the head?' Platon wanted to know. "I told him that this is something the Orthodox Jews are required to wear.'

"He asked: 'Do you wear this too?'

"'No, I do not. I am different.'

"He says: 'Since they cannot do any other work let them pray.'"

These variations do not change the basic fact that General Platon had paid a visit to the Bielski otriad and was impressed with the important work they were doing. From then on he protected Tuvia Bielski and through this the entire otriad.

62. Hersh Smolar, Personal Interview, was amazed at the calm with which Tuvia received the news about the threats against his life.

63. Yisrael Gutman and Shmuel Krakowski, *Unequal Victims: Poles and Jews During World War II* (New York: Holocaust Library, 1986), pp. 109–127; "National Democrats," in Wielka, *Encyclopedia Powszechna* (*The Great Popular Encyclopedia*) (Warszawa: Wydawnictwo Naukowe, 1964), p. 622.

64. Shmuel Geler, Yad Vashem Testimony.

65. Killing of Jews by Poles was reported independently in several sources. Cholawski writes that in 1943, Poles in the woods received orders to attack Jews and Russians. He specifically refers to an attack on a Jewish group of partisans by Polish partisans in the summer of 1943. See Shalom Cholawski, *Soldiers from the Ghetto* (New York: The Herzel Press, 1980), p. 162. Killings of Jewish partisans

by Polish partisans in the Nalibocka forest were also mentioned in Gutman and Krakowski, *Unequal Victims,* p. 131; Josef Marchwinski, Yad Vashem Testimony, No. 03/3568; Jashke Mazowi, Personal Interview, Tel Aviv, Israel, 1989.

66. Tadeusz Bor-Komorowski, *The Secret Army* (London: Victor Gollancz, Ltd. 1951), pp. 171–172.

67. Kazimierz Iranek-Osmecki, *He Who Saves One Life* (New York: Crown Publishers, Inc., 1955), pp. 260–262.

68. These killings were independently reported in several sources. Shmuel Geler, Yad Vashem Testimony; Jacov Greenstein, Personal Interview, Tel Aviv, Israel, 1984–1990; Josef Marchwinski, Yad Vashem Testimony; Hersh Smolar, Personal Interview. Also, see some discussion on these issues: Tec, *In the Lion's Den,* pp. 182–184.

69. Tuvia Bielski, Personal Interview, Brooklyn, New York, 1987.

70. Norman Davies, *God's Playground: A History of Poland* (*1795 to the Present*) (New York: Columbia University Press, 1984), Vol. II, pp. 471–475; Norman Davies, *Heart of Europe: A Short History of Poland* (Oxford: Clarendon Press, 1984), pp. 76–100; Jan Karski, *The Great Powers and Poland, 1919–1945, From Versailles to Yalta* (New York: University Press of America, Inc., 1985), pp. 488–489. The 1944 Warsaw uprising and the destruction that followed is a reflection of Stalin's determination to dominate Poland. For an excellent account of this historical chapter see Janusz Z. Zawodny, *Nothing But Honor: The Story of the Warsaw Uprising, 1944* (Stanford, California: Hoover Institute Press, 1979); To humiliate and destroy the leadership of the Polish underground, the Soviets would sometimes put Nazi criminals and AK (Home Army) leaders into the same prison cell. This happened to the arrested Polish leader Kazimierz Moczarski. He was placed in a cell with Juergen Stroop, the man in charge of the destruction of the Warsaw ghetto uprising in 1943. Moczarski wrote a book based on his imprisonment with Stroop. See Kaziemierz Moczarski, *Rozmowy Z Katem* (*Talks with a Hangman*) (Warszaawa: Państwowy Instytut Wydawniczy, 1978).

71. Jacov Greenstein, Personal Interview; see also Tec, *In the Lion's Den,* pp. 182–184.

72. Hersh Smolar, Personal Interview; the Soviet partisan authorities, in an effort to substitute for these Polish partisans, promoted the Polish communist Marchwinski; Josef Marchwinski, Yad Vashem Testimony.

73. Hersh Smolar, Personal Interview, talks about the effectiveness of the White Poles, Polish partisans loyal to the Polish Government in Exile in London. Their activities in parts of Western Belorussia are described in Janusz Prawdzic-Szlański, *Nowogródczyzna W Walce, 1940–1945* (*Fighting Nowogródek Region*) (London: Oficyna Poetów I Malarzy, 1976). This book contains copies of Soviet documents ordering the destruction of the Kościuszko detachment headed by Miłaszewski. See pp. 111–112.

74. These unanticipated positive consequences of the detachment Ordzonikidze are mentioned by quite a number of people. Some are Moshe Bairach, Personal Interview, Tel Aviv, Israel, 1988–1989; Sonia Bielski, Personal Interview, Brooklyn, New York, 1989; Zus Bielski, Personal Interview, Brooklyn, New York, 1989; Luba Rudnicki, Personal Interview.

Chapter 12

1. A partisan in the Bielski otriad, this woman asked me not to mention her name in connection with the rape. I interviewed her for about three hours.

2. Hersh Smolar, Personal Interview, Tel Aviv, Israel, 1989–1990.

3. Ibid.

4. Ibid.

5. Porter says that the literature about the partisan movement does not admit that women were treated in a sexist fashion. See Jack N. Porter, "Jewish Women in the Resistance," in Isaac Kowalski (ed.), *Anthology of Armed Resistance to the Nazis, 1939–1945* (New York: Jewish Combatants Publishing House, 1986), Vol. I, p. 292; Earl Ziemke, "Composition and Morale of the Partisan Movement," in John A. Armstrong (ed.), *Soviet Partisans in World War II* (Madison: The University of Wisconsin Press, 1964), p. 147;

6. Nechama Tec, *In the Lion's Den: The Life of Oswald Rufeisen* (New York: Oxford University Press, 1990), p. 195.

7. Ziemke, "Composition and Morale of the Partisan Movement," pp. 147–148.

8. Hersh Smolar, Personal Interview.

9. *Sefer Hapartisanim Hajehudim* (*The Jewish Partisan Book*) (Merchavia: Sifriat Poalim, Hashomer Hatzair, 1958), Vol. 1, p. 442; Fanny Salamia-Loc, *Woman Facing the Gallows* (Amherst, Massachusetts: Word Pro , Inc., 1981). First published in Hebrew in 1972, the book is a personal account of a Jewish partisan in a Soviet otriad and the discrimination she had to face as a woman.

10. Chaja Bielski, Personal Interview, Haifa, Israel, 1987–1991; Raja Kaplinski, Personal Interview, Tel Aviv, Israel, 1987–1989; Lili Krawitz, Personal Interview, Tel Aviv, Israel, 1989.

11. Hersh Smolar, Personal Interview; it is interesting that some women used these expressions as well. See Chapter 10.

12. Particularly for the last year of the war an estimated ten to twenty percent of the entire Soviet partisan movement were former Nazi collaborators. See Earl Ziemke, "Composition and Morale of the Partisan Movement," p. 147. Several Jewish partisans have reported stepped-up anti-Semitism in the Soviet otriads. A few examples are Zorach Arluk, Personal Interview, Tel Aviv, Israel, 1988; Jashke Mazowi, Personal Interview, Tel Aviv, Israel, 1989; Mordechai Ginsburg, Yad Vashem, No. 3682/270; Itzyk Mendelson, Yad Vashem Testimony, No. 3355/186;

13. A few such cases are described in Chapter 8.

14. Zorach Arluk kept his wife in Iskra, but had to send his wife's parents to the Bielski otriad.

15. Doctors in particular were in short supply, as were nurses, and would be welcomed into an otriad.

16. In deference to her wishes, when using this quote I am omitting the woman's name.

17. Sulia Rubin, Personal Interview, Saddle River, New Jersey, 1988.

18. Lili Krawitz, Personal Interview. The presence of many old people attests to this policy.

19. No clear-cut figures are available. These estimates were offered by several

partisans. Moshe Bairach, Personal Interview, Tel Aviv, Israel, 1988–1989; Raja Kaplinski, Personal Interview; Lazar Malbin, Personal Interview, Tel Aviv, Israel, 1987–1988.

20. Riva Reich, Personal Interview, Tel Aviv, Israel, 1989, emphasizes how helpful the extra food was. As I have noted in the previous chapter some women did guard duty and this gave them extra food.

21. A few examples of men who acquired guns after they came to the otriad are Bairach, Geler, Sawicki, Kopold. Herzel Nachumowski after he received a gun became an excellent fighter and scout.

22. Pesia Bairach, Personal Interview, Tel Aviv, Israel, 1990.

23. Raja Kaplinski, Personal Interview.

24. Moshe Bairach, Personal Interview; Shmuel Geler, Yad Vashem Testimony, No. 1556/112; Raja Kaplinski, Personal Interview.

25. Sulia Rubin, Personal Interview.

26. This view has been expressed by practically everyone. A few examples are Chaja Bielski, Personal Interview; Eljezer Engelstern, Yad Vashem Testimony, No. 3249/233; Shmuel Geler, Yad Vashem Testimony.

27. Chaja Bielski, Personal Interview; Raja Kaplinski, Personal Interview. Most people say that the "marriages" were stable and lasted a lifetime.

28. Shmuel Amarant, "The Tuvia Bielski Partisan Company," a chapter in *Nvo Shel Adam* (*Expressions of a Man*) (Jerusalem: Misrad Hahinuch v Tarbut, 1973), translated by R. Goodman; Chaja Bielski, Personal Interview.

29. Cila Sawicki, Personal Interview, Tel Aviv, Israel, 1989.

30. Sulia Rubin, Personal Interview.

31. Ibid.

32. Ibid.

33. Ibid.

34. Ibid. In a sense, the mother was right because Sulia continues to be married to the same man.

35. Ibid.

36. Ibid.

37. Ibid.

38. Pesia Bairach, Personal Interview.

39. Lili Krawitz, Personal Interview. She described many such marriages that had lasted a lifetime. She is still groping for explanations and answers.

40. Tamara Rabinowicz, Personal Interview, Haifa, Israel, 1990.

41. Ibid.

42. Raja Kaplinski, Personal Interview.

43. Lili Krawitz, Personal Interview, is particularly adamant when she talks about their sexual escapades. Toward the end, however, she says that, after all, men now sleep with their secretaries and women who are dependent on them.

44. Chaja Bielski, Personal Interview.

45. Pesia Bairach, Personal Interview.

46. Abraham Viner, Personal Interview, Haifa, Israel, 1990.

47. Pinchas Boldo, Personal Interview, Haifa, Israel, 1990.

48. Most Bielski partisans I spoke to were very much aware that Lilka suffered but overlooked Tuvia's infidelities.

49. Riva Reich, Personal Interview; Cila Sawicki, Personal Interview.

50. Pinchas Boldo, Personal Interview.

51. Lilka Bielski, Personal Interview, Brooklyn, New York, 1989.

52. The person who expressed these views asked for anonymity.
53. Tamara Rabinowicz, Personal Interview.
54. Lilka Bielski, Personal Interview.
55. Ibid.
56. Chaja Bielski, Personal Interview.
57. Practically all I interviewed or spoke to informally expressed this view.
58. Chaja Bielski, Personal Interview.
59. Pinchas Boldo, Personal Interview; Raja Kaplinski, Personal Interview; Lili Krawitz, Personal Interview; and practically everyone else.
60. Chaja Bielski, Personal Interview.
61. Sonia Bielski, Personal Interview, Brooklyn, New York, 1989; Sulia Rubin, Personal Interview.
62. Chaja Bielski, Personal Interview.
63. Pinchas Boldo, Personal Interview. Cila Sawicki, Personal Interview, says that Chaja was for the people like valium.
64. There are also differences in the reported figures. Shmuel Amarant, for example, says that there were "tens of children." Chaja Bielski and Raja Kaplinski give the figure of twenty. Others bring it up to thirty or even eighty; Chaim Basist, Personal Interview, Tel Aviv, Israel, 1988. No doubt the number of children fluctuated with time.
65. Jacques Bloch, "Jewish Child Care, Its Organisation and Problems" (A report presented at the Geneva Council of the International Save the Children Union) in When Winter Comes . . . Special Issue of the Information Bulletin of the OSE Union (Geneva, December 1946), p. 12; Z.H Wachsman, "The Rehabilitation of Jewish Children by the 'OSE' (New York: American Committee of OSE Inc., New York, 1947), p. 3. These and other sources appear in Debora Dwork, *Children with a Star: Jewish Youth in Nazi Europe* (New Haven: Yale University Press, 1991), p. 274, note 27.
66. Ibid. Dwork's book deals with the fate of children in general and the obstacles they encountered in staying alive. Special problems faced by children who wanted to live in the forbidden Christian world are also dealt with by Nechama Tec, *When Light Pierced the Darkness: Christian Rescue of Jews in Nazi-Occupied Poland* (New York: Oxford University Press, 1986), pp. 137–149.
67. Lili Krawitz, Personal Interview, tells how her husband carried an orphaned boy on his shoulder. Motl and Fruma Berger also took care of an orphan during the big hunt. Motl Berger, Personal Interview, Brooklyn, New York, 1989; Fruma Berger, Personal Interview, Brooklyn, New York, 1989.
68. Amarant, "The Tuvia Bielski Partisan Company."
69. Luba Garfunk, Personal Interview, Tel Aviv, Israel, 1989; Cila Sawicki, Personal Interview, had brought some sugar from the ghetto that she saved especially for this boy.
70. Amarant, "The Tuvia Bielski Partisan Company."
71. Ibid. Several other partisans have mentioned Czesia's school with pride. Among them was Chaja Bielski, Personal Interview; Shmuel Geler, Yad Vashem Testimony; Raja Kaplinski, Personal Interview.
72. Amarant, "The Tuvia Bielski Partisan Company."
73. Ibid.
74. Ibid.; Shmuel Geler, Personal Interview, Tel Aviv, Israel, 1989.
75. Amarant, "The Tuvia Bielski Partisan Company."
76. Shmuel Geler, Yad Vashem Testimony, says that two or three chil-

dren were born in the forest. Eljezer Engelstern, Yad Vashem Testimony.

77. Quite a number of women expressed these sentiments. Not to betray their trust, I am not including their names.

78. Luba Garfunk, Personal Interview.

79. When Arkie Lubczanski was banned from the otriad his "wife" was not included in this order. In fact, she voluntarily left with him but then returned because no Russian otriad would take her.

80. Riva Kaganowicz-Bernstein, Personal Interview, New York, 1988.

81. Riva Reich, Personal Interview.

82. Lili Krawitz, Personal Interview; Sulia Rubin, Personal Interview.

83. Ibid. Sulia Rubin had this story secondhand and the statement is based on one observation. Whether Sulia's observation applies to many more cases is not important. What matters are her perceptions of reality and not reality itself.

84. Ibid. Instead of balanced facts, Sulia's observations and comments reflect a negative attitude toward the opposite sex.

85. Two examples of the many women who share these attitudes are Ester Krynicki Gorodejski Berkowitz, "Sichrojnes Fun Der Deitscher Okupacje" ("Memoirs from the German Occupation"), pp. 587–602, in *Mir,* N. Blumenthal (ed.) (Jerusalem: Memorial Books, Encyclopedia of the Diaspora, 1962). Cila Kapelowicz reached the Bielski otriad after she escaped from the Mir ghetto and after the few relatives and friends with her were murdered by Russian partisans. She now lives in South Africa and I interviewed her when she was on a visit to Israel in 1987.

Chapter 13

1. Lili Krawitz, Personal Interview, Tel Aviv, Israel, 1989.

2. Chaja Bielski, Personal Interview, Haifa, Israel, 1987–1991.

3. Riva Reich, Personal Interview, Tel Aviv, Israel, 1989.

4. Other women active in the inspection of the otriad who spoke about it are Tamara Rabinowicz, Personal Interview, Haifa, Israel, 1990; Lili Krawitz, Personal Interview; Rosalia Gierszonowski-Wodakow, Personal Interview, New York, 1988 (she will be referred to as Wodakow).

5. Shmuel Amarant, "The Tuvia Bielski Partisan Company," a chapter in *Nvo Shel Adam* (*Expressions of a Man*) (Jerusalem: Misrad Hahinuch v Tarbut, 1973), translated by R. Goodman; Chaja Bielski, Personal Interview.

6. Daniel Ostaszynski, Personal Interview, Tel Aviv, Israel, 1989.

7. Lili Krawitz, Personal Interview.

8. Riva Reich, Personal Interview.

9. Amarant, "The Tuvia Bielski Partisan Company"; Eljezer Engelstern, Yad Vashem Testimony, No. 3249/233.

10. Chaim Basist, Personal Interview, Tel Aviv, Israel, 1988; Shmuel Geler, Yad Vashem Testimony, No. 1556/112.

11. Amarant, "Tuvia Bielski Partisan Company"; Tamara Rabinowicz, Personal Interview; Eljezer Engelstern, Yad Vashem Testimony, Riva Reich, Personal Interview.

12. Shmuel Geler, Yad Vashem Testimony.

13. Eljezer Engelstern, Yad Vashem Testimony.

14. Amarant, "The Tuvia Bielski Partisan Company."
15. Luba Garfunk, Personal Interview, Tel Aviv, Israel, 1989.
16. Riva Reich, Personal Interview.
17. Lili Krawitz, Personal Interview.
18. Amarant, "The Tuvia Bielski Partisan Company"; Tuvia Bielski in *Yehudei Yaar* (*Forest Jews*) (Tel Aviv: Am Oved, 1946).
19. Lili Krawitz, Personal Interview.
20. Pesia Bairach, Personal Interview, Tel Aviv, Israel, 1990.
21. Moshe Bairach, Personal Interview, Tel Aviv, Israel, 1988–1989.
22. Ibid.
23. There are differences in the estimated number of typhoid cases. Luba Garfunk, Personal Interview, says that 140 people had typhoid fever. Shmuel Geler, Yad Vashem Testimony, thinks that there were sixty cases. Dora Shubert, Personal Interview, Tel Aviv, Israel, 1989, estimates that there were ninety cases.
24. Amarant, "The Tuvia Bielski Partisan Company."
25. Tamara Rabinowicz, Personal Interview; Riva Reich, Personal Interview.
26. Luba Garfunk, Personal Interview; Shmuel Geler, Yad Vashem Testimony; Rosalia Wodakow, Personal Interview.
27. Hersh Smolar, Personal Interview, Tel Aviv, Israel, 1989–1990.
28. Jacov Greenstein, Personal Interview, Tel Aviv, Israel, 1984–1990.
29. Berl Chafetz, Personal Interview, Boston, Massachusetts, 1989.
30. Tuvia Bielski in *Forest Jews*.
31. Alter Titkin was killed during a food expedition. Chaja Bielski, Personal Interview, quotes Titkin.
32. Ibid.
33. Sonia Bielski, Personal Interview, Brooklyn, New York, 1989.
34. Chaja Bielski, Personal Interview; Abraham Viner, Personal Interview, Haifa, 1990.
35. Daniel Ostaszynski did not mention this incident. Nor did he speak about Sonia's protection. I heard about it independently from Sonia and Zus. Sonia Bielski, Personal Interview; Zus Bielski, Personal Interview, Brooklyn, New York, 1989.
36. Chaja Bielski, Personal Interview; Abraham Viner, Personal Interview.
37. Chaja Bielski, Personal Interview; Eljezer Engelstern, Yad Vashem Testimony.
38. For a description of this early rebellion see Chapter 7.
39. Raja Kaplinski, Personal Interview, Tel Aviv, Israel, 1987–1989.
40. Ibid.; Tuvia Bielski in *Forest Jews*.
41. Abraham Viner, Personal Interview.
42. Ibid.
43. Lili Krawitz, Personal Interview; Perale S. Szlossberg, "Zichrojnes Un Iberlebungen, in *Arajateinu Nalibok: Hajeja Ve Churbana* ("Memories and Experiences" in *Our Town Naliboki: Its Life and Destruction*) (Tel Aviv: Irgun Yozei Nalibok, 1987), p. 124; *The Jewish Partisan Book,* pp. 461–462.
44. Many were convinced that Kesler wanted Tuvia's position. Among them were Chaja Bielski, Personal Interview; Lazar Malbin, Personal Interview, Tel Aviv, Israel, 1987–1988; Raja Kaplinski, Personal Interview.
45. Tuvia was certain that Kesler wanted to take over his position. See Tuvia Bielski in *Forest Jews;* also, Tuvia Bielski, Personal Interview, Brooklyn, New York, 1987.

46. Pinchas Boldo, Personal Interview, Haifa, Israel, 1990, points out that only with time, after he witnessed the continuous destruction of small family groups, was he convinced by Tuvia's arguments. Many individuals who had reached the Bielski otriad alone were the sole survivors of such unprotected small groups. Among them were Ester Krynicki Gorodejski Berkowicz, "Sichrojnes Fun Der Deitscher Okupacje" (Memoirs from the German Occupation, pp. 587–602 in *Mir,* ed., N. Blumenthal (Jerusalem: Memorial Books, Encyclopedia of the Diaspora, 1962); Cila Kopelowicz, Personal Interview, Jerusalem, Israel, 1987; Rojak, Yad Vashem Testimony, No. 3597/240; Hana Stołowicki, Yad Vashem Testimony, No. 3439/191.

47. Tuvia Bielski in *Forest Jews.*

48. In the summer of 1942 the government defense committee of the USSR, headed by Stalin, started a drive aimed at collecting contributions from the workers' salaries. Later in Western Belorussia this request for voluntary contributions was changed into a mandatory contribution of gold. The new order threatened people with the death sentence. For a discussion of this order and its implications, see Nachum Alpert, *The Destruction of Słonim Jewry: The Story of the Jews of Słonim During the Holocaust* (New York: Holocaust Library, 1989), pp. 297–298; the difficulties in enforcing this rule are also discussed by Tuvia Bielski in *Forest Jews* and by Pinchas Boldo, Personal Interview.

49. Lazar Malbin, Personal Interview.

50. Moshe Bairach, Personal Interview.

51. Chaja Bielski, Personal Interview, Tel Aviv, Israel, 1988–1989.

52. Luba Rudnicki, Personal Interview, Tel Aviv, Israel, 1989. Aware of the otriad's need for funds, she and her husband gave Tuvia some money for the purchase of arms. She insists that no one was forced into giving. This was reiterated by Lilka Bielski, Personal Interview, Brooklyn, New York, 1989.

53. Luba Garfunk, Personal Interview.

54. Cila Sawicki, Personal Interview, Tel Aviv, Israel, 1989.

55. Rosalia Wodakow, Personal Interview.

56. This is reconfirmed by practically all the partisans from whom I have information.

57. Kesler sought and received support from General Sokolov who was in charge of the partisans in the Lida region. See *The Jewish Partisan Book,* p. 454; Tuvia Bielski in *Forest Jews.*

58. Sulia Rubin, Personal Interview, Saddle River, New Jersey, 1988.

59. Chaja Bielski, Personal Interview.

60. Eljezer Engelstern, Yad Vashem Testimony.

61. Tuvia Bielski in *Forest Jews.*

62. Chaja Bielski, Personal Interview; Shmuel Geler, Yad Vashem Testimony.

63. Riva Reich, Personal Interview.

64. Lili Krawitz, Personal Interview; Szlossberg, "Memories & Experiences."

65. Baruch Kopold, Personal Interview, Haifa, Israel, 1990.

66. Daniel Ostaszynski, Personal Interview.

67. Shmuel Geler, Yad Vashem Testimony; Tuvia Bielski in *Forest Jews.*

68. Abraham Viner, Personal Interview.

69. Hersh Smolar, Personal Interview.

70. Tuvia Bielski in *Forest Jews;* Tuvia Bielski, Personal Interview. Tuvia had the support of many of his people. The entire headquarters was behind him,

particularly Asael Bielski and Chief of Staff Lazar Malbin. Toward the end of Malbin's life, when I visited him in an old-age home, he was adamant about it, insisting that Kesler's death averted a disaster.

Chapter 14

1. Filip Friedman, "Zagłada Żydow Polskich W Latach 1939–1945" (The Destruction of Polish Jewery, 1939–1945) *Biuletyn Głównej Komisji Badania Zbrodni Niemieckiej W Polsce,* No. 6, 1946, pp. 165–206.

2. Chaim Basist, Personal Interview, Tel Aviv, Israel, 1988; these were his family, parents, and several children.

3. Just before the big hunt twenty-two partisans from the Orlanski otriad in the Lipiczańska forest joined the Bielski group. Also, with a group of other Jewish partisans, Izyk Mendelson, Yad Vashem Testimony, No. 3355/186, moved from a Russian otraid to the Bielski detachment.

4. The official name of the Bielski detachment was Kalinin, but no one used it. Particularly in the last phase, after the return to Nalibocka forest, the Bielski camp had been transformed into a "shtetl." These changes were discussed by Shmuel Amarant, "The Tuvia Bielski Partisan Company," a chapter in *Nvo Shel Adam* (*Expressions of a Man*) (Jerusalem: Misrad Hahinuch v Tarbut, 1973) translated by R. Goodman; Eljezer Engelstern, Yad Vashem Testimony, No. 3249/233; *Sefer Hapartisanim Hajehudim* (*The Jewish Partisan Book*) (Merchavia: Sifriat Poalim, Hashomer Hatzair, 1958), Vol. 1, pp. 454–460.

5. Seidel Kushner, Yad Vashem Testimony, No. 3276/280; "Novogrudok," *Encyclopedia Judaica* (Jerusalem: Keter Publishing House, Ltd., 1971), Vol. 12, pp. 1237–1238.

6. Other leaders as well as the existence of two underground groups are mentioned by Sholem Cholawski, *Yahadut Habelorussia Hadromit Memilhemet Hashnija* (*The Jews of Belorussia During World War II*) (Tel Aviv: Sifriat Hapoalim, 1982), p. 178. See also *The Jewish Partisan Book,* pp. 450–453.

7. Daniel Ostaszynski, Personal Interview, Tel Aviv, Israel, 1989, quoted this letter to me. In contrast, Eliachu Berkowitz, Yad Vashem Testimony, No. 2053/182, says that "A part of the otriad was for coming and liberating us and a part against it. There was a Christian otriad, even though anti-Semitic, that was willing to help Bielski free us. But there was a group of Jews that said 'They sit there and keep warm. Let them continue to stay there. We will not die for them.' The decision was not to come and take us out. We therefore decided to get organized." It is hard to say how he got this information.

8. Cholawski, *The Jews of Belorussia During World War II,* p. 178.

9. Eliachu Berkowitz, Yad Vashem Testimony.

10. Seidel Kushner, Yad Vashem Testimony; "Novogrudok," *Encyclopedia Judaica.*

11. The figures offered by Berkowitz, Cholawski and Ostaszynski range from 233 to 300. The number of survivors of the May 7, 1943, Aktion is given as 230 in "Novogrudok," in *Encyclopedia of the Holocaust* (London: Macmillan Publishing Co., 1990), Vol. 3, p. 1072.

12. Eliachu Berkowitz, Yad Vashem Testimony; "Organizacja Bojowa W Nowgródku: Ucieczka Żydow Przez Tunnel" ("Fighting Organization in Nowogródek: Jewish Escape Through the Tunnel) in Betti Ajzenstein (ed.), *Ruch*

Podziemny W Ghettach I Obozach: Materiały I Dokumenty (Warszawa: Centralna Żydowska Komisja Żydowska W Polsce, 1946), pp. 182–185.

13. Pnina Hirschprung, Personal Interview, Tel Aviv, Israel, 1988.

14. Eliachu Berkowitz, Yad Vashem Testimony.

15. Ibid.

16. Ibid. As a digger, Berkowitz was familiar with the project and describes it in great detail. Basically, he reconfirms what others have reported; Pnina Hirschprung, Personal Interview, saw how they measured the tunnel to check if it was straight. Daniel Ostaszynski, Personal Interview, as one of the leaders who helped organize the project, is familiar with the different problems. Rosalia Wodakow, Personal Interview, New York, 1988, participated in the building of the tunnel. She was particularly involved with the disposal of soil.

17. A teenager at the time, Riva Kaganowicz-Bernstein, Personal Interview, New York, 1988, though not directly involved in the building of the tunnel, remembers that she and most others knew that some escape was being prepared.

18. Eliachu Berkowitz, Yad Vashem Testimony, says that the voting took place on a Sunday and that out of 230 people 165 voted in favor of a breakout; Daniel Ostaszynski, Personal Interview, feels that just the awareness that a ghetto escape was being prepared lifted the spirits of the inmates.

19. There are variations in the reported length of the tunnel, ranging from 80 meters to 270. Cholowski, *The Jews of Belorussia During World War II,* mentions 200 meters; "Fighting Organization in Nowogrodek," gives a 270-meter length, while Ostaszynski, Personal Interview, estimates it to be 80 to 100 meters long.

20. Pnina Hirschprung, Personal Interview.

21. Eliachu Berkowitz, Yad Vashem Testimony, mentions this incident without, however, telling how it was resolved.

22. Ibid.; Riva Kaganowicz-Bernstein, Personal Interview; Pnina Hirschprung, Personal Interview. Riva and Pnina were not directly involved in the project. Each was impressed by the mutual care and helpfulness present during the actual exit and each emphasized the special concern shown by the organizers of the breakout.

23. Those who spoke about the fire admit that no one knew for sure what exactly happened. While all argue that some ghetto inmates were killed after the escape, there were variations in the number of survivors. For example, Shmuel Amarant, "The Tuvia Bielski Partisan Company," says that 150 runaways reached the Bielski otriad. This 150 figure is repeated by Eljezer Engelstern, Yad Vashem Testimony, and by Daniel Ostaszynski; in contrast, the article "Novogrudok," *Encyclopedia Judaica,* says that only 100 people survived the breakout.

24. Riva Kaganowicz-Bernstein, Personal Interview.

25. Seidel Kushner, Yad Vashem Testimony.

26. Ibid.

27. For the fate of some Christians in Kołdyczewo see Nechama Tec, *In the Lion's Den: The Life of Oswald Rufeisen* (New York: Oxford University Press, 1990), p. 99.

28. Szymon Berkowski, "Der Ojfshtandt In Lager Kołdyczewe" ("The Revolt in Kołdyczewo Camp"), pp. 4–5 in Joseph Foxman (ed.), Baranowicz In Umkum Un Viderstandt (*Baranowicz In Martyrdom And Resistance*) (New York: Baranowicher Farband of America, Inc., 1964); Joseph M. Foxman, "The Escape from Kołdyczewo Camp," pp. 172–175 in Yuri Suhl (ed.), *They Fought Back:*

The Story of The Jewish Resistance in Nazi Europe (New York: Schocken Books, 1975).

29. A. Sroganowicz, "Mir antlojfn Fun Baranowiczer Lager," Mir Antolojfn Fun Koldyczew Lager," p. 12 in *Baranowicze in Martyrdom and Resistance*. The original testimony of Sroganowicz is reprinted in Polish in Ajzenstein's *Underground Movements in Ghettoes and Camps,* pp. 178–180; *The Jewish Partisan Book,* p. 453.

30. Pinchas Markowski, "Mir Antlojfn Fun Kołdyczewer Lager Zum Di Partisaner" ("We Escaped from Kołdyczewo Camp to the Partisans"), pp. 3–4, in *Baranowicze in Martyrdom and Resistance.*

31. Berkowski, "The Revolt in Kołdyczewo Camp," pp. 4–5; Foxman, "The Escape from Kołdyczewo Camp," pp. 172–175; *The Jewish Partisan Book,* p. 453; Lea Estreich, Personal Interview, Queens, New York, 1992; Itzyk Estreich, Personal Interview, Queens, New York, 1992. Of the five sources cited, four mention that seventeen lost their way and were caught. Only the Jewish Partisan book states that twenty-six were caught. Itzyk Estreich, one of the leaders of this breakout, who seems very well informed, says that of the seventeen caught two saved themselves.

Berkowski and Foxman say that when the exodus ended with all the inmates on the other side, two men, Berkowski and Romek, went back. They went to shut the doors of the barracks and to mine them. The mines were reserved for the guards who would come to wake the Jews next morning. Indeed, as planned, there was an explosion. Ten Germans died and the barracks burned down.

When I asked Estreich about this he categorically denied that such an explosion took place. He says that once they left the place no one turned back.

32. Yehuda Szimszonowicz, Yad Vashem Testimony, No. 3312/296, was rejected by several Russian otriads. He eventually reached the Bielski camp where he survived the war. Several Bielski partisans emphasized that their otriad benefited from the arrival of the runaways from Kołdyczewo camp. Among them were Tamara Rabinowicz, Personal Interview, Haifa, Israel, 1990; Chaja Bielski, Personal Interview, Haifa, Israel, 1987–1991; Raja Kaplinski, Personal Interview, Tel Aviv, Israel, 1987–1989.

33. Amarant, "The Tuvia Bielski Partisan Company."

34. Tuvia Bielski in *Yehudei Yaar* (*Forest Jews*) (Tel Aviv: Am Oved, 1946).

35. Among those who were active in theatrical productions were Riva Kaganowich-Bernstein, Personal Interview; Sulia Rubin, Personal Interview, Saddle River, New Jersey, 1988; Rosalia Wodakow, Personal Interview.

36. Shmuel Geler, Yad Vashem Testimony, No. 1556/112. Przepiórka was the same man who during the big hunt asked Tuvia for his gun so that he could shoot himself. Tuvia reassured him that though ill he would be protected like everyone else.

37. Ibid.; Eljezer Engelstern, Yad Vashem Testimony; Raja Kaplinski, Personal Interview.

38. Tuvia Bielski in *Forest Jews.*

39. Shmuel Geler, Yad Vashem Testimony; the basic aim of the Zorin otriad was to save Jews. Like the Bielski detachment it had an open-door policy admitting every Jewish fugitive. Although the Bielski otriad was a purely Bielski creation, the Zorin detachment was from the start sponsored by the Soviets. It was smaller, not as well organized, and poorer than the Bielski detachment. For interesting descriptions of the Zorin group, officially known as the otriad 106, see

Yitzhak Arad, "Jewish Family Camps in the Forests," pp. 333–353 in *Rescue Attempts During the Holocaust,* Proceedings of the Second Yad Vashem International Historical Conference (Jerusalem: Yad Vashem, 1977); Jacov Greenstein, "An Officer and a Savior," pp. 228–230 in Lester Eckman and Chaim Lazar, *The Jewish Resistance: The History of the Jewish Partisans in Lithuania and White Russia During the Nazi Occupation: 1940–1945* (New York: Shengold Publishers, Inc., 1977); Hersh Smolar, *The Minsk Ghetto: Soviet-Jewish Partisans Against the Nazis* (New York: Holocaust Library, 1989), pp. 116–118; J. Tennenbaum, *Underground: The Story of a People* (New York: Philosophical Society, 1952), pp. 408–409. In these and other sources the stated size of the Zorin otriad ranges from 600 to 800.

40. Eljezer Engelstern, Yad Vashem Testimony.

41. Dr. Amarant's political lectures are mentioned in *The Jewish Partisan Book,* p. 444; and by Tuvia Bielski in *Forest Jews;* as the official historian of the Bielski otriad, Dr. Amarant had collected a great deal of material.

42. Tuvia Bielski in *Forest Jews* says that "There was no place for Zionism in our official educational system for the children. The school was established primarily to separate the children from the atmosphere of fighting. Of the two teachers one was a Zionist. There was no Hebrew education per se. . . . The first of May was celebrated as well as all the Jewish holidays." To the Soviets religious observances were more acceptable than Zionism.

43. Shmuel Geler, Yad Vashem Testimony; actually, most partisans deny that they discussed recent events, particularly as these referred to Jewish suffering. For example, Amarant, "The Tuvia Bielski Partisan Company"; Chaja Bielski, Personal Interview; and many others. Indeed, from personal experience, I know that for quite some time, after the war, there was a tacit agreement among survivors of the Holocaust not to mention their wartime experiences. See Nechama Tec, *Dry Tears: The Story of a Lost Childhood* (New York: Oxford University Press, 1984).

44. Martin Gilbert, *The Holocaust: A History of the Jews of Europe During the Second World War* (New York: Holt Rinehart and Winston, 1985), p. 694.

45. Amarant, "The Bielski Partisan Company."

46. Eljezer Engelstern, Yad Vashem Testimony.

47. Shmuel Geler, Yad Vashem Testimony.

48. Eljezer Engelstern, Yad Vashem Testimony.

49. This version is given by Tuvia Bielski in *Forest Jews;* and Shmuel Geler, Yad Vashem Testimony. There are other slightly different accounts. For example, Dr. Amarant, in "The Tuvia Bielski Partisan Company," writes that "Four prisoners of war were brought to headquarters. Three were young, the fourth was older and weeping bitterly. Two of the young ones claimed that they were communists, accusing the other two of being Nazis. Only one admitted his identity, abused and reviled the Jews, cursing and threatening. They were shot in front of headquarters, everyone witnessing the 'event.'" Eljezer Engelstern, Yad Vashem Testimony, says that the three SS men were lynched by the entire otriad.

50. Although the quote comes from Amarant, "The Tuvia Bielski Partisan Company," others support what he says. Among them was Chaja Bielski, Personal Interview; Raja Kaplinski, Personal Interview. Different sources offer different dates for this event. I have settled on July 9. This date was given by Rosalia Wodakow; for her, it was most memorable because her mother was one of the victims.

51. There are variations in the reported number of German attackers. Amarant, "The Tuvia Bielski Partisan Company," says that there were 100 Germans; Luba Garfunk, Personal Interview, Tel Aviv, Israel, 1989, reduces the

figure to thirty-nine. Eljezer Engelstern, Yad Vashem Tertimony, and Shmuel Geler, Yad Vashem Testimony, both give the figure of thirty; Riva Kaganowicz-Bernstein, Personal Interview, thinks that there were only twenty.

52. Similarly, the number of people lost that day varies with the particular source, ranging from seven to sixteen. Thus, for example, Chaim Bassist, Personal Interview, says that sixteen partisans lost their lives but Chaja Bielski, Personal Interview, puts the number at nine as does Tuvia in *Forest Jews*. Even though a few sources mention nine dead, rather than rely on consistent answers, I have selected the figure eleven because it was mentioned by Rosalia Wodakow, Personal Interview, who lost her mother that day.

53. Amarant, "The Tuvia Bielski Partisan Company."

54. Shmuel Geler, Yad Vashem Testimony; Babi Yar is a ravine outside of the city of Kiev. This is where at the end of September 1941 over 30,000 Jews were massacred. The murder was committed by the German Einsatzkommandos who were assisted by Ukrainian militiamen. See Gilbert, *The Holocaust,* p. 206.

55. Riva Reich, Personal Interview, Tel Aviv, Israel, 1989, talks about people's anxiety and marvels at the fact that they could return to the places that were earlier attacked by Germans.

56. Pinchas Boldo, Personal Interview, Haifa, Israel, 1990; Raja Kaplinski, Personal Interview.

57. Pinchas Boldo, Personal Interview.

58. Raja Kaplinski, Personal Interview, saw Tuvia as someone who felt very responsible for his people even at this last stage.

59. Amarant, "The Tuvia Bielski Partisan Company."

60. Among the majority that strictly obeyed this order was the brave fighter Herzl Nachumowski, Personal Interview, Tel Aviv, Israel, 1987. He approved of it.

61. Pinchas Boldo, Personal Interview.

62. Chaim Basist, Personal Interview.

63. Tuvia Bielski in *Forest Jews*.

64. All agree that in the end the Bielski otriad included more than 1200 individuals. More precise figures vary with the particular source. In *Forest Jews* the stated number is 1230. Tamara Rabinowicz says that there were 1234, Personal Interview; Abraham Viner, Personal Interview, Haifa, Israel, 1990, who worked as a clerk at the otriad's headquarters and had access to the records, says that at the time of liberation the otriad grew to 1291 people.

65. Some of those who see Polonecki in a very negative light are Moshe Bairach, Personal Interview, Tel Aviv, Israel, 1988–1989; Chaim Basist, Personal Interview; Raja Kaplinski, Personal Interview.

66. Chaim Basist, Personal Interview.

67. Moshe Bairach, Personal Interview.

68. Tamara Rabinowicz, Personal Interview; Shmuel Geler, Personal Interview, Tel Aviv, Israel, 1989.

69. Moshe Bairach, Personal Interview; Raja Kaplinski, Personal Interview.

70. Shmuel Geler, Yad Vashem Testimony.

71. Moshe Bairach, Personal Interview, says that he had heard it from others.

72. Most people tell the story secondhand. Among those who doubt that Tuvia was responsible for the shooting are Moshe Bairach, Personal Interview; Chaim Basist, Personal Interview; Eliachu Berkowitz, Yad Vashem Testimony; Herzel Nachumowski, Personal Interview; Lazar Malbin, Personal Interview, Tel Aviv, Israel, 1987–1988; Riva Reich, Personal Interview.

73. Tamara Rabinowicz, Personal Interview.

74. Pinchas Boldo, Personal Interview.

75. It seems that quite a few people tried to cope with the unexpected death of Polonecki by denying Tuvia's responsibility for it. Given the large size of the group, only a few could have witnessed the shooting. Of those who say that they did not see, some made up farfetched stories. One of the Bielski partisans, Chaim Basist, Personal Interview, says, "I was not at the spot. I heard all kinds of versions. They told that Szematovietz, the Russian commissar, shot him after Polonecki refused to remove his stuff. I did not see it. This is what I had heard. . . . As I look back, if I were Tuvia, I would have been much harder on the people. After all, what was at stake was the fate of more than 1000 people." Eliachu Berkowitz, Yad Vashem Testimony, places the responsibility for the shooting with the Russian general Dubov, who supposedly ordered Tuvia to shoot the man for disobedience. His conclusion is that Tuvia did what he was ordered to do.

76. Luba Garfunk, Personal Interview. This is essentially restated by many others.

77. Moshe Bairach, Personal Interview, quotes his friend Baruch Lewin.

78. Shmuel Geler, Personal Interview, Tel Aviv, Israel, 1989. When I interviewed Tuvia in 1987, I was not familiar with many of the historical facts. I asked him in a general way if he had had discipline problems and if he had to shoot anyone for disobedience. Tuvia talked about Kesler but did not mention Polonecki. Also the book *Forest Jews* fails to mention the incident with Polonecki.

79. Amarant, "The Tuvia Bielski Partisan Company."

80. Tuvia Bielski in *Forest Jews*.

81. Tamara Rabinowicz, Personal Interview; Raja Kaplinski, Personal Interview. Several Bielski partisans reported that the Soviet authorities invited Tuvia and Malbin to the city of Minsk. There they were asked to testify about their experiences in the forest. It has been assumed that the material became a part of the historical archives in Minsk. In an effort to gain access to this data I wrote to the (1) Academy of Science in Moscow, (2) Central State Historical Archives, Minsk, Belorussia, (3) Institute of History of CPB, Minsk, Belorussia, (4) Yuri Dubrinin, Russian Ambassador in Washington, D.C., and (5) Jack Matlock Jr., American Ambassador in Moscow. Not all my letters were answered. Those that were denied the existence of any archival material about the Bielski otriad or any other Jewish partisan group in Western Belorussia. During my recent contacts with Yad Vashem in Jerusalem, I was told that they hope to be more successful in obtaining some of this material. As of the beginning of 1993 this has not yet happened.

82. Pinchas Boldo, Personal Interview.

83. This place was mentioned by Chaja Bielski, Personal Interview. I assume that as a wife she knew where Asael died. Königsberg is mentioned in *Forest Jews* and other sources.

84. What happened to Tuvia Bielski fits well into Max Weber's theory about the charismatic leader. Weber says that "The charismatic lord has to prove his being sent 'by grace of God' by performing miracles and being successful in securing the good life for his followers or subjects. Only as long as he can do so will he be recognized. If success fails him, his authority falters." See Max Weber, *The Theory of Social and Economic Organization* (Glencoe, Illinois: The Free Press, 1947), pp. 359–369.

85. Pnina Hirschprung, Personal Interview, cried when she spoke about the fact that, after the war, the Bielski partisans did nothing for Tuvia. Others echo her

concern, among them Moshe Bairach, Personal Interview; Luba Dworecki, Personal Interview, Tel Aviv, Israel, 1988; Raja Kaplinski, Personal Interview; Riva Reich, Personal Interview. At the same time almost all of them said that they were not in a position to help, particularly right after the war.

86. Daniel Ostaszynski, Personal Interview.

87. Pinchas Boldo, Personal Interview.

88. Hersh Smolar, Personal Interview, Tel Aviv, Israel, 1989–1990.

89. Ibid.

90. Most of the Bielski partisans see Tuvia's move to America as a grave mistake that interfered with his career. Some of those who felt this way were Motl Berger, Personal Interview, Brooklyn, New York, 1989; Pinchas Boldo, Personal Interview; Luba Garfunk, Personal Interview; Raja Kaplinski, Personal Interview.

Chapter 15

1. *Sefer Hapartisanim Hajehudim* (*The Jewish Partisan Book*) (Merchavia: Sifriat Poalim, Hashomer Hatzair, 1958), Vol. 1, pp. 337, 346.

2. I describe the move of the Kesler group to the Bielski otriad in Chapter 6. The literature contains several accounts of Jewish partisans who were disarmed by Soviet partisans. For one example see Dov Levin, *Fighting Back: Lithuanian Jewry's Armed Resistance to the Nazis, 1941–1945* (New York: Holmes & Meier, Publishers, Inc., 1985), pp. 184–186.

3. The two Bielski partisans who offered the estimate of fifty dead were Chaja Bielski, Personal Interview, Haifa, Israel, 1987–1991, and Baruch Kopold, Personal Interview, Haifa, Israel, 1990.

4. When I asked Chaja to be more specific she sent me the following letter.

"Dear Nechama: You asked me to complete a very emotional and difficult task. This is why it took me such a long time to fulfill it. I found it extremely difficult to remember all the names of all the people who died in our partisan camp. 48 years went by trying to do justice is hard on my memory, as I said before. But I appreciate them very much because they gave their lives to bring a piece of bread for an old man, child and woman. They fought against the Nazis making ambushes on the roads and put explosives under the railway lines. They remain in my memory as holy men.

1/5/43 *Chrapiniewo:* Sonia Bielski; Regina Titkin; Grisha Meitis; Izchak Leibovicz; Elijahu Bakshit; Herzl Efroimski; Bernstein; Lova Wolkin; one more, I don't remember his name.

11/4/43 *Zabiełowo:* Lansman.

The end of January: Aba Wolfowicz.

3/43 (*by an ambush*): Motl Dworzecki; Josef Zelikowicz; Szumanski, the son; David Sztein.

4/43 in a hutor, near the village Dobre-Pole: Alter Titkin; Abraham Polonski; Rubin Polonski; Joshua Ostaszynski; Jehuda Kowalski; Leibel Gimpilowski; Leizer Chaitowicz; Szumanski (the second son); Israel, from Delatitz; one more.

Jasinowo: Miriam Cipilewicz; Mrs. Białabroda; Haim Bloch; a baby; and two guards.

Abraham Kalmonowicz (he hanged himself).

8/43 *Nalibocka Forest:* Kaplan; Gwenofelski, drowned.

9/43 *Żurawielnik:* Ida Bielski; Jacob Słucki; Abraham Mowszowicz; Otminova.
Nalibocka Forest: Israel Kesler; Mr Bialobroda; Josef Szmulowicz.
7/12/44 Nalibocka Forest: Elijachu Ostaszynski; Pacowski; Mark Epsztein; Zvi Leibowicz; Gordon; Sara Gierszonowski; Szumanski (the father); Luk; one more.
Elijahu Jewnowicz (was killed by the Nazis in a hutor).
Chana (from Nowojelnie).
Hirsh Feldman; Shmuel Lisman.
Polonecki.

Nechama, please look if the list is O.K. Maybe you'll have to ask other witnesses too." Note that this attack had happened on July 9, 1944. See Note 50 in chapter 14.

5. John A. Armstrong (ed.), *Soviet Partisans In World War II* (Madison: The University of Wisconsin Press, 1964), p. 151.

6. Professor Herbert Spirer, a statistician and friend, helped me with these calculations.

7. Jacov Greenstein, Personal Interview, Tel Aviv, Israel, 1984–1990.

8. "Russia," *Encyclopedia Judaica* (Jerusalem: Keter Publishing House Ltd., 1971), Vol. 15, p. 478.

9. Dov Levin, "Baltic Jewry's Armed Resistance to the Nazis," in Isaac Kowalski (ed.), *Anthology of Armed Resistance to the Nazis, 1939–1945,* 3 volumes (New York: Jewish Combatants Publishing House, 1986), Vol. 3, pp. 42–48.

10. Yitzhak Arad, *The Partisan from the Valley of Death to Mount Zion* (New York: Holocaust Library, 1979), pp. 115–134; for additional descriptions of the plight of different family camps in different areas see Yehuda Merin and Jack Nusan Porter, "Three Jewish Family-Camps in the Forests of Volyn, Ukraine During the Holocaust," *Jewish Social Science,* Vol. 156, No. 1 (1984), pp. 83–92.

11. Shmuel Krakowski, *The War of the Doomed: Jewish Armed Resistance in Poland, 1942–1944* (New York: Holmes & Meier Publishers, Inc., 1984), p. 59. This study, pp. 80–100, shows that south of Lublin the survival rate of Jews in the forest was much lower, in some cases non-existent. The precarious situation of Jewish partisans is also described by Harold Werner, *Fighting Back: A Memoir of Jewish Resistance* (New York: Columbia University Press, 1992), pp. 104, 141, 155–56.

12. Nechama Tec, *When Light Pierced the Darkness: Christian Rescue of Jews in Nazi-Occupied Poland* (New York: Oxford University Press, 1986), pp. 150–183. Only after I finished writing this book did I read the recently published book by Raul Hilberg, *Perpetrators, Victims, Bystanders: The Jewish Catastrophe, 1933–1945* (New York: HarperCollins, 1992). Particularly relevant are Hilberg's discussions of Survival and the determination of people to live. See pp. 186–191.

Biographical Appendix

This is a brief selective description of characters who appear and reappear in this book. I hope that these short remarks will help eliminate some of the confusion that might have been created by the many foreign names. In addition, by putting these names into their proper places I wish to clarify some of the historical events reflected in their activities. Moreover, by telling, albeit briefly, how the lives of these principal characters unfolded following their time in the forest, I might satisfy the reader's natural curiosity.

The individuals mentioned represent only a fraction of those whose life histories form the foundation for this book. To make the material more accessible I have arranged the names in alphabetical order.

Zorach Arluk, born in 1914 in Lida. Early on, during the Nazi occupation, he became involved in ghetto underground activities. He ran away armed from Lida ghetto, but is not sure if it was in December 1942 or January 1943. Zorach joined the Russian partisan detachment Iskra. He distinguished himself as an excellent fighter. Arluk married a girl whom he helped save from the ghetto and was allowed to keep her in his otriad.

After the war the couple settled in Israel where Zorach became a businessman. He is now retired. The Arluks had two daughters and a son; one of the daughters died. They have six grandchildren.

Moshe Bairach was born in 1918, in the Polish town of Pabianice. He escaped from the Germans and came to the Russian-occupied part of Poland. After many hardships, helped by Jewish strangers, he settled in the town Żołudek in Western Belorussia. During the 1942 liquidation of the Żołudek ghetto, he was among the small group of Jewish men spared. His papers offered protection to a wife. Although single, Moshe pretended to be married to Pesia Lewit. The two moved to Lida ghetto and from there, with a guide sent from the Bielski otriad, as a part of a large group of ghetto inmates, they reached the forest in the spring of 1943. In the otriad Moshe went frequently on food expeditions, did guard duty at the airports, and, in general, never declined any job.

Pesia and Moshe Bairach, were married and after the war moved to Israel. Moshe became the owner of a clothing department store. He is now retired. The couple has two children, and three grandchildren.

Pesia Lewit-Bairach was born in Żołudek in 1922. During the liquidation of the Żołudek ghetto Pesia hid in a bunker. With the help of a Belorussian policeman she contacted Moshe Bairach whose papers saved her (see above). Very energetic, always ready to help, as a partisan Pesia worked on all kinds of jobs. She learned sewing and worked as a seamstress.

In Israel, until very recently, Pesia was employed as a guide in the Israeli Museum of the Diaspora. Right now she is busy helping many of the new Russian immigrants.

Motl Berger was born in 1912 in Wsielub. From the start of the German occupation he had been trying to stay out of the enemy's reach. After several unsuccessful attempts to remain outside the ghetto, he left the Nowogródek ghetto in the summer of 1942. During this last escape Berger was a part of the first group of young men who were invited to come to the Bielski otriad. A special guide brought this group to the forest. As a member of the Bielski otriad Motl acquired a gun, became an active fighter, and very often went on food expeditions. In the forest he met Fruma Gulkowitz; they became a couple and were liberated by the Red Army.

After the war, Motl and Fruma came to the United States and settled in Brooklyn, New York. Motl worked as a printer; now retired, he continues to live in Brooklyn. Over the years both Motl and Fruma have been speaking and writing about the Holocaust. They have two sons and two grandchildren.

Riva Kaganowicz-Bernstein escaped from the Nowogródek ghetto through the tunnel in the fall of 1943. She was fourteen years old and an orphan. In the forest she joined two other fugitives, mother and daughter, Sarah and Rosalia Gierszonowski. The three ended up in the Bielski otriad where they lived together like a family. Although comforted by her adoptive mother, Riva felt lonely and out of place. She found her position as a malbush particularly humiliating. A day before liberation her adopted mother, Sarah Gierszonowski, became a victim of the last German attack.

After the war Riva was determined to get an education. She moved to the United States where she became a teacher. She married a commercial artist and lives in New York City. She continued to teach. She has two children and is expecting a grandchild.

Asael Bielski was born in 1908, in the village Stankiewicze. From the start of the German occupation he refused to submit to the Nazi terror, roaming the countryside, supporting himself by doing odd jobs. At the beginning of 1942, at the head of less than twenty Jewish fugitives, he formed his own partisan group.

This group was soon joined by his wife Chaja and her relatives, Asael's older brother, and other family members. This enlargement led to a reorganization of the group into a more formal partisan detachment, an otriad. Tuvia Bielski became its official commander. Asael became the second in command and was also in charge of the fighters. Zus Bielski became head of reconnaissance. The youngest brother, Arczyk, a teenager was also a member of this group. He would work as a scout.

An excellent, brave fighter, Asael was devoted to the Jewish people and was in full agreement with Tuvia's policy that aimed at saving as many Jews as possible.

In 1944 Asael was drafted into the Red Army. He died in battle, in Marienbad, Germany, leaving behind him his pregnant wife, Chaja.

Chaja Bielski, born in 1918, in a village close to town of Nowogródek. Chaja came from a prosperous family. Unlike Asael, she was well educated and deeply concerned about the plight of the underprivileged. She spent much of her life in the town of Nowogródek. From the start of the German occupation Chaja concentrated on eluding Nazi authorities. With members of her family she moved illegally into the forbidden Christian world where she was protected by Belorussian friends. In the early spring of 1942, with several members of her family, Chaja joined Asael's partisan group and officially became his wife.

During her stay in the forest Chaja would occasionally join her husband for food expeditions. Most of her time she devoted to helping people inside the camp. She was particularly concerned with the welfare of women and children.

Chaja remarried and settled in Israel. She has two daughters, one from each marriage: Asaela and Tova. A grandmother of five, Chaja continues to live in Israel where she devotes all her energies to helping the new Israeli immigrants from the Soviet Union.

Lilka Bielski was born in 1926 to a prosperous family, the Titkins, in the small town Mołodeczno. During the Soviet occupation, Lilka with her father, Alter Titkin, and his wife Regina, moved to Lida. In Lida, in 1941, she met and fell in love with Tuvia Bielski.

In 1942 Lilka joined Asael's partisan group in the forest. She was a part of a group that consisted of her father, Alter Titkin, his wife and her son, Grisha Meitis, her sister Sonia who was married to Tuvia Beilski, and Bielski himself.

After Sonia Bielski was killed during a German attack in Chrapiniewo at the start of 1943, Lilka became Tuvia Bielski's wife. A beautiful and retiring girl, in the forest Lilka remained in her husband's shadow.

In 1945 Lilka and Tuvia Bielski settled in Israel. There Lilka gave birth to two children, a daughter and a son. In 1956 the Bielskis moved to the United States. In Brooklyn, New York, Lilka gave birth to another son. In June 1987 Lilka became a widow. She has seven grandchildren, continues to live in the same house in Brooklyn, and pays frequent visits to her children and grandchildren.

Sonia Bielski, the *second wife of Tuvia Bielski,* was the sister of Regina Titkin. Regina Titkin was the stepmother of Lilka Titkin who became the third wife of Tuvia Bielski. Sonia Bielski was killed during a German attack, in January 1943, near Chrapiniewo.

Sonia Bielski, born Boldo, comes from a very prosperous family in Nowogródek. In the summer of 1942 Sonia ran away from the Nowogródek ghetto in a group led by a special guide sent by the Bielski otriad. In the forest she met Zus Bielski and became his wife. Through Zus she arranged for her parents' transfer to the forest. In 1943 Sonia moved with Zus to the newly established detachment Ordzonikidze. In 1945 Sonia moved to Israel with her husband. In 1956 they settled in the United States in Brooklyn, New York. A homemaker, Sonia has three sons and five grandchildren.

Tuvia Bielski was born in 1906 in the village Stankiewicze. After Tuvia married Rifka he moved to the small town of Subotniki. During the Soviet occupation he lived in Lida. In Lida Tuvia fell in love with Sonia, the sister of Regina Titkin. Tuvia divorced his wife, Rifka, and married Sonia. During the German occupation Tuvia refused to submit to the Nazi terror. He eluded the Germans by staying illegally in the countryside. Tuvia found a place for his wife, Sonia, with his friends, a Belorussian peasant family. In May 1942 Tuvia and Sonia were joined by Sonia's relatives, her sister Regina Titkin, Regina's husband, Alter Titkin, Regina's son, Grisha Meitis, and Regina's stepdaughter, Lilka Titkin. Tuvia and this group of relatives joined Asael's partisan unit in the forest.

Together in the summer of 1942 they formed a new partisan detachment with Tuvia as its commander. From the very beginning Tuvia was in favor of the policy of expansion. Not only would he accept all Jewish fugitives into their unit but he would also actively look for them. He sent guides into the ghettos to bring out Jews. He made the surrounding villages safer for escaping Jews by punishing informers.

Despite internal opposition by those who pointed to the dangers of large size, Tuvia never abandoned his policy of taking in all Jewish fugitives. On the contrary, the more determined the Germans were to annihilate the Jews, the more he insisted on saving them. With time Tuvia gained more converts to his determination to rescue Jews. Tuvia also managed to elude the Germans with a minimum of losses. Eventually he established a permanent community in the forest that became useful as a supplier of goods to the Russian partisan movement.

After the death of his second wife, Sonia, Tuvia married the young and beautiful Lilka Titkin. In 1945 Tuvia and Lilka settled in Israel. Tuvia joined the Israeli Army during the war of independence. In Israel Tuvia became a cab owner and cab driver. Disillusioned, in 1956 Tuvia and his family settled in the United States, in Brooklyn, New York. In the United States Tuvia drove a truck for his older brother who was a factory owner. Later he became an owner of one truck and then two trucks. In 1987 Tuvia died and a year later was re-buried in Israel in Jerusalem in Haar Hmnuchot.

Tuvia and Lilka had one daughter and two sons. Now there are seven grandchildren.

Zus Bielski, born in 1912 in Stankiewicze, moved to Nowogródek after his marriage to Cyrl. At the start of the German occupation Cyrl gave birth to a girl. Like his brothers, Zus refused to submit to the Nazis and stayed illegally in the countryside. Cyrl refused to join Zus. She and the baby were killed during an early ghetto Aktion.

When the Bielski brothers organized the otriad in 1942, Zus was appointed head of intelligence operations. Shortly after that he married Sonia Boldo, a new forest arrival from Nowogródek ghetto. In the fall of 1943 Zus and Sonia transferred to the new Ordzonikidze otriad.

In this unit Zus was also in charge of intelligence operations. Zus and Sonia settled in Israel in 1945. In 1948 Zus served in the Israeli Army. In 1956 the Bielskis moved to the United States where Zus owned a fleet of cabs. Zus and Sonia have three sons and five grandchildren. They continue to live in Brooklyn, New York.

Pinchas Boldo at seventeen, with members of his family, moved to the forest and became a partisan fighter. First he was in the Bielski otriad and later in the new Ordzonikidze unit. Pinchas' mother, Rachel Boldo, and Chaja Bielski were sisters. Quite a number of the Boldo family, including aunts and cousins, joined the Bielski otriad and survived the war.

After the war Pinchas became a physician and a radiology specialist. Dr. Pinchas Boldo lives in Israel. He is married, has two children, and seven grandchildren.

Cila Dworecki (Jopan) and her sister, *Luba Dworecki* (Ram), joined the Bielski otriad as young teenagers with their parents. They came to the forest at the personal invitation of Tuvia Bielski who through a letter urged their father to come. The family arrived at the forest led by a special guide sent by the Bielski otriad to the ghetto. The Dworecki family survived the war in the forest. They settled in Israel where each of the sisters married.

Cila, a homemaker, has two children and nine grandchildren. Luba, a secretary, has two children and three grandchildren.

Luba Garfunk, born in 1921 in Lida, ran away to the forest with a little boy and husband, during the liquidation of the ghetto, in the fall of 1943.

The three were accepted into the Bielski otriad. After the war they settled in Israel. Luba was a homemaker while her husband became the head of a publishing house. Now widowed, Luba has two children and seven grandchildren.

Jacov Greenstein as a teenager ran away from Nazi-occupied Poland and settled in Minsk. From the Minsk ghetto he moved to the forest where he became a partisan. With his present wife Bela, he had to look for a Russian unit that would accept her. After several refusals Jacov and Bela joined the Soviet detachment, Ponomarenko. Jacov distinguished himself as a fighter; both survived in the same otriad.

After the war the couple settled in Israel. Jacov became an owner of an automotive supply store. He has produced several publications and continues to write about different aspects of the war. He and his wife have two children and five grandchildren.

Before *Dr. Hirsh* came to the forest he worked in a hospital in Iwieniec. There the Germans had forced him to poison two Russian pilots. This matter was investigated in the Bielski otriad. Dr. Hirsh admitted that he had poisoned the pilots but he was acquitted because he was forced to do it. After the war the Soviet authorities accused him of the same crime. There was a trial in Smolensk and Dr. Hirsh was sentenced to twelve years in a Siberian prison. At the trial one of the witnesses was General Dubov. After that Dr. Hirsh lived in Novosibirsk.

Raja Kaplinski (Kaganowicz) born in 1922, in Nowogródek, reached the forest in the summer of 1942, in a group headed by a guide sent by the Bielski otriad. In the otriad she had many close friends, among them Sonia Bielski, the wife of Zus Bielski. Raja became the official secretary of the Bielski otriad, a position she held until the liberation.

After the war Raja settled in Israel where she met and married her present husband, an engineer.

Baruch Kopold, born in Iwje in 1923, became very active in the Jewish underground right after the German occupation. With other underground members he escaped during the liquidation of the Iwje ghetto in January 1943. At that time the Bielski otriad was under attack and some fugitives from Iwje ghetto returned to the ghetto, Baruch among them. He was moved to Lida ghetto and from there escaped to the forest during the liquidation of Lida ghetto in 1943. In the forest he joined the Bielski otriad and stayed there until the liberation.

After the war Baruch settled in Israel where he became a president of a bank. He has three children and six grandchildren.

Hanan Lefkowitz, born in 1925, in Traby, was forced by the Germans into Iwje ghetto. In the winter of 1942 he ran away and searched for partisans. After initial difficulties he was accepted into the Russian otriad, Stalin. Hanan distinguished himself as a fighter and was severely wounded while fighting the enemy.

After the war he settled in Israel and became an owner of an insurance company. A widower, Hanan has two children and four grandchildren.

Lazar Malbin was born in Stołpce in 1903. During the German occupation he lived with his wife and children in Nowogródek ghetto. When he lost his entire family during an Aktion in the summer of 1942, he escaped into the forest. There he met Tuvia Bielski who appointed him Chief of Staff of the Bielski otriad. He retained this position until the very end.

After the war Malbin moved to Israel. Although an engineer, he worked as an administrator. He never remarried and died in 1989.

Jashke Mazowi, born in 1920, ran away from ghetto Lida in 1942. After many hardships and close calls he succeeded in joining the Soviet partisan unit, Iskra. He distinguished himself as a fighter and, except for a short time when, for safety, Jashke moved to the Bielski otriad, he stayed in Iskra until the arrival of the Red Army in 1944. Jashke settled in Israel where he worked as an engineer. Retired, he is married, has two children and five grandchildren.

Herzl Nachumowski was born in 1920, in Iwje. From Iwje ghetto he ran away to the forest in the summer of 1942. He joined the Bielski otriad and distinguished himself as a fighter who went frequently on food expeditions.

After the war Herzl settled in Israel where he worked as a policeman. Now retired, he has two children and four grandchildren.

Daniel Ostaszynski was born in Nowogródek in 1921. During the German occupation he moved up from working for the ghetto administration to the position of Judenrat head. Some people viewed with suspicion his ghetto cooperation with the Germans. Others had defended him, pointing to his 1943 participation in the ghetto underground. Ostaszynski came to the forest in 1943 during the ghetto breakout through the tunnel. He stayed in the Bielski otriad until the liberation by the Red Army.

After the war he settled in Israel where he worked as an accountant. He married, has three children and six grandchildren.

Tamara Rabinowicz, though born in Stołpce, lived in Vilna. From the very beginning Tamara refused to work for the Germans and together with her husband was planning to escape. Forced into the Vilna ghetto the couple escaped to Lida. From there they succeeded in running away only during the liquidation of the ghetto, in 1943. In the forest they were accepted into the Bielski otriad.

After the war Tamara and her husband settled in Israel, where both worked as educators. Retired and widowed, Tamara has two children and five grandchildren.

Riva Reich was born in 1914 in Stołpce into the prosperous and educated family, Kantorowicz. Shortly after the Germans came they executed her husband. Later Riva lost her little daughter when the Belorussian peasant to whom she entrusted the child delivered her to the police.

From the start of the German occupation Riva concentrated on eluding the authorities. Eventually she succeeded in running away during the liquidation of the Stołpce ghetto, September 1942. While roaming the countryside, she was picked up by Bielski scouts who brought her to the camp in the forest. There Riva was very active working as a nurse and supervisor of the camp's hygiene.

After the war Riva married a lamp designer. They settled in Israel. Riva Reich lives in Israel where she continues to work as a private nurse, taking care of older people. Once more a widow, she has one daughter and two grandchildren.

Sulia Wołożhinski-Rubin was born in 1924 into an upper-class family in Nowogródek. She was one of those young people constantly and actively searching for a way out. After many hurdles, in response to an invitation, she ended up in the Bielski otriad, toward the end of 1942.

Sulia lives in New Jersey with her husband Boris, whom she married in the forest. He is a retired owner of a knitwear factory where she was employed as an officer of the company. They have two daughters and three grandsons.

Luba Rudnicki, young and just married, escaped from Nowogródek ghetto in the fall of 1942 to the countryside where together with her husband, Janek Rudnicki, and three more ghetto fugitives she was protected by Poles. The group was treacherously attacked by Russian partisans. Two were killed, and the rest ran away to the forest and joined the Bielski otriad. Later, the Rudnickis were transferred to the new otriad Ordzenikidze. A cousin of Tuvia Bielski's first wife, Rifka, Luba and Tuvia continued their friendship.

After the war the Rudnickis settled in Israel. There Luba worked as a secretary of a school, and her husband became a bank president. Now both are retired. They have two children and five grandchildren.

Cila Sawicki, born in 1922, in Lida, joined the Bielski otriad with her husband after the liquidation of the ghetto, in the fall of 1943. After the war the couple settled in Israel where Cila worked as a teacher. Her husband was employed by the Israeli Department of Defense. The couple has two children and four grandchildren.

Alter Titkin was the father of Lilka Bielski, wife of Tuvia Bielski. Alter Titkin was married to Regina, the sister of Sonia Bielski, the second wife of Tuvia Bielski. At his age Alter was not expected to participate in food missions, but he insisted on going. In the spring of 1943, at the age of forty-nine, Alter was killed during a food expedition. This group was denounced by a Belorussian peasant.

Regina Titkin, the wife of Alter Titkin, was the stepmother of Lilka Bielski. In January 1943, together with her sister Sonia and her son Grisha Meitis, she was killed during an attack in Chrapiniewo.

Esia Lewin-Schor, a cousin of the Bielskis, reached the forest in 1942. She escaped from Nowogródek ghetto in a group with a guide sent by the Bielski otriad. Later she was able to bring her father. Esia stayed in the Bielski otriad until the liberation. After the war she settled in the United States, became a teacher, and married a store designer. The couple has two children and three grandchildren.

Hersh Smolar, born in 1905, in Zambrów, Poland, became an active member of the Communist Party as a young teenager. For underground Communist activities Smolar spent four years in Polish prisons. In 1939 he escaped from prison and arrived in Russian-occupied Poland. Left behind by the retreating Red Army, he reached Minsk and became a ghetto inmate and a founder of the first Jewish underground, November 1941. A journalist-writer, from the Minsk ghetto Smolar established contacts with the non-Jewish underground outside the ghetto. He initiated massive departures of Jews into the Belorussian forests. As a result, 10,000 Jews left the ghetto. In 1942 Smolar moved to the Nalibocka forest, fought, and helped organize many Soviet partisan units.

Smolar was very helpful to Tuvia Bielski. His high position in the Soviet partisan movement gave Tuvia protection, particularly with General Platon. After the war, in Communist Poland, Smolar became the head of the Cultural Department of the Jewish Central Committee in Poland. Disappointed because of government-sponsored anti-Semitism Smolar left for Israel in 1971. Separated from his wife for many years, he has two sons.

Abraham Viner was born in Naliboki; from there the Germans moved him to camp Dworzec. He ran away in 1942 and unsuccessfully tried to join partisans. Bielski scouts brought him to the otriad in 1943. As a member of the Bielski otriad he worked in the office as a clerk.

After the war he settled in Israel and became a high school principal. I interviewed him in 1990 about a year before his death. Viner had two children and three grandchildren.

Rosalia Gierszonowski-Wodakow reached the forest in 1943, during the tunnel breakout in Nowogródek ghetto. With her mother and Riva

Kaganowicz-Bernstein she ended up in the Bielski otriad. In the forest she performed all kinds of duties, guarding the base and inspecting the cleanliness of the otriad.

After the war she moved to the United States. Rosalia was married twice. For years she worked at the United Nations in New York as an interpreter and supervisor of interpreters. She is a widow and has now retired. She has two children and four grandchildren.

Organization of the Bielski Otriad (last phase in the Nalibocka Forest, 1943–1944)[a]

HEADQUARTERS[b]

Commander: Tuvia Bielski[c]
Head of the Fighting Force and Second in Command: Asael Bielski
Head of Reconnaissance: Zus Bielski
Chief of Staff: Lazar Malbin
Chief of Special Operations: Salomon Wolkovisky
Commissar: Shematovietz
Assistant Commissar: Gordon, Ivan Vasilewicz
Quartermaster: Pesach Friedberg
Chief of Sabotage: Leibush Feldman
Platoon Heads: Haim Abramovicz, Novicki
Section Heads: Yehuda Bielski, Shlosberg, Elijahu Ostaszynski, Greenkowsky
Engineer: Ribinski
Historian and Lecturer: Dr. Shmuel Amarant
Secretary: Raja Kaplinski
Clerks: Abraham Viner, Bedzowski
Head Doctor: Dr. Hirsh
Nurses: Lili Kravitz, Hana Reebuk, Riva (Kantorovicz) Reich, Ita Grenau, Mrs. Hirsh (Dr. Hirsh's wife)
Workshops Manager: Matatiahu Kaback
Kitchen Manager/Chemist: Itzhak Rosenhaus

[a] This list was prepared and supplied by Chaja Bielski.

[b] While the list of the workshops seems complete, here and there a first name or second name is missing. For example, the head leather worker is not mentioned by name.

[c] Unless asked not to, I have tried to keep the spellings consistent. For example I chose the Polish spellings of "i" at the end of a name rather than "y"; similarly, instead of "tz" I chose "cz." Some names are spelled differently depending on the source. For example, Oppenheim and Openheim; Volkoviski, Wolkoviski and Wolkowiski, and so on.

WORKSHOPS

Head Shoemaker: Kolaczek
Head Tailors: Shmuel Kagan, Ester Gorodinski
Head Barbers: Bialobroda, Igelnik
Head Leather worker:
Head Watchmaker: Pinczuk
Head Carpenter: Neta Huberman
Head Hatmaker: Haim Leibovicz
Bakery: Mordechai Gershowicz
Bathhouse: Berkowicz
Soapmaker: Sioma Pupko
Slaughterhouse: Mordechai Shevachovicz
Blacksmith: Mordechai Berkowitz
Cattleshed: Aharon Dzienciolski
Tannery: Orkovicz
Kindergarten and School: Czesia
Bread Distribution: Batia Rabinovicz
Metal Workshop and Arms: Shmuel and Itzhak Openheim
Slaughterer: David Bruk
Head Sausage Shop: Monis
Head Corps: Potasznik
Manager of the Mill: Reznik
Chemical Products: Shmuel Mikulicki
Manager of Transportation: Michal Mechlis
In Charge of Potato Distribution: Akiva

A Selected Glossary of Foreign Words That Appear in the Text

Aktion a term used by the Germans describing a raid on a ghetto, usually resulting in mass killings or forced transfers to concentration camps.

Bambioshka a Russian term used to describe a partisan food mission.

Gebietskommissar a high German official in charge of a geographical district.

Hasid a term used in rabbinical literature to designate those who maintain a higher standard in observing religious and moral commandments.

Hutor a farm that stands alone away from a village.

Judenrat a German term used for a Jewish council, appointed by the Germans and sometimes including prewar Jewish community leaders. The stated purpose of this body was to serve as an intermediary between the German authorities and the Jewish community. In reality, members of the council were expected to fulfill all Nazi demands, regardless of how they or those they were supposed to represent felt about them. Noncompliance led to severe punishment, usually death.

Kozachok a Russian folk dance.

Kneidlach a pasta dish usually served during Passover; dumplings made out of matzoh or matzoh balls.

Legalshchyk a peasant who supplied the partisans with information about the availability of food in his village. Himself poor, this peasant knew who had hidden food and where. The partisans would often reward him with some of the confiscated food.

Malbush a derogatory term used for members in the Bielski unit who lacked arms and who were at the bottom social level; a Hebrew term for clothes. No one seems to know how the term acquired its negative meaning.

Matzoh unleavened bread eaten during Passover.

Obława a raid.

Otriad a partisan detachment that varied in size.

Politruk a person in a Russian partisan unit in charge of propaganda and political education.

Razwiedka Reconnaissance group, patrol, scouts.

Shochet a Jewish ritual slaughterer.

Tavo a Hebrew word meaning come; a masculine address that within this context could be translated into "come here"; a term applied to a steady boyfriend in the Bielski detachment.

Yeshiva a school for higher Jewish religious learning.

Zhidi a Belorussian term for Jews.

Ziemlanka a bunker in the forest, part of which was built underground.

Index